Environmental Impact Assessment

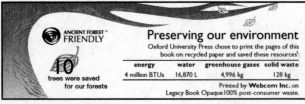

Environmental Impact Assessment

Practice and Participation

Edited by **Kevin S. Hanna**

OXFORD
UNIVERSITY PRESS

OXFORD
UNIVERSITY PRESS

8 Sampson Mews, Suite 204, Don Mills, Ontario M3C 0H5

www.oupcanada.com

Oxford University Press is a department of the University of Oxford.
It furthers the University's objective of excellence in research, scholarship,
and education by publishing worldwide in

Oxford   New York

Auckland   Cape Town   Dar es Salaam   Hong Kong   Karachi
Kuala Lumpur   Madrid   Melbourne   Mexico City   Nairobi
New Delhi   Shanghai   Taipei   Toronto

With offices in

Argentina   Austria   Brazil   Chile   Czech Republic   France   Greece
Guatemala   Hungary   Italy   Japan   Poland   Portugal   Singapore
South Korea   Switzerland   Thailand   Turkey   Ukraine   Vietnam

Oxford is a trade mark of Oxford University Press
in the UK and in certain other countries

Published in Canada by Oxford University Press

Library and Archives Canada Cataloguing in Publication

Environmental impact assessment : practice and participation / edited by
Kevin S. Hanna. — 2nd ed.

Includes bibliographical references and index.
ISBN 978-0-19-543022-6

1. Environmental impact analysis—Canada—Textbooks.
I. Hanna, Kevin S. (Kevin Stuart), 1961–

TD194.68.C3 E55 2009   333.71'40971   C2009-900878-5

Cover images: iStockphoto

The pages of this book have been printed on paper which has been certified
by the Forest Stewardship Council® (©1996 FSC®, Cert. no. C004071),
and which contains 100% post-consumer waste.

Printed and bound in Canada.

3 4 — 14 13 12

Contents

Foreword

Why a book on environmental impact assessment (EIA)? Why a focus on Canada? At a global scale, environmental impact assessment was formally introduced in 1969 through the National Environmental Policy Act in the United States. In December 1973, the Canadian federal Cabinet created a policy-based environmental assessment (EA) process, and in 1975 the Ontario Environmental Assessment Act established the first legislation-based process in Canada. As a result, Canada has acquired over 35 years of experience with EIA. Policy and practice in Canada have both reflected concepts and approaches in place around the world and contributed to their development. Thus, it is timely to pause and reflect on the progress made, the lessons learned, and the opportunities available. Stated another way, the second edition of this book provides a valuable inventory and assessment related to the philosophy, process, and results of EIA in Canada.

However, there are at least two further reasons for us to pause and reflect. First, the cumulative effects of an influential neo-conservative ideology have resulted in challenges for environmental management and protection. Key aspects of this ideology include the creation of more 'business friendly' regulatory environments, a greater reliance on market mechanisms than on public regulation, and a reduction of the role of government. Second, and likely to be significant in at least the short and medium term, is the uncertainty caused by the economic downturn that started in the last quarter of 2008. With those who are working concerned about losing jobs and those who are retired or approaching retirement worried about the viability of pension funds, it may be difficult to have our society engaged in environmental issues or concerned about the negative environmental impacts of 'development'. One obvious outcome is that advances in environmental stewardship are under serious risk of being eroded or lost. Thus, it is appropriate to consider how what we have learned about EIA can be used in a changing and turbulent political and economic context to ensure that our development decisions are more sustainable than they have been in the past.

Readers will have their own views about what constitutes 'best practice' with respect to EIA and how this has evolved. In the mid-1970s, most attention focused on *biophysical impacts* of projects. Subsequently, *social impacts* were recognized as important, leading to what became known as 'social impact assessment'. In parallel, various approaches to *public participation* as well as to *monitoring* were applied. Related to public participation, the incorporation of 'traditional knowledge' with scientific understanding has gained increasing attention. Later, there emerged an interest in *cumulative effects* and *strategic impact assessment* (extending beyond projects, to consider policies, plans, and programs). Over the decades, once provinces had introduced policies and processes for EIA, consideration was directed towards the coordination, or *harmonization*, of federal and provincial approaches. And now harmonization is also needed with respect to territorial approaches.

This book presents the reader with a thorough discussion and assessment of the attributes of best practice across the country, from British Columbia to Atlantic Canada and from the more densely populated southern areas to the less-populated North. In this manner, it provides readers with the basic information and helpful insights they need to reflect about best practice—and especially about what has been accomplished and what still needs to be done.

In this second edition, several notable additions will be found. In the first chapter, Hanna highlights the need to give attention to the efficacy or effectiveness, as well as to the efficiency, of impact assessment. He argues persuasively that too often attention has been focused on 'efficiency', or how to ensure that resources are allocated to minimize wastefulness, with less or no attention given to 'efficacy', or how well basic goals or expectations are achieved. He is quite right to draw attention to the distinction between effectiveness and efficiency. And, if a third edition is contemplated, a third 'E' might be added: equity (who gains and who loses).

In addition to the insight from Hanna in chapter 1, there are two new chapters. Chapter 10, by Kellar and Hanna, focuses on how EIA was used for the 18 projects and two large-scale linear facilities for the 2010 Winter Olympics in Vancouver and Whistler. Five of the projects were assessed through a harmonized process that drew upon federal and provincial EA arrangements. Kellar and Hanna explicitly consider the efficacy of the EA processes to deal with projects characterized by non-negotiable timelines, national pride, and global attention. Given such characteristics, their findings, although discouraging, are not surprising: EA is viewed by many as a 'perfunctory green stamp of approval', the EA process is not permitted to impede development, and there are difficulties in addressing cumulative effects.

Chapter 13, the other new chapter, written by Rusk, Granchinho, and Barry, reviews experience with EA in the Nunavut Settlement Area and thus provides a complement to chapters addressing experience in the Northwest Territories and the Yukon. Nunavut, an area of two million square kilometres and containing about 30,000 residents living in about two dozen communities, typifies much of the northern environment of Canada—very large area, low population densities, significant biodiversity, and striking cultural differences. By reviewing the mandate for and activities of EA under the auspices of the Nunavut Impact Review Board, which has many attributes of a co-management organization, the authors identify challenges regarding (1) connecting regional land-use plans (only two are in place in Nunavut) and EIAs, (2) developing EAs for some types of marine protected areas, (3) maintaining necessary human and other capacity and collaboration for various territorial organizations, and (4) incorporating traditional knowledge into EA processes, especially when Inuktitut lacks comparable words for certain scientific terms, and vice versa.

The distinctive contributions by Canadians in the field of environmental impact assessment have not changed since the first edition. Still appropriately identified prominently in this book are (1) the incorporation of Aboriginal and other lo-cal peoples' traditional and scientific knowledge into environmental impact state-ments (Thomas Berger's 1977 report, *Northern Frontier, Northern Homeland*, es-tablished a benchmark in this regard, not only for Canada, but globally); (2) the use of partnerships or alliances with communities and local groups to assist in the design and implementation of the assessment process; (3) the handling of issues specific to the context and distinctive physical and cultural environment of the Canadian North; (4) the frequent success in dealing with inter-jurisdictional is-sues arising through the shared and overlapping responsibilities and authority of federal, provincial, and territorial agencies and the legal frameworks introduced by federal and provincial governments; and (5) the design and implementation of a 'package' (policy, administrative arrangements, legal framework) for EIA.

Noting such distinctive contributions is not the same as concluding that Canada has been a global leader as a consequence. Much remains to be done. For example, if Berger established a benchmark in the late 1970s, what can be said about how the bar has been raised since then in regard to the incorporation of traditional knowledge into the EIA process? Or, if Canada has accumulated experience in harmonizing inter-jurisdictional issues, which aspects need attention to lift best practice to a higher level? Both of these aspects are addressed in the new chapters in this edition.

As always in such an undertaking, not all themes or topics are or can be ad-dressed, and this creates opportunities for future inventory and assessment. For example, how should *adaptive environmental management* be used along with more conventional impact assessment practices? How should harmonization with impact arrangements in the United States for *transnational development initiatives* be improved? How might EIA be better connected to other planning approaches, such as *watershed management* and *biodiversity management*. Often, no single planning approach is sufficient, and a productive avenue could be to determine more systematically how various approaches can be used together. To what extent has the Canadian experience been directly applied to, or been an influence on, the design of EIA processes in other countries? There is evidence that assessment pro-cesses introduced elsewhere have been modelled on Canadian approaches. For ex-ample, the Canadian experience was drawn upon explicitly in Indonesia. It would be interesting to know more about which aspects of the Canadian experience have been the most effective, efficient, and equitable when adapted and adopted else-where. In terms of resource sectors or regions, are there specific issues related to marine or estuarine environments? Finally, and as mentioned earlier, when we evaluate how EAs are conducted, how should we systematically consider questions related to effectiveness, efficiency, and equity? Such issues provide opportunities for a third edition.

Reading the updated chapters from the first edition and the new chapters in this second edition has given me a clearer and more systematic understanding of where environmental impact assessment came from and where it is now. The chapters have also provided insight about what challenges and future opportuni-ties remain. Some of the issues addressed in the book reflect distinctive contextual conditions in Canada. Many, however, are not specifically tied to a region and thus should be of interest to readers in many countries.

As a final comment, I commend Kevin Hanna for dedicating the second edition to Len Gertler. When I joined the University of Waterloo in 1969, Len was director of the planning school, and over the subsequent decades, until his death in 2005, I had the pleasure of participating with him in events as diverse as a field trip to New Brunswick and Nova Scotia and a development project in Bali, Indonesia. Throughout, I had a front row seat to watch Len's involvement in environmental planning and impact assessment, and had the privilege of seeing close-up his vision, creativity, commitment, and perseverance. I benefited immensely from being able to work with and learn from him. Len certainly was a leader in both planning and impact assessment.

Given Len's remarkable career and many accomplishments, it is especially fitting that in November 2006 the Ontario government established the Len Gertler Memorial Loree Forest, an area of 339 hectares located on the Niagara Escarpment at about the midpoint between Collingwood and Thornbury. This memorial park overlooks Georgian Bay, offering wonderful views and a mix of outdoor activities. The dedication of this book to Len is further tribute to a remarkable man and his legacy.

Bruce Mitchell
University of Waterloo
Waterloo, Ontario
February 2009

PART I

Environmental Impact Assessment: Process, Practice, and Critique

Chapter 1

Environmental Impact Assessment: Process, Setting, and Efficacy

Kevin S. Hanna

> Environmental impact assessment is, in its simplest form, a planning tool that is now gener-
> ally regarded as an integral component of sound decision making. . . . As a planning tool
> it has both an information gathering and decision making component which provides the
> decision maker with an objective basis for granting or denying approval for a proposed
> development.
>
> —Mr Justice La Forest, *Friends of the Oldman River Society v. Canada* (1992)

This is a book about the processes and practice of environmental impact assessment
in Canada. Environmental impact assessment (EIA or EA) is arguably one of the most
influential and consistent aspects of environmental regulation and policy in North
America. It is worth noting here that various terms have become synonymous with
EIA—for example, *impact assessment* and *environmental assessment*—and they are
used somewhat interchangeably throughout this book. *Environmental impact assessment* is best defined as a process for identifying and considering the impacts of
an action. As the quote above suggests, EIA is about making better-informed decisions. Environmental impact assessment is not about rejecting development; rather
it is about making sure that development proceeds with full knowledge of the environmental consequences. The Canadian Environmental Assessment Agency defines
environmental assessment as a process that

1. identifies possible environmental effects,
2. proposes measures to mitigate adverse effects, and
3. predicts whether there will be significant adverse environmental effects, even
    after the mitigation is implemented.

As an approach to environmental management, as a system, and as practice, EIA
has evolved in significant ways over the last few decades. The Canadian setting in

particular represents the lasting influence of EIA, the diversity of issues now addressed by EIA processes, and the more complex definition of environment that
many public agencies must now consider. The federal and provincial governments
in Canada each have their own EIA system, and as the later chapters in this book
illustrate, despite some common ingredients these systems can vary substantially
in what they cover and to whom and what they apply.

Before we explore the detailed discussions of process and application as well as
the jurisdictional case studies, a brief review of the context and basic process of
environmental impact assessment is in order. This is what this chapter provides.
In the Foreword, Bruce Mitchell notes the contributions made by Canadians to
the field of impact assessment and situates the book in the context of EIA teaching
and research. This introductory chapter provides fundamental information about
what EIA is and what an EIA system consists of; here I discuss the ideal EIA process,
the planning setting, and the pervasive question of EIA's efficacy as environmental
management. The succeeding chapters provide detailed explorations of process, the
evolution of Canadian EIA, analyses of methods and approaches, case studies, and
illustrations of EIA as practised by Canada's provincial and federal governments.

EIA initially garnered a flourish of scholarly interest in the 1970s, an interest that
lasted into the early 1990s, when other environmental issues and scholarly trends
came to overshadow impact assessment. Since the 1980s, EIA has quietly evolved
into one of the more consistent and unquestionably powerful instruments for environmental management in Canada. It now informs virtually all public project
and program development at the federal and provincial levels, and in some Canadian jurisdictions, major private-sector undertakings have also been subject to it.
Moreover, environmental impact assessment has become an increasingly complex
policy area. Beyond its initial focus on technical exercises and studies, EIA now
seeks to incorporate consideration of cumulative impacts, health, social, and economic impacts, and public participation as requisite elements in its application.
This is evidenced in the chapters that follow.

The Setting

It is best to begin with an understanding of where EIA is situated in policy- and
decision-making processes. Environmental impact assessment is ideally embedded in planning. Planning might be seen as having several faces—the process of
preparing a program or policy, of determining or deciding a course of action,
or of simply implementing development. Development actions are a product of
planning, and it is the action that has been most commonly subject to EIA. In the
EIA lexicon, an action is referred to as an undertaking or a project. Actions can
be physical projects, such as constructing a hydroelectric dam or a highway, or
non-physical projects and policies, such as creating a social welfare program or
raising the price of postage. Planning suggests a proactive process for addressing
a goal or need; it suggests actions that involve forward or strategic thinking. But
as Gibson and Hanna (in chapter 2) note, EIA has not always been applied pro-

actively in Canada. In the not-so-distant past, EIA has tended to be applied as an afterthought—when it was applied at all. But this has changed, and we can now say that EIA, in the Canadian context, is part of proactive policy and planning for numerous public and private entities.

If planning is an ongoing process, or continuum, that includes not just goal setting, evaluation, and plan formulation, but also fine-tuning the resulting action, then there is an evident role for EIA. Impact assessment is preferably a planning tool, one that fits into a larger process or model of decision making and environmental management. But this is not always the case, and in some jurisdictions EIA may only come into play after the development decision is made.

Perhaps the planning process can best be conceptualized through the use of the *rational comprehensive planning* (RCP) *model*—despite the discomfort of some planning theorists and even the debates within EIA scholarship about EIA's relationship to the RCP model. While there has been evolution and innovation in the ways that planning theorists describe planning both as a practice and as a process, the RCP model remains of enduring importance. It has been my experience in government that RCP represents, albeit in many modified forms, the way that many public and private entities like to think they plan, or at least the way they like others to think they plan. The rational comprehensive planning model, or approach, has four basic elements:

1. Goal setting
2. Identification of alternatives
3. Evaluation of means
4. Implementation of the decision (Hudson 1979)

Other models and theories have evolved around the RCP model. In many respects, these have largely sought to make the RCP model more comprehensive, more detailed, and more capable of dealing with issues of greater complexity. This evolution represents the growing recognition that environmental, social, and economic issues are inherently complex and interwoven. Friedmann (1987), for example, presents the RCP model in terms of an expanded decision-making model with seven stages largely attuned to the evaluation of policy alternatives:

1. Identifying goals and objectives
2. Identifying alternatives for meeting the goals and objectives
3. Predicting consequences and impacts that could be reasonably expected to flow from each alternative
4. Evaluating and considering the consequences and impacts with respect to the goals and objectives
5. Making the decision
6. Implementing the decision
7. Monitoring the impacts and consequences of the decision and responding to them

If it is indeed to be comprehensive, the RCP model requires substantial information, time, and resources. The process would also seem to require objectivity and broad consultation with those likely to be affected by the undertaking. Forester (1987) notes that the application of the RCP model assumes that practitioners, planners, and decision makers have access to six basic ingredients:

1. A well-defined problem
2. A full array of alternatives to consider
3. Full baseline information
4. Complete information about the consequences and impacts of each alternative
5. Full information about the values and preferences of those affected
6. Adequate time, skill, and resources to analyse and consider the above

At first glance, the RCP model imparts an image of simplicity—the stages provide a rational, deliberate process, one that is easily understood and standard in its application (Hudson 1979). The model also seems to have wide applicability and the potential for the consideration of all the essential issues, alternatives, compromises, and approaches for meeting whatever goals and objectives it is employed to address. Although RCP may at first appear to be an ideal process, it has several limitations. One criticism centres on the key word *comprehensive*. In his early critique of the RCP approach, Lindblom (1973) suggests that the cost of comprehensiveness may often exceed the benefits. In practice, there are limitations on time, on agency resources, and on the information available; in addition, there may not be a concise understanding of the nature of the problem or even a clear indication of goals and objectives. Comprehensiveness requires considerably more of everything than most agencies can realistically provide.

Another criticism flows from the apparent simplicity of the RCP model. Planning as a process or as a set of practice activities in urban, environment, or resource agencies is now seen as being more complex than something limited to the implementation of a rational approach. Planning has been variously described as embodying communicative or deliberative practice—notions that centre understanding on how planners or environment and resource managers interact with one another or with the public, and on the power that they and their agencies have and how they use and express it. Planning has also come to be seen in more activist tones, where the planner seeks to achieve economic, social, or environmental equity—a somewhat halcyon perspective on practice (Hanna 2005). But planning— despite the orations of theorists and even though it may be an institutional, political, and bureaucratic process—ultimately yields a physical outcome. Where a planning process exists, whatever it is composed of, it is the means by which we decide how land will be used and how development will modify our biophysical and social/cultural environment.

So why should we think about the rational comprehensive approach with respect to EIA? There are four basic reasons:

1.   The RCP model remains a valid representation of practice because in many modified forms it represents the way most agencies seem to approach planning. The stages outlined by Friedmann can be observed in practice, and each of these stages has come to be composed of many activities or even sub-stages. This increasing complexity reflects agency responsibility, the issues or projects under consideration, and the political social setting within which planning occurs.

2.   While criticisms of the unworkable nature of comprehensiveness are valid, the response in practice has been to bound issues—in other words, to limit what constitutes comprehensiveness and to make do with ingredients that may be less perfect than those that Forester (1987) outlines. This can make the model more workable.

3.   The RCP model is relevant to EIA because it represents the framework within which EIA is often used as a tool for planning and decision making.

4.   And finally, the model is significant because the EIA process itself is based on it in part—as the discussion below illustrates.

Fundamental Principles of EIA

One of the key values of EIA is the chance it gives proponents and decision makers to design and implement an action with the best available knowledge of its impacts and likely performance. The capacity for EIA to provide such information depends largely on the principles and values that inform and guide it both as a system and as part of the policy process. Barry Sadler's (1996) work on evaluating practice and performance in EIA provides a key discussion of the principles and core values of impact assessment, or at least what they should be. Sadler (1996) writes of EIA as having five main guiding principles:

1.   *A strong legislative foundation.* EIA should be based on legislation that provides clarity with respect to objectives, purpose, and responsibilities. Application of EIA should be codified, based in law rather than in discretionary guidelines.

2.   *Suitable procedures.* The quality, consistency, and outcomes of EIA should reflect the environmental, political, and social context within which EIA operates, and should demonstrate the ability to respond to divergent issues.

3.   *Public involvement.* Meaningful and effective public involvement must be present. Not only must those affected and interested be consulted, but their concerns should be able to affect the decision. As Healey (1997) notes, the power of public involvement is in whether or not such involvement has the capacity to affect the decision.

4.   *Orientation towards problem solving and decision making.* The context of EIA is inherently practical and applied. Thus, the EIA system should have relevance to issues of importance, it should generate needed information,

and it must influence, and be connected to, the settings where conditions of approval are set and decisions are made.

5. ***Monitoring and feedback capability.*** The consideration of impacts should not end with approval and implementation; rather the process must have some capacity for ensuring compliance, accuracy of impact prediction, and evaluation of project performance. Not only does such a role strengthen EIA, it provides information that can fine-tune the EIA process, provide knowledge of what impacts actually do occur, and measure project performance.

We can expand upon these principles. Senécal and colleagues (1999)[1] developed a list of 'Principles of EIA Best Practice', part of which is a framework for the basic principles that should guide the design, operation, and practice of EIA; these hold that an EIA system should be

- ***purposive***—the process should inform decision making and effect environmental protection and community well-being;
- ***rigorous***—it should apply the *best practicable* science, employing methodologies and techniques appropriate to the problems under consideration;
- ***practical***—it should result in information and suggestions that not only assist with problem solving but can also be reasonably implemented by proponents;
- ***relevant***—it should provide sufficient, reliable, and usable information for planning and decision making;
- ***cost-effective***—it should achieve the objectives of EIA within the limits of available information, time, resources, and assessment techniques;
- ***efficient***—it should impose the minimum cost burdens in terms of time and finance on proponents and participants consistent with meeting the requirements and objectives of EIA;
- ***focused***—it should concentrate on significant environmental effects and key issues, that is, those that need to be taken into account in making decisions;
- ***adaptive***—it should be adjusted to the realities, issues, and circumstances of the proposals under review without compromising the integrity of the process, and be iterative, incorporating lessons learned throughout a project's life cycle;
- ***participative***—it should provide appropriate opportunities to inform and involve the interested and affected publics, and their inputs and concerns should be addressed explicitly in the documentation and decision making;
- ***interdisciplinary***—it should ensure that the appropriate techniques and experts in the relevant biophysical and socio-economic disciplines are employed and integrated, including traditional knowledge;
- ***credible***—it should be carried out with professionalism, rigour, fairness, objectivity, impartiality, and balance, and be subject to independent checks and verification;

- *integrated*—it should address the interrelationships of social, economic, and biophysical aspects;
- *transparent*—it should have clear, easily understood requirements for EIA content, ensure public access to information, identify the factors that are to be taken into account in decision making, and acknowledge limitations and difficulties; and
- *systematic*—it should result in full consideration of all relevant information on the affected environment, of proposed alternatives and their impacts, and of the measures necessary to monitor and investigate residual effects.

Of course, these principles may seem idealistic, and—like the rational comprehensive model—in practice their application is variable across Canada and indeed internationally. Within the idealistic, hopeful nature of the principles that Senécal and his colleagues articulated, there are also important applied values. First, as I commented above, these principles provide a guide for the design, practice, and operation of EIA, even if hesitantly or partially applied. But equally as important, especially within the context of this book, they provide a framework for the evaluation and critique of EIA practice.

Stages in the EIA Process

The practice of EIA centres on a process for considering an action and its likely impacts and outcomes. The process of conducting an EIA can best be conceptualized as involving stages or procedures, and as with the discussions above, I will present these stages here in a somewhat idealistic framework. The following chapters illustrate the variable ways that procedures are defined in different jurisdictions, the methods used, and the considerations and applications actually involved with EIA. Preferably, EIA begins as early as possible in the planning, project, and decision-making process; it is applied to all development proposals that may cause significant environmental effects; it considers a range of biophysical impacts and relevant socio-economic factors, including health, culture, gender, lifestyle, age, and cumulative effects consistent with the concept and principles of sustainable development; and it provides for the involvement and input of communities and industries affected by a proposal, as well as the larger interested public (Senécal et al. 1999). Effective EIA provides opportunities for public involvement throughout the assessment process, not just at one or two stages, and the results of such consultation should have the capacity to affect the recommendations or decisions of the EIA agency and the proponent. In terms of the specific components, an EIA system will generally have seven stages, each composed of various steps. Different jurisdictions will describe, label, and blend the stages in different ways.

Stage 1

The first stage is the *proposal* itself. This is the basic concept of the undertaking. It may be articulated as a need, such as more electricity or more water, and it then outlines the options for meeting the need, such as the damming of a river. In some jurisdictions, there may be an assumption that when the EIA application is tendered, alternatives have already been considered or will be addressed during other stages of the impact assessment. The proponent, however, may be required to show that alternative means for achieving the project's goals have been considered. The nature of the project, the extent to which the proposal has been developed, and how thoughtfully it has been conceived will have great bearing on the way the EIA evolves. EIA should occur as early as possible in the project life. Some systems have an option whereby proponents can consult with the EIA agency as they develop their proposal. This helps ensure that a more responsive and ultimately acceptable proposal will be tendered to the EIA agency.

Stage 2

Screening occurs in the second stage. It answers the basic question, is an EIA required? Screening is used to determine whether or not a proposal should be subject to EIA and, if so, at what level of detail (Senécal et al. 1999). It is here that it is determined whether the review will entail larger public hearings, an internal agency-based panel review, or a small-scale administrative assessment. EIA systems apply to a broad range of actions, many of which are routine and their environmental impacts negligible. At the screening stage, such undertakings might be quickly reviewed and approved, or given a cursive review to ensure that no larger impact issues are likely. This is a practical need, since in some jurisdictions the majority of actions subject to EIA generally do not require comprehensive assessment and do not warrant the expenditure of EIA resources. When well designed and made conceptually sound, the screening stage can ensure that important and relevant proposals are subject to the assessment scrutiny they require, without subjecting small projects to needless delay and EIA costs. Screening criteria typically include legal requirements (is the undertaking subject to EIA legislation?), scale (does it fall within a size or cost threshold?), the nature of the proponent of the project (is it public or private [in some places all public projects are subject to EIA]? are certain permits required?), the nature of the project (e.g., it may be that all hydroelectric or chemical facilities are subject to EIA within a respective jurisdiction), or a combination of these.

Stage 3

Once it is determined that an EIA will be conducted, *scoping* begins. Scoping is where it is decided what the EIA will address. The issues and impacts that are likely to be important are identified, and the terms of reference for the EIA are established. Since the EIA may be conducted under considerable time and resource limitations, this stage can take on particular importance (Harrop and Nixon 1999). Scoping frames the attention of the impact assessment. Existing baseline data supports scoping, but the scoping stage can also help planners and resource managers decide what additional or new baseline information is needed. Decisions about stakeholder consultation, methods of assessing and predicting impacts, and additional consideration of alternatives begin with scoping. Some jurisdictions will provide relatively precise lists of what the scope of an EIA will be; others may provide more fluid and discretionary advice, allowing the EIA to be tailored to the circumstances, which for some projects may be largely biophysical and for others mostly social. Public participation should be an integral part of determining the scope of the EIA. It is through such consultation that the EIA system can identify what is important to those who may be affected by the proposed undertaking. As with other elements of EIA practice, there is variation in how the scope of assessment is decided and applied.

Stage 4

After scoping is complete, *assessment of the proposal* begins. It is here that data collection, impact prediction, and evaluation occur. Baseline data may already exist in some form, although not uncommonly it must be expanded and new data collected. Baseline information describes the current environmental (physical, social, and economic) conditions of the area that would be affected by the proposal. It provides the foundation for the assessment and prediction of impacts. Impact prediction also occurs at this stage, and as the term implies, it involves the forecasting of the likely impacts and outcomes of the proposal. Such prediction may address a range of project design and operating scenarios. Likely impacts are also assessed for their significance. As Baker and Rapaport outline in chapter 3, specific methods that have been refined within EIA have become synonymous with the science of assessing impacts.

Significance is a subjective notion determined by the importance that the stakeholders—the proponent, the regulators, and the decision makers—attach to specific impacts. It is also during the assessment stage that mitigation measures are identified and a monitoring or compliance program is outlined. The process of mitigation involves outlining the measures that can be taken to reduce or eliminate the impacts identified. It also provides the proponent with the opportunity to make the project better, to respond to the concerns of those affected, and to improve the likelihood that the proposal will be favourably received by the EIA and other approval agencies. Effective mitigation measures can make a project more likely to be accepted and perhaps even ensure that it is more efficiently implemented.

Stage 5

The task of *preparation, submission, and review* follows assessment, although in practice preparation of the submission should occur throughout the EIA process. At this point, the information that has been collected and analysed is brought together and placed in the EIA report; in essence, this is where the findings of assessment are presented. The contents of the report are usually determined by the regulating EIA agency. In some jurisdictions there will be clear expectations about what the report will contain and how this information will be organized; these expectations may be communicated through agency publications, such as a guide to the EIA process, through pre-consultation with the proponent where the agency's expectations are made clear, or though the formal provision of terms of reference. The report is then tendered to the EIA agency for review and a decision.

Stage 6

While the *decision* may appear to be a simple matter, in practice the decision-making process is complex. The decision in EIA might be better seen as a recommendation. The recommendation might be to approve a proposal as it is or with conditions, reject it in its present form, or reject the concept outright. The decision/recommendation flows from the review, and in some instances it may appear to be the last part of the review. As outlined above, EIA is a tool in the planning process. It contributes knowledge that is used in decision making. Initially, environmental impact assessment was not intended to be the point at which the formal decision about whether or not to proceed with a proposal would occur; rather, it was meant to entail assessing impacts and communicating such knowledge to decision makers. The issue of whether a decision to approve a proposed project is seen to be part of or separate from EIA is problematic, but perhaps it is no longer terribly relevant—some would now hold that EIA has become the place where the decision is in fact made, and some EIA processes now provide formal approvals.

The context for considering the proposal will depend on the jurisdiction, the scale of the undertaking, and the results of screening and scoping. In many instances of lesser public concern, the EIA agency will conduct a review and formulate recommendations within an internal administrative setting; in instances where the undertaking is considered substantial, a public hearing may become the setting for review. As Gertler shows in chapter 5, the hearing process in EIA can be an important place for the interpretation of the range of regulations and acts relevant to a proposal, and the boards that hear EIA applications have played vital roles in the environmental policy and regulatory process, sometimes well beyond their EIA consideration capacity.

A glance at EIA legislation across Canada shows that the product of an impact assessment tends to be a recommendation; the power to make a decision based on the knowledge provided by an EIA more commonly resides at the political level. EIA legislation may give the 'minister' (of environment or whatever agency is responsible for the EIA system or the subject of the proposal) the power to decide. This power is usually exercised as a formal acceptance of the EIA agency's recommendation. In other words, the responsible minister *signs off* on what the agency recommends. For the great majority of proposals reviewed by EIA systems in Canada, the arbiter of whether or not the subject undertaking proceeds, in what form or under what conditions, is the EIA process.

Stage 7

The final stage, and a relatively recent one in the evolution of EIA, is *monitoring and compliance follow-up*. It is one thing to render a recommendation and attach conditions, but quite another to ensure that the proponent complies with the conditions. In some jurisdictions the EIA systems have poor linkages to the agencies that actually enforce conditions of approval and monitor compliance, while in other jurisdictions these links have become stronger. Monitoring not only has an obvious function in supporting compliance, it also provides information that can be used in further assessments to improve EIA efficiency or enhance baseline information.

The Progression of EIA

The last few points I would like to make in this introductory chapter relate to the non-static nature of EIA. There has been overall progress in the application and elements of EIA in Canada. Aspects of this progress include the specialized types of impact assessment, enhancement of public participation and consultation, inter-jurisdictional EIA processes, implementation of sustainability objectives, and the recognition that EIA provides a venue for public learning about environment, community, and governance.

Types of EIA

Environmental impact assessment has evolved to include consideration of a range of impacts beyond those that affect just the biophysical environment. This evolution of interests reflects recognition of the complex meaning that the *environment* has and the integrated nature of the impacts of development and of the responses needed to mitigate them. Part of this progression has been the development of distinct impact assessment forms that now address issues such as cumulative, social, and economic impacts, as well as assessment at a strategic level. These forms of impact assessment have developed under the larger rubric of EIA. In some instances they might be applied independently, but in many settings they have emerged as steps or components of a comprehensive EIA system.

Cumulative impact assessment can be described as the analysis of effects that are additive or interactive and result from the recurrence of actions over time. Cumulative impacts are incremental and result when undertakings build on or add to the impacts of previous impacts. In chapter 8, Creasey and Ross examine this form of impact assessment from a case study perspective, based on their experience with the Cheviot mine project in Alberta.

Social and economic assessment is concerned with the impacts of an action on the social and economic constructs of human society. In chapter 7, Pushchak and Farrugia-Uhalde also use a case study approach, specifically a Canadian experience with high-level radioactive waste disposal, not only to look at this type of impact assessment, but also to explore unique issues surrounding the impacts of development on First Nations communities and the participation of such stakeholders in EIA.

Strategic impact assessment is applied to policies, programs, or plans rather than to the physical action itself. As Noble and Harriman-Gunn outline in chapter 6, strategic EIA is a conceptually and methodologically difficult form of assessment to apply, but it is an increasingly essential form of impact assessment. Strategic assessment also holds promise for advancing sustainability assessment by possibly ensuring that sustainability criteria are considered at the level of conceptual planning (see Gibson and Hanna, chapter 2). Other jurisdictions, notably the European Union, have rapidly advanced the application of strategic assessment, while in Canada its advancement has been hesitant and the benefits not well recognized.

These types of impact assessment represent a progression and evolution in EIA that flows from the complexity of our understanding and recognition of environment, as well as from an increasing sophistication in the way that impacts are envisaged and addressed under the EIA rubric.

Participation and Consultation

One of the tenets of a good EIA system is public participation and broad consultation with those likely to be impacted by the proposal. Participation, like other aspects of EIA, has evolved. We can say that, with a few exceptions, the Canadian tendency has been to strengthen and expand the participatory elements of EIA. Sinclair and Diduck, in chapter 4, provide an overall image of participation in Canadian EIA, noting the progression and hesitancy inherent in making participation meaningful. In chapter 11, using the Mackenzie Valley as a case study, Armitage looks at another aspect of participation—collaboration among affected parties and the need to realize integration in planning. Armitage's focus on the theme of integration brings to light an important, and not always well acknowledged, dynamic in EIA: it is a process that at a very fundamental level requires integrated efforts, knowledge, and application to be effective and inclusive.

Effective EIA

Even though EIA is an established environmental management tool, little is known about its effectiveness. Effectiveness is a long-standing issue in EIA, fundamental to its theoretical development and essential to enhancing its contribution to sustainability. However, systematic evaluations of the actual impacts and influence of EIA on Canadian environmental management are rare indeed. A good portion of the EIA reform we have seen in Canada has focused on making EIA systems function more efficiently. But there is little indication that these efforts have improved EIA as a form of environmental management. Indeed, injecting timelines and reducing budgets and opportunities for participation all in the name of 'efficiency' may inevitably weaken EIA. Regulators, industries, and the public alike increasingly challenge the value of EIA as a tool for development decision making and, even more so, for better environmental management. And in this respect one of the greatest challenges that EIA may face in the next few years is the argument over the validity and efficacy of EIA as a tool for environmental management.

Arguably, the most significant measure of the effectiveness of EIA is the extent to which it achieves its goal of environmental management (Morrison-Saunders and Bailey 1999; Doyle and Sadler 1996). Cashmore et al. (2004) point out that EIA's indirect influence on environmental management, by stimulating changes in policy and practice, may be more important than its direct role in the decision about a proposed development (see also Caldwell 1993; Bartlett 1986). In other words, EIA does have the potential to contribute to better environmental management through 'a multiplicity of additional, and often interlinked, transformative potentialities' (Cashmore et al. 2008, 1246). Identifying such potential, however, as well as the underlying criteria of what constitutes effectiveness in EIA when examined from a broader environmental management paradigm, is easier said than done. Effectiveness simply means 'to have an effect' (Emmelin 2006), but what is effective in one context and under one regulatory system or resource sector may not be considered so under another. Furthermore, understandings and interpretations of effectiveness may vary from the proponent to the regulator to the general public. We can certainly observe these variations in the Canadian context. In chapter 2, Gibson and Hanna note that the efficiency mantra may limit the efficacy of EIA as a sustainability tool. And Kellar and Hanna (in chapter 10) describe a case where efficiency may well have won over effective EIA in the push to get projects approved in time for the 2010 Winter Olympics.

Inter-jurisdictional Application

In Canada, EIA is applied at the federal and provincial levels, but many environmental issues cross jurisdictional boundaries, both in terms of the location of the undertaking and hence the impacts and in terms of the nature of the impacts. As the nature of EIA and the impacts it considers have become more complex, EIA has in some cases become inter-jurisdictional in its application. This is part of the progression of EIA—the simple reality that two EIA stages may sometimes examine the impacts of the same undertaking and not always reach the same conclusions or recommendations. The Red Hill Creek Expressway (in Ontario), the earlier Oldman River Dam (Alberta), and the Rafferty-Alameda Dam (Saskatchewan) each had substantial provincial involvement and inter-jurisdictional implications, and as Gibson and Hanna show in chapter 2, these projects have been influential in the evolution of Canadian EIA. Fitzpatrick and Sinclair's chapter 9 deals with the Sable Gas Panel Review, a more recent and successful application of multi-jurisdictional EIA within a relatively complex biophysical and political/social setting. While the Sable Gas case study illustrates a relative outlier in federal-provincial EIA cooperation, it does symbolize the hesitant progression towards more jurisdictionally integrated approaches to EIA.

The challenge in advancing inter-jurisdictional EIA systems is ensuring that such models, when they do appear, act to strengthen the application of impact assessment rather than weaken it. In chapter 12, Slocombe, Hartley, and Noonan discuss the influence that Native land claims have had on the application and evolution of EIA in Canada's North and the trend towards devolution of EIA responsibility. These have inter-jurisdictional implications, though with a uniquely Canadian twist, as indeed does the case described by Armitage (chapter 11). In Canada the course of land claims has seen new interpretations of environment and resource use and new visions of local governance. In many key regions of the country, EIA will have to adapt to these changing perceptions. Indeed, in Nunavut, as Rusk and colleagues note in chapter 13, EIA is emerging as the primary venue for natural resource–use planning, not as a parallel process. And in many respects Nunavut may well emerge as an illustration of the ideal application of EIA.

The progress of Canadian EIA has been gradual, complex, and often hesitant. Even so, EIA has become an influential and relatively stable environmental policy tool, even during times when environmental protection has been weakened by governments. Environmental impact assessment has often served as the locus for considering not just the environmental impacts of major developments, but also their broader implications. The chapters that follow provide a current guide to Canadian EIA, the methods employed, the types of assessment practised, and the history and legal evolution. Each of Canada's EIA systems, provincial and federal, is described in terms of how it works, its attributes, and the practices unique to it.

The contributors to this book provide critical and pragmatic illustrations of the practice and process of Canadian environmental impact assessment. While they certainly provide matter-of-fact images of EIA practice and process, they also offer critical insight into the challenges facing EIA in Canada. Together the authors draw on a wide range of knowledge based on their own experience, research, and practice to create a unique analysis of EIA as practised in Canada.

Note

1.   The authors developed the 'Principles of EIA Best Practice' for the International Association for Impact Assessment and the Institute of Environmental Assessment, UK.

Acknowledgements

I would like to thank the peer reviewers who provided comments and suggestions and who gave their time to help in the development of this second edition. Their contribution is gratefully acknowledged. And of course I would like to recognize the help and hard work of each of the chapter authors. Thanks to Bram Noble, too, for his helpful notes regarding efficacy; these have been used in this chapter.

References

Bartlett, R. 1986. Rationality and the logic of the National Environmental Policy Act. *Environmental Professional* 8:105–11.

Caldwell, L. 1993. Achieving the NEPA intent: New directions in politics, science and law. In J. Cannon and S. Hildebrand (Eds), *Environmental analysis: The NEPA experience.* London: Lewis Publishers.

Cashmore, M., A. Bond, and D. Cobb. 2008. The role and functioning of environmental assessment: Theoretical reflections upon an empirical investigation of causation. *Journal of Environmental Management* 88:1233–48.

Cashmore, M., R. Gwilliam, R. Morgan, D. Cobb, and A. Bond. 2004. The indeterminable issue of effectiveness: Substantive purposes, outcomes and research challenges in the advancement of EIA theory. *Impact Assessment and Project Appraisal* 22:295–310.

Doyle, D., and B. Sadler. 1996. *Environmental assessment in Canada: Frameworks, procedures, and attributes of effectiveness.* Ottawa: Canadian Environmental Assessment Agency.

Emmelin, L. 2006. Tools for environmental assessment in strategic decision making—Reflections in the conceptual basis for a research programme. In L. Emmelin (Ed.), *Effective environmental assessment tools—Critical reflections in concepts and practice.* Research report no. 2006:03. Blekinge Institute of Technology.

Forester, J. 1987. *Planning in the face of power.* Berkeley: University of California Press.

Friedmann, J. 1987. *Planning in the public domain: From knowledge to action.* Princeton: Princeton University Press.

Friends of the Oldman River Society v. Canada. 1992. S.C.R. 3. Ottawa: Supreme Court of Canada.

Hanna, K.S. 2005. Planning for sustainability, two contrasting communities. *Journal of the American Planning Association* 71 (1): 27–40.

Harrop, D.O., and J.A Nixon. 1999. *Environmental assessment in practice.* London: Routledge.

Healey, P. 1997. *Collaborative planning: Shaping places in fragmented societies.* Hong Kong: Macmillan Press.

Hudson, B. 1979. Comparison of current planning theories: Counterparts and contradictions. *Journal of the American Planning Association* 45 (4): 387–98.

Lindblom, C.E. 1973. The science of 'muddling through'. In A. Faludi, *A reader in planning theory.* Oxford: Pergamon Press.

Morrison-Saunders, A., and J. Bailey. 1999. Exploring the EIA/environmental management relationship. *Environmental Management* 24 (3): 281–95.

Sadler, B. 1996. *International study of the effectiveness of environmental assessment. Environmental assessment in a changing world: Evaluating practice to improve performance.* Final report. Ottawa: Ministry of Supply and Services Canada.

Senécal, P., B. Sadler, B. Goldsmith, K. Brown, and S. Conover. 1999. *Principles of environmental impact assessment, best practice.* Fargo, ND: International Association for Impact Assessment; Lincoln, UK: Institute of Environmental Assessment.

Chapter 2

Progress and Uncertainty: The Evolution of Federal Environmental Assessment in Canada

Robert B. Gibson and Kevin S. Hanna

This chapter provides a discussion of environmental assessment (EA or EIA) at the federal level, examining its evolution and advancement by federal agencies over the past 30 years. We outline a range of factors and competing interpretations concerning the progress of the federal EA process. Overall, the idea of progressive evolution towards stronger contributions to sustainability serves as the basic foundation for our treatment of federal environmental assessment. Although this book generally uses the term *environmental impact assessment*, Canada's federal process is referred to as *environmental assessment*; here we use the two terms interchangeably.

The Evolution of Environmental Assessment as a Concept and in Practice: Some General Themes

Environmental assessment is the collective term for a host of quite different activities and processes. The common assumption is that these processes centre on efforts to anticipate the environmental effects of new undertakings and to make better decisions about them. Across Canada there is still great variation in what are considered to be environmental effects, undertakings, and decisions, and great variation in several other key aspects of EA as well, as our colleagues show in subsequent chapters in this book. But after some 30 years of deliberation and experience, it is possible to delineate, at least generally, the main lines of change in the concept and practice of environmental assessment.

Environmental assessment is now broadly recognized as an approach to planning and decision making, but in most jurisdictions it first emerged from environmental regulations. This evolution, from environmental regulations to advanced environmental assessment, can be divided into four stages (Gibson 2002b):

- ***Stage 1.*** Reactive pollution control through measures responding to locally identified problems (most often air, water, or soil pollution), with technical solutions considered and issues addressed through closed negotiation of abatement requirements between government officials and the polluters
- ***Stage 2.*** Proactive impact identification and mitigation through relatively formal impact assessment and project approval/licensing, but still focused on biophysical concerns (though now integrating consideration of various receptors) and still treated mostly as a technical issue with no serious public role (but perhaps expert review)
- ***Stage 3.*** Integration of broader environmental considerations in project selection and planning through environmental assessment processes that include
  - consideration of cultural, historical, and economic impacts as well as bio-physical effects;
  - required examination of alternatives, aiming to identify the best options environmentally as well as socially and economically; and
  - public reviews, which would reveal conflicts and uncertainties among experts and consequently the significance of public choice
- ***Stage 4.*** Integrated planning and decision making for sustainability, addressing policies and programs as well as projects and cumulative local, regional, and global effects, with review and decision-making processes that
  - are devoted to empowering the public;
  - recognize uncertainties and favour precaution, diversity, reversibility, and adaptability;
  - are strategic; and
  - integrate sustainability into planning and decision-making processes

The fourth stage does not exist to any degree anywhere in the world. As we shall see, in Canada the federal assessment process has not wholly reached the third stage, and many provincial EA processes are further behind. As in other matters of civic responsibility, in environmental assessment the province of Ontario moved backwards in the late 1990s. A few years later, British Columbia also weakened its EA system. Nevertheless, as noted by Hanna in this volume, EA has been one of the more consistent and stable aspects of environmental policy in Canada and overall there seems to be a more or less consistent pattern of movement in the direction of stage 4. There are, however, no guarantees that the pattern will continue, for in times of economic difficulty or when government ideology swings to 'small government, business friendliness', the environment is seen as a cost of doing business, an amenity to be sacrificed for the sake of development and growth. When this happens, the integrity of an EA system can be compromised.

Framing the argument somewhat differently, and more positively, we can say that the evolution of Canadian environmental assessment over the last 30 years has involved changes in 12 positive trend lines (Gibson 2002b; Gibson et al. 2005); specifically, environmental assessment in concept and practice has moved towards being

- more mandatory and codified (increased adoption of law-based processes, further specification of requirements, reduction of discretionary provisions);
- more widely applied (covering small as well as large capital projects, continuing as well as new initiatives, sectoral and area developments as well as single proposals, strategic- as well as project-level undertakings);
- more often initiated early in planning (beginning with purposes and broad alternatives, sometimes beginning with the driving policies, programs, and plans);
- more open and participatory (not just proponents, government officials, and technical experts, but also the broader public and interest groups);
- more comprehensive of environmental concerns (socio-economic, cultural, and community effects as well as biophysical and ecological effects; regional and global as well as local effects; life cycle as well as immediate effects);
- more integrative (considering cumulative and systemic effects rather than just individual impacts);
- more accepting of different kinds of knowledge and analysis (informal and traditional knowledge as well as conventional science, preferences as well as 'facts');
- more closely monitored (by the courts, informed civil society bodies, and government auditors and enforcement agencies watching for compliance with assessment obligations, and by stakeholders watching for actual effects of approved undertakings);
- more cautious, perhaps even humble (recognizing and addressing uncertainties and applying the notion of precaution);
- more sensitive to efficiency concerns (questions about process emphases, costs, and relations with other evaluation and decision-making processes);
- more often adopted beyond formal environmental assessment processes (through sectoral law at various levels, but also in land-use planning and through voluntary corporate initiatives, etc.); and
- more ambitious (aiming for positive contributions to overall sustainability rather than just for mitigation of adverse effects for individually 'acceptable' undertakings).

None of this movement has been accidental (Gibson 2002b; Gibson et al. 2005). Most of the positive steps can be seen as responses to rising public concerns about environmental realities that have become more evident in recent decades and are even more pressing now. At the same time, progress in each of the categories has been the product of concerted efforts, often in the face of resistance from industry, government agencies, and the provinces. Some of the gains have been modest and tentative, and as may be expected in struggles among competing interests, there have been retreats as well as advances (Gibson 2002b). These three elements—the gradual movements in response to evident problems, the persistent resistance, and the continuing conflicts and tensions—are evident in the chronology of change in Canadian federal environmental assessment.

An Environmental Impact Assessment Chronology

In the United States, the key components of environmental assessment emerged almost fully formed through court rulings on the deceptively few lines on environmental assessment in the 1969 National Environmental Policy Act (NEPA). At that time in Canada, especially at the federal level, rising public awareness of environmental damage, deepening skepticism about government and corporate reassurances, and the visibility of the NEPA precedent led to expectations that we, too, should introduce environmental assessment requirements. But the NEPA example also warned authorities of what such requirements might entail.

The US experience showed that NEPA was giving the environment a powerful presence in policy and program processes; the act had become disruptive to business-as-usual in US agencies. At the beginning of the 1970s, not long after NEPA came into play in the United States, environmental assessment entered the Canadian federal agenda. But the prospect of a similarly strong and somewhat autonomous environmental assessment process made federal agencies in Canada highly uncomfortable. While EA was one of the first issues to be considered in the newly formed federal Department of the Environment, it quickly became clear that there was no federal willingness to follow the American model (Gibson 2002b). The first steps in Canada were hesitant and guidelines based. Guidelines, normally, are suggestions for good behaviour, generally lacking the force of law. Agencies might see guidelines as being discretionary as opposed to obligatory and required. But as we outline below, the progress of environmental assessment in Canada has shown that guidelines are not always easy for agencies to ignore.

The following chronology identifies the main steps that were taken in creating and amending the federal process over the 30 years from the early 1970s to, roughly, the present. These steps include a selection of the most significant, illustrative, and influential developments in related Canadian environmental assessment work.

8 June 1972 The federal Cabinet decides that all new federally initiated projects and those under federal jurisdiction must be screened for 'potential pollution effects' (FEARO, undated, 3) and that proposed projects found likely to have significant effects be referred to the Department of the Environment for further assessment. The requirement is presented in policy only. The idea of following the American (NEPA) model of a broadly scoped legislated process is rejected for three main reasons:

- Fear that legislation would open the floodgates to litigation and shift authority to the courts, thus delaying decision making
- Concern that mandatory obligations would reduce governmental and ministerial discretion and flexibility
- Discomfort with a broad scope that would invite critical public examination of conventional projects, policies, and agency programs, and would

expand the purview of the new Department of the Environment at the expense of more established departments.

20 December 1973 Cabinet approves a more formal and detailed directive establishing a policy-based federal environmental assessment process. A wide range of federal government projects and projects requiring federal money, land, or approval is potentially covered. *Environment* is defined broadly to include social as well as biophysical considerations. Proposed projects that have potentially significant environmental effects or that raise significant public concern may be subject to public panel review with informal hearings.

Departments currently responsible for the relevant planning and/or approvals retain that responsibility. The responsible authorities are expected to ensure that environmental assessments are done where appropriate, but they are left to decide whether, how, and when assessment would be appropriate. This approach, called *self-assessment*, is officially intended to diffuse environmental responsibility throughout government. In effect, it makes serious attention to environmental assessment requirements essentially voluntary.

1975. The province of Ontario's Environmental Assessment Act establishes Canada's first legislated assessment process. Loosely based on the NEPA approach but also including an enforceable decision, the Ontario process covers social, economic, and cultural as well as biophysical effects, requires examination of alternatives, and provides for public hearings. The bill introduced to the Ontario legislature anticipated that the process would be applied only to specially designated undertakings. However, in a minority government situation, the law is amended during the legislative process to apply automatically to all provincial government undertakings not expressly exempted.

1975. The governments of Canada and Quebec and the Cree and Inuit peoples of Northern Quebec sign the James Bay and Northern Quebec Agreement (Quebec et al. 1998), which includes the first of many claims-based environmental assessment processes (Northeastern Quebec Agreement with the Nascapi in 1978, Inuvialuit Final Agreement in 1984, Gwich'in Agreement in 1992, Nunavut Land Claims Agreement and the Umbrella Final Agreement with the Council for Yukon Indians in 1993, and the Sahtu Dene and Metis Agreement in 1994).

1974–7 The Mackenzie Valley Pipeline Inquiry, led by Mr Justice Thomas Berger, sets an international standard for critical and cross-cultural public assessment of proposed development options (see Figure 2.1 for the location of key assessments mentioned in this chapter). The Berger (1977) report assumes a prominent place in the Canadian environmental and resource management literature, indeed in the lexicon of environmental and resource policy (see chapter 11 by Armitage for a detailed discussion of the Mackenzie Valley setting). While the model established by the Berger inquiry will never be used again in Canada, it influences subsequent deliberations and practice in environmental assessment. In essence, it creates expectations about what an assessment process should be.

Figure 2.1   Locations of some key events in Canadian environmental impact assessment

1974–7 The federal Environmental Assessment and Review Process (EARP) is gradually given detail and substance through interdepartmental negotiation within the federal government (FEARO 1977). Many federal departments move slowly or not at all to ensure effective implementation. Major early cases involving public panel reviews are heavily criticized. Requirements for the first EARP review (of Point Lepreau nuclear power station in New Brunswick) are watered down to avoid conflict with the financing and construction schedules (Emond 1978, 236–51). The second EARP review (of the Wreck Cove hydroelectric power project in Nova Scotia) is not initiated until after the project has been approved. Findings that would have affected project design are not available until after a good portion of the project work has been completed (Emond 1978, 251–9).

15 February 1977 The EARP process is strengthened slightly in response to public criticism. Early public participation is encouraged, and public review panels are no longer to include a representative from the proponent department (FEARO 1977).

1981–2 The environment minister asks Cabinet to consider the further strengthening of the process, including the addition of a legislated foundation. Instead Cabinet orders a review of the desirability and effectiveness of the existing process.

The review, undertaken largely by independent management consultants, covers most departments and agencies responsible for implementing the EARP. The consultants' reports, dated July 1982, document uneven implementation and, within federal agencies, a broad disinterest in and disregard for the policy obligations—even in some parts of the Department of the Environment (Lavalin 1982). After the review findings are leaked to the press, pressure for improvement mounts.

1983 Gordon Beanlands and Peter Duinker (1983) publish a report on the quality of environmental assessment work and how to improve it. They propose putting a greater emphasis on ecosystem effects rather than on effects on individual species and other receptors, and they recognize the importance of local knowledge in identifying significant and valued ecosystem components. In the same year, Charles Caccia is appointed Canada's minister of the environment and begins to press for legislated strengthening of federal environmental assessment.

22 June 1984 Caccia wins a compromise. The EARP is adjusted and registered formally as the 'Environmental Assessment Review Process Guidelines Order' under the Government Organization Act of 1979. The order-in-council includes firmer language of obligation ('the initiating department . . . shall ensure . . .'), but self-assessment remains, as does considerable flexibility on when and how to apply the requirements (Canada 1984). While the notion of a guidelines order appears to be oxymoronic—guidelines are flexible, while orders are mandatory—the approach reflects the federal government's conflicting desires to reassure proponent departments while at the same time assuaging the public.

1984–9 The new Guidelines Order fails to secure, to any significant extent, more effective commitment to environmental assessment on the part of federal authorities. Assessment obligations are often avoided where federal assessment would challenge provincial agendas and autonomy—even in high-profile controversial cases.

September 1987 Another federal review of the environmental assessment process leads to the release of a green paper, *Reforming Federal Environmental Assessment: A Discussion Paper* (FEARO 1987). The paper considers broadening the scope of assessment, specifying categories of projects requiring assessment, providing intervenor funding to facilitate effective public involvement, and requiring post-approval monitoring plans, but is silent on the question of enshrining the process in legislation. The environment minister promises to seek Cabinet approval for a reform package following public consultations on the green paper, but no reform package emerges in the following year.

April 1989 In an unexpected decision on the application of the Guidelines Order to a water management project in Saskatchewan (the Rafferty-Alameda Dam), Mr Justice Cullen of the Federal Court of Canada rules that the Guidelines Order is legally binding. This ruling is later upheld by the Federal Court of Appeal (1990) and then by the Supreme Court of Canada (in the *Oldman River Dam* case, 1992).[1] Suddenly, implementation of the Guidelines Order is mandatory and the discretionary element that agencies counted on is lost. Federal authorities immediately begin to take environmental assessment more seriously, and the federal government begins work on an intentionally legislated process.

10 June 1990 The federal government introduces a bill to establish a Canadian environmental assessment act. The proposed law, a product of deliberations dominated by proponent departments, is less ambitious than the Guidelines Order and full of openings for ministerial discretion. The law does not include application to the strategic level (policies and programs). Instead, the government announces that it will initiate a non-legislated process for policies and programs submitted for Cabinet approval.

1990–4 Environmental groups, especially public interest environmental law associations, play a major role in the parliamentary committee review of the proposed legislation. Significant improvements are incorporated, and a large portion of the discretion is excised. The law is passed in 1992 but not proclaimed in force. A new government makes further amendments in 1994.

February 1993 The Cabinet directive for assessments at the strategic level (policies, programs, and plans) is published. The requirements involved rest on mere policy, and the strategic-level assessments are to be shrouded in Cabinet secrecy. Only summaries of findings are to be released (FEARO 1993). Implementation proves to be weak, paralleling the earlier experience with implementation of the non-legislated EARP at the project level.

January 1995 The Canadian Environmental Assessment Act, which had received legislative approval in 1992, is proclaimed in force along with a set of key regulations governing its application (Canada 1992). The new law retains an apparently restrictive definition of *environment* that omits direct socio-economic and cultural effects. It also leaves environmental assessment as a largely advisory exercise. However, the law provides openings for evaluation of needs, alternatives, and broadly environmental considerations; promises intervenor funding; adds attention to cumulative effects; and encourages follow-up plans.

The preamble to the Act emphasizes the expectation that EA will foster sustainable development:

> The Government of Canada seeks to achieve sustainable development by conserving and enhancing environmental quality and by encouraging and promoting economic development that conserves and enhances environmental quality. . . . Environmental assessment provides an effective means of integrating environmental factors into planning and decision-making processes in a manner that promotes sustainable development.

In contrast to these opening sentiments, most of the Act remains focused on mitigating serious adverse environmental effects. The result is an apparent inconsistency between the Act's purposes and its main provisions. Openings for sustainability-centred assessments are nonetheless available in the new law.

January 1997 In one of the few significant reversals in the strengthening of the environmental assessment process, the Government of Ontario weakens its Environmental Assessment Act, chiefly by eliminating intervenor funding and introducing a mechanism for allowing proponents to propose and carry out narrowly scoped assessments.

20 June 1997 The public review panel responsible for assessing the proposed Voisey's Bay Mine and Mill on the north coast of Labrador issues guidelines for the review that adopt a higher-sustainability test. The mine's proponent is required to discuss 'the extent to which the Undertaking may make a positive overall contribution towards the attainment of ecological and community sustainability, both at the local and regional levels' (Voisey's Bay Panel 1997).

The Voisey's Bay case is noteworthy for at least two reasons. First, it is a joint assessment run with surprising accord under the authority of four sometimes-conflicting jurisdictions—it is carried out under both the Newfoundland and Labrador provincial Environmental Assessment Act and the Canadian Environmental Assessment Act (CEAA), and is subject to a memorandum of understanding with the Labrador Inuit Association and the Innu Nation. And second, it introduces a new and more demanding test for environmental assessment in Canada. The public review panel requires the proponents to show not just that they had taken environmental factors into account in their planning and were prepared to mitigate any serious adverse effects, but also that the proposed activities would contribute positively to local and regional sustainability (Voisey's Bay Panel 1997; Gibson 2000).

29 January 1998 To address persistent concerns about duplication and inefficiency, federal and provincial environment ministers sign an accord on environmental assessment harmonization. The accord provides guidance on basic assessment provisions but mostly promotes cooperative use of existing processes. The federal government subsequently signs several bilateral agreements with individual provinces, setting out arrangements for joint assessments where jurisdictions overlap (CCME 1998).

July 1999 The Canadian Environmental Assessment Agency (CEA Agency) releases a revised Cabinet Directive on the Environmental Assessment of Policies, Plans, and Programs, linking these strategic-level assessments to legislated requirements obliging departments to prepare sustainable development strategies. The broadly stated objective of the directive, like that of the Act for project-level assessments, is to provide a means of integrating environmental factors into federal planning and decision making in support of sustainable development. The new directive continues to avoid enforceable obligations and public scrutiny, but does include a conception of environmental assessment that goes beyond some aspects of the project-level process in the Canadian Environmental Assessment Act. In particular, it stresses assessment of alternatives and attention to enhancing positive effects as well as to mitigating negative ones (CEA Agency 1999).

15 October 1999 The public review panel in the Red Hill Creek Expressway case issues environmental impact statement guidelines that adopt and elaborate on the higher-sustainability test introduced earlier by the Voisey's Bay panel. The review panel, convened under the CEAA, sets out the test as a means of promoting sustainable development, which is identified in the law as a fundamental purpose of environmental assessment. The panel interprets progress towards sustainable development as meeting three objectives: (1) the preservation of ecosystem integrity, including the capability of natural systems, local and regional, to maintain their structure and functions and to support biological diversity; (2) respect for the right of future generations to the sustainable use of renewable and non-renewable resources; and (3) the attainment of durable social and economic benefits. The panel requires the proponent to demonstrate how the project met these three goals.

The federal review of the Red Hill Creek Expressway proposal had actually begun in 1998, when it was determined that the project required a federal Fisheries Act permit, and thus an environmental assessment was required. A lower-level assessment began in 1998, but this was moved up in 1999 to a full panel with public hearings. The Region of Hamilton-Wentworth boycotts the process and launches eventually successful legal action against the federal government (see April 2001 below). The main focus of the region's lawsuit is that the case is not legally subject to federal review under the CEAA.

1999–2001 A five-year review of the Canadian Environmental Assessment Act, required in the legislation, is undertaken with extensive consultations. The examination focuses on matters on which many disparate interests might agree. Efficiency, assessment quality, and public participation issues get the most attention. Bigger issues of scope and decision criteria, including the matters raised by the Voisey's Bay and Red Hill cases, are not addressed.

March 2001 Bill C-19 to amend the Canadian Environmental Assessment Act is introduced in Parliament. No major improvements are proposed, but the bill would strengthen public involvement in modest assessments (the vast majority of cases), streamline some decision making, and focus more attention on follow-up monitoring (Gibson 2001).

April 2001 The Federal Court of Canada hands down its decision in the *Red Hill Creek* case. The court decides that the Canadian Environmental Assessment Act 'does not apply to the completion of the Red Hill Creek Expressway' and 'that no review was required under the CEAA.' The court rules that the expressway is exempt from the CEAA under a grandfathering clause in the Act that states: 'Where the construction or operation of a physical work . . . was initiated before 22 June 1984 [the date on which the CEAA came into force], this Act shall not apply in respect of the issuance or renewal of a licence, permit, approval.' The court decides that the term *construction* used in the Act includes 'a whole series of events such as acquiring and clearing land, imposing building restrictions, and securing funding and approvals, all dedicated to, and prerequisites to, the actual physical step of construction.' In essence, the court accepts the region's contention that the project began before 1984. While the decision may be a blow against an impact assessment of the Red Hill Creek Expressway, the sustainability criterion is not discernibly affected.

July 2001 The province of Manitoba introduces a sustainable development code of practice under its Sustainable Development Act (1997). Among other things, it requires public officials to 'strive towards' ensuring that assessments of proposed programs and projects are carried out 'to determine and address their sustainability impacts' (Manitoba 2001).

June–October 2003 The amended Canadian Environmental Assessment Act receives royal assent in June 2003 and is declared in force in October along with associated regulatory adjustments. While the revisions have modest objectives, some substantive changes are included. Several focus on efficiencies, including those to be achieved through more frequent use of class screenings, additions to the exclusions list (of projects not typically subject to EA), elimination of public hearings for projects assigned to the comprehensive study stream, and the introduction of a 'federal environmental assessment coordinator' to improve communication among agencies and levels in assessment deliberations. As well, the revised Act requires more attention to post-approval monitoring, strengthens some compliance provisions, and promises its broader application to federal lands, including Indian reserves, airports, all Crown corporations, and projects with a transboundary dimension (international or provincial). The amendments also seek to enhance public participation, mostly through earlier notification of assessments, consultation throughout the process rather than only at a specific stage, public electronic access to EA information, and more direct recognition of community knowledge and Aboriginal traditional knowledge in EA work (also see Herring, in this volume; Canada 2003; CEA Agency 2003; Benevides 2004).

July 2005 The Joint Review Panel for the Mackenzie Gas Project issues its 'determination on sufficiency', announcing its finding that enough information has been submitted to justify moving on to the hearings phase of its review of the $16.8 billion Mackenzie gas gathering and pipeline project in the Northwest Territories (JRP 2005). Elaborating on earlier terms of reference, the panel advises hearing participants to anticipate the panel's application of a 'contribution to sustainability' test:

> In preparing for public hearings, the Proponent, Interveners and other participants should
> be aware that the Panel will evaluate the specific and overall sustainability effects of the
> proposed project and whether the proposed project will bring lasting net gains and wheth-
> er the trade-offs made to ensure these gains are acceptable in the circumstances.

September–October 2007 Two federal-provincial review panels recommend
against proposed mining projects that they had judged likely to have adverse
long-term effects. In September, the panel reviewing the proposed Kemess North
copper-gold mining project in British Columbia concludes that the project is not
likely to bring lasting gains, especially because of the long-term effects of the mine
wastes on ecological and Aboriginal interests. The project as proposed would
dump 700 million tonnes of acid-generating mine tailings and waste rock into
nearby Duncan (Amazay) Lake. One month later, the panel reviewing a basalt
quarry and marine terminal project proposed for Whites Point, Nova Scotia, also
finds that the undertaking would not make a positive contribution to sustainability. The panel concludes that economic gains during the project's life would accrue
mostly to the proponent and would undermine longer-term qualities and sustainable community economic development opportunities. The recommendations of
both panels are accepted by the relevant federal and provincial authorities.

Characteristics and Implications of the Evolution of EA in Canada

Based on our discussion of the evolution of Canadian EA and the chronology provided above, we would describe the progression of Canada's federal EA as having
seven main characteristics (Gibson 2002b).

1. Canadian environmental assessment policies and laws have evolved slowly over
the past 30 years. While this evolution has been hesitant and uneven, it has, with a
few exceptions, been positive and progressive overall.

2. The chronology notes the introduction of a few of the provincial and Aboriginal claim agreement processes. There are many more, and the processes adopted
in other Canadian jurisdictions—municipal as well as provincial, territorial, and
Aboriginal—vary widely in application, scope, and process (Couch 1988). Even in
the design of territorial and Aboriginal processes, where the federal government
has played a powerful role, there are substantial differences. Some of the variation
is the result of changes in government thinking about environmental assessment
over time. But as is illustrated by the contrast between Ontario's EA law and the
policy-based federal process, both adopted in the mid-1970s, different philosophies, political strategies, and practical circumstances are also involved.

Harmonization efforts, which have now been underway for many years, have
produced little more than agreements to cooperate on assessments involving two
or more jurisdictions. One difficulty has been that the different interests involved
have had conflicting objectives. Proponents of undertakings subject to different

Canadian processes have generally advocated a broad simplification of assessment requirements. Environmentalist organizations have favoured an upward harmonization, which in essence means bringing all federal, provincial, and other assessment requirements to the same high standard of rigorous public process. In contrast, the provinces have seen harmonization largely as a means of minimizing federal involvement in provincial matters. As a result, federal harmonization objectives have centred on the negotiation of agreements with individual provinces for cooperation in specific cases of overlapping process application.

Between 1997 and 2000, the federal government also funded a multi-stakeholder initiative of the Canadian Standards Association to develop a standard (effectively a consistent set of guidelines) for environmental assessment best practice. The consensus standard development process led eventually to a relatively progressive package in the fourteenth draft, but then the provincial representatives withdrew and the initiative was suspended.

3. The non-legislated approaches used for the first 20 years of environmental assessment at the federal level in Canada generally failed to ensure consistent and sincere implementation efforts. Law-based obligations were required, and this marked a transition from discretionary to required application and from guidelines to codified environmental assessment. In this vein, while the courts have not been commonly used either to advance or to dispute the application and practice of EA in Canada, when they have played a role, their decisions have been forceful, not always anticipated, and ultimately influential.

4. Despite the improvements included in the recent amendments to the CEAA, not all of the key elements of effective environmental assessment are yet reliably part of the federal process. For example, the Act does not clearly cover the full range of interrelated environmental, social, economic, cultural, or ecological effects. Nor does the Act impose on the stakeholders consistent requirements to consider the purposes of and alternatives to undertakings. It also leaves strategic-level assessment to a non-legislated process (a process required by Cabinet but not embedded in the legislation) and contains only weak provisions for enforcing compliance with the law and with terms and conditions of approvals (Gibson 2001; Benevides 2004).

5. All important moves along the nine positive trend lines listed above have been resisted by both federal and provincial agencies. A surprising number of the biggest advances have resulted from accidents, litigation, or unanticipated political difficulties that in turn led or obliged the federal government to take steps that it might otherwise have avoided. Politicians have been generally more positive about environmental assessment and more willing to strengthen assessment provisions than the bureaucrats who face meeting the requirements or whose mandates centre on economic development objectives, which they see as being at odds with EA.

This resistance within government is hardly surprising, but it has not been merely, or even chiefly, a matter of the desire to avoid additional work. Rather it is a product of the propensity that bureaucracies have for avoiding actions that involve surrender of, or a constraint on, their power. Even modest EA processes require some relinquishing of authority over the consideration of options and impacts, and have the potential to inject uncertainty into even the most resilient agency status quo. Next to officials in federal departments and agencies with economic mandates, the strongest voices against more ambitious federal environmental assessment requirements have been provincial authorities who patrol the boundaries of provincial jurisdiction; these provincial voices consistently favour autonomous provincial authority to set the assessment rules and do the necessary work (Provincial/Territorial Working Group 2000).

It seems that the key motivation for resistance has been proponent unhappiness with the basic premise of environmental impact assessment. Effective environmental assessment requires decision makers to accept a much broader and largely unfamiliar set of obligations and objectives, to subject themselves to greater public scrutiny, and in the course of this to cede some of their authority to an external process. None of these elements is comfortable for established agencies. Effective assessment requirements should challenge conventional development assumptions and conventional planning practices—and for that reason they are resisted.

6. The evolution towards more comprehensive, demanding, and mandatory assessment has been accompanied by persistent, and evidently growing, expressions of concern about the inefficiencies of EA processes. Throughout the evolution of federal assessment policy and law, efforts to extend or tighten requirements have been accompanied by steps to streamline, harmonize, restrict the scope of, or otherwise smooth the path of deliberation and decision.

Efficiency concerns are often couched in terms of resource limitations, delayed decision making, duplication with other processes, and undue attention to matters that are unimportant or unfamiliar to the agency and perhaps better addressed elsewhere. These concerns have been tied to the common argument that unreasonable environmental demands and, by proxy, inefficient processes will discourage economic development and growth. In this context, efficiency has taken on a simple economic imperative—short term in perspective. Unfortunately, it has not always been easy to distinguish between process requirements that are genuinely wasteful and those that open avenues for broader and more critical reconsideration of proposals that deserve careful attention. The Provincial/Territorial Working Group (2000) lists many cases of allegedly unnecessary federal incursions into provincial jurisdiction, where the federal process introduced considerations that the provincial exercise had not, or would not have, addressed.

7. Environmental assessment at the federal level has generally moved towards being broader, more critical, and more difficult to avoid. The recent changes to the Canadian Environmental Assessment Act support this observation. Environmental assessment has also become more challenging for decision makers, even though many agencies have adapted to EA and incorporated it into their planning processes over the last 30 years.

It seems likely that the evolution of federal environmental assessment will continue in ways that promise greater benefits and influence, but will also bring further challenges. Canada is sinking deeper into unsustainability, as is much of the rest of the world (WWF 2007). While it does not follow that we need to reverse direction on all fronts, we do need to ensure that all of our policies, programs, and projects offer net gains for sustainability. A transformation from environmental assessment to sustainability assessment seems to be one good way of encouraging this. However, while contribution to sustainability is frequently presented as an official objective of environmental assessment, serious application of sustainability-based evaluation criteria is not yet common in environmental assessments and ancillary decision making. In a host of related practical circumstances—urban planning, corporate responsibility reporting, regional growth management initiatives, new versions of progress indicators, and so on—reasonably comprehensive lists of sustainability considerations have been adopted. But carefully designed and intentionally influential sustainability assessments remain rare (Gibson 2002a).

Interpreting and Evaluating the History of Environmental Assessment in Canada

This review of the evolution of and experience with environmental assessment at the federal level in Canada suggests that the story is not just one of progressive change. Viewed from a sustainability perspective, the overall record does involve some improvements, although there have been backward as well as forward steps. But we can also read this history as a story of a still-unresolved succession of struggles over power and influence—over what voices are heard in EA deliberations, over who makes the decisions and which interests might be favoured in the outcomes. Moreover, we can see the whole EA experience as reflecting a fundamental tension between what is practical and what is a genuine need to enhance, support, and sustain social equity and ecological complexity.

Similarly, there are different ways of evaluating the changes that have occurred. Given the global as well as the national environmental realities we face, perhaps the most suitable basic foundation for evaluation is a set of criteria for determining what seems to be needed to reverse the current trend towards ever greater unsustainability and ultimately to construct more durable and responsive socio-ecological relations. At a very rough level, it is not too difficult to identify the key requisites for recognizing, and thus achieving, sustainability or the characteristics of good environmental assessment processes (Gibson 2002a; Gibson et al. 2005).

We can say that the current federal environmental assessment process sits roughly at stage 3 in the evolution of environmental assessment, but there is movement towards federal EA being a contributor to the practical implementation of sustainability policies. Whether and how quickly we might see a general move to stage 4—effectively the transformation of environmental assessment to sustainability assessment in Canada—is a matter for speculation (Gibson 2002b).

A shift to sustainability assessment would be demanding. It would entail, among other things,

- more direct integration of social, economic, biophysical, and other considerations;
- adopting strategic approaches that integrate sustainability into planning and decision-making processes;
- greater attention to systemic complexity and uncertainty; and
- a higher test for approvals (enhancements and net gains for sustainability rather than the simple mitigation of serious adverse effects).

There are good arguments for each of these individually, and the overall shift would focus environmental assessment and overall decision making more directly on the major issues before us. Formal commitment to sustainability assessment is already present in the existing law, and exploratory applications, such as in the Voisey's Bay and Kemess North and Whites Point cases, established that the current provisions can be used as a foundation for assessments focused on sustainability. Sustainability-centred assessment would also fit well with leading international initiatives in EA and the broader sustainability purposes of the Canadian government, as is represented in the mandate of the federal commissioner of the environment and sustainable development and the legislated requirement that all federal departments and agencies have regularly audited sustainable development strategies. However, a transition to sustainability-based assessment involves at least two significant difficulties.

The first problem is that a poorly conceived shift to sustainability-based assessment could reduce the attention given to ecological considerations. Environmental assessment law, indeed environmental law generally, has struggled for three decades to bring serious attention to ecological concerns, and that battle is not yet won. In the business, political, and bureaucratic communities, there remain many proponents of narrowly conceived economic development who would welcome an opening to slide back to a time of environmental disregard, development without the encumbrance of environmental concerns, and growth unfettered by the nuisance of having to consider sustainability. Rigorous sustainability-based assessment is, however, no friend of narrow economic priorities. Moreover, its place is at the core of decision making. While the current forms of environmental assessment do give concentrated public attention to ecological issues, they frequently play a marginal end role, making recommendations to closed, economically driven approval processes.

Proper sustainability-based assessment would forcibly draw public attention to the full suite of social/economic/ecological interdependencies, demand open evaluation of the trade-offs and compromises often proposed in EA processes, and apply a higher test for identifying and measuring durable benefits. If such were achieved, the effective influence of ecological considerations would be considerably enhanced. But it all depends on ensuring that sustainability assessment is designed and situated to be powerful—that is to say, actually implemented.

The second challenge is efficiency. The demands of sustainability assessment bring the practical challenges of keeping processes manageable and practicable. They disturb entrenched agency interests as well as facilitate crucial innovations.

It does not necessarily follow that more time and resources are required—better integration often permits greater efficiency in planning and decision making, quicker implementation, and the sharing of limited resources. But achieving integration, introducing new criteria and analytical methods, and fostering new habits of mind and practice are rarely easy. Governments will be inclined to resist more demanding expectations. Advocates of sustainability assessment will be tempted to seek more time and resources, while its enemies will be pleased to complain of waste and delay.[2] Indeed, the efficiency mantra can become a way of reducing the effectiveness of EIA, by limiting the scope and scale of attention through imposed timelines and limiting the resources needed to fully evaluate EIA applications. The route around this peril requires careful and thoughtful design, ensuring process efficiency as well as effectiveness in core decision making.

This is all consistent with the experience of environmental assessment in Canada. Like any innovation in policy and law that demands new thinking and new practice, EA suggests trouble to entrenched interests and conservative agencies. To address or silence its critics, EA must be powerful, efficient, and effective. Without both elements, environmental assessment has been weak in the past. Without them in the future, sustainability assessment will be hesitant and thus unsuccessful. The recent changes made to the Canadian Environmental Assessment Act are modest and by some measures disappointing. But they do reflect a basic recognition of the dual need to enhance efficiency and strengthen application.

Over the last three decades, the Canadian experience with process reform suggests that the continued transformation and advance of environmental assessment will be gradual, sometimes hesitant, and certainly resisted by some. Progress will also be achieved as much through chance as through pressure from those who support strong environmental policies. But we can say that Canada has been following a discernible path and Canadians are perhaps now in a better position to see what further efforts are needed and how successful those underway and recently initiated are likely to be.

Notes

1. The key rulings were in *Canadian Wildlife Federation Inc. v. Canada Minister of the Environment*, 99 N.R. 72, 27 F.T.R. 159 nt. [1990] 2 W.W.R. 69 (Fed.C.A.); and *Friends of the Oldman River v. Canada Minister of Transport*, [1992] S.C.R. 3, 132 N.R. 321, [1992] 2 W.W.R. 193 (S.C.C.).

2. It is probably not entirely a coincidence that while the Kemess North and Whites Point panels were completing their sustainability-focused reviews, the federal government was preparing to launch a Major Projects Management Office to streamline regulatory approvals of resource industry projects (Canada 2007).

References

Beanlands, G.E., and P.N. Duinker. 1983. *An ecological framework for environmental impact assessment in Canada.* Halifax: Institute for Resource and Environmental Studies, Dalhousie University.

Benevides, H. 2004. Real reform deferred: Analysis of recent amendments to the Canadian Environmental Assessment Act. *Journal of Environmental Law and Practice* 13:2.

Berger, T.R. 1977. *Northern frontier, northern homeland: The report of the Mackenzie Valley Pipeline Inquiry.* Vol. 1. Ottawa: Supply and Services Canada.

Canada, Government of. 1984. *Environmental Assessment and Review Process Guidelines Order* (1984), S.O.R./84–467, 22 June 1984.

———. 1992. *Canadian Environmental Assessment Act,* S.C. 1992, c. 37.

———. 2003. *Canadian Environmental Assessment Act,* S.C. 2003, c. 9. http://www.ceaa.gc.ca/013/index_e.htm.

———. 2007. Canada's new government launches Major Projects Management Office (press release). 1 October. http://www.mpmo-bggp.gc.ca/directive-e.php.

CCME (Canadian Council of Ministers of the Environment). 1998. *Canada-wide accord on environmental harmonization, environmental assessment sub-agreement.* 29 January 1998. http://www.ccme.ca/initiatives/environment.html.

CEA Agency (Canadian Environmental Assessment Agency). 1999. *The 1999 Cabinet Directive on the Environmental Assessment of Policy, Plan and Program Proposals.* Ottawa/Hull: CEA Agency. http://www.ceaa.gc.ca/016/index_e.htm.

———. 2003. *Strengthening environmental assessment in Canada: Implementing changes to the Canadian Environmental Assessment Act* (information package). Gatineau: CEA Agency, October. http://www.ceaa.gc.ca/013/whatsnew_e.htm.

Couch, W.J. 1988. *Environmental assessment in Canada: 1988 summary of current practices.* Ottawa/Hull: Canadian Council of Resource and Environment Ministers.

Emond, D.P. 1978. *Environmental assessment law.* Toronto: Emond-Montgomery.

FEARO (Federal Environmental Assessment Review Office). N.d. *Guide for environmental screening.* Ottawa/Hull: FEARO.

———. 1977. *A guide to the federal environmental assessment and review process.* Ottawa/Hull: FEARO.

———. 1987. *Reforming federal environmental assessment: A discussion paper.* Ottawa/Hull: FEARO.

———. 1993. *The environmental assessment process for policy and program proposals.* Ottawa/Hull: FEARO.

Gibson, R.B. 2000. Favouring the higher test: Contribution to sustainability as the central criterion for reviews and decisions under the Canadian Environmental Assessment Act. *Journal of Environmental Law and Practice* 10 (1): 39–55.

———. 2001. The major deficiencies remain: A review of the provisions and limitations of Bill C-19, an Act to amend the Canadian Environmental Assessment Act. *Journal of Environmental Law and Practice* 11:83–103.

———. 2002a. *Specification of sustainability-based environmental assessment decision criteria and implications for determining 'significance' in environmental assessment.* Monograph prepared under a contribution agreement with the Canadian Environmental Assessment Agency Research and Development Programme. http://www.ceaa-acee.gc.ca/0010/0001/0002/index_e.htm.

———. 2002b. From Wreck Cove to Voisey's Bay: The evolution of federal environmental assessment in Canada. *Impact Assessment and Project Appraisal* 20 (2): 151–9.

Gibson, R.B., with S. Hassan, S. Holtz, J. Tansey, and G. Whitelaw. 2005. *Sustainability assessment: Criteria and processes.* London: Earthscan.

JRP (Joint Review Panel) for the Mackenzie Gas Project. 2005. Joint Review Panel determination on sufficiency, Inuvik, 18 July.

Lavalin Econosult. 1982. Program evaluation study (EARP) final report: Environment Canada. July [unpublished].

Manitoba, Government of. 2001. *Manitoba's provincial sustainable development code of practice.* Winnipeg: Government of Manitoba.

Provincial/Territorial Working Group. 2000. *Canadian Environmental Assessment Act five year review: Provincial and territorial input.* April. http://www.eao.gov.bc.ca/ceaa/background.htm.

Quebec, Government of, Sociètè d'ènergie de la Baie James, Sociètè de dèveloppement de la Baie James, Commission hydroèlectrique de Quèbec, Grand Council of the Crees of Quebec, James Bay Crees, Northern Quebec Inuit Association, Inuit of Quebec, Inuit of Port Burwell, and Government of Canada. 1998. *James Bay and Northern Quebec Agreement and Complementary Agreements.* Sainte-Foy: Les Publications du Quèbec. http://www.ainc-inac.gc.ca/pr/agr/que/jbnq_e.html.

Voisey's Bay Mine and Mill Environmental Assessment Panel. 1997. *Environmental Impact Statement (EIS): Guidelines for the review of the Voisey's Bay Mine and Mill undertaking.* 20 June.

WWF (World Wildlife Fund). 2007. *Living planet report 2006.* Geneva: WWF International. http://www.panda.org/news_facts/publications/living_planet_report/index.cfm.

Chapter 3

The Science of Assessment: Identifying and Predicting Environmental Impacts

Douglas Baker and Eric Rapaport

Introduction

The historical challenge of environmental impact assessment (EIA) has been to predict project-based impacts accurately. Both EIA legislation and EIA practice have evolved over the last three decades in Canada, and the development of the discipline and science of environmental assessment (EA) has improved how we apply EA to complex projects. The practice of environmental assessment integrates the social and natural sciences and relies on an eclectic knowledge base from a wide range of sources. EIA methods and tools provide a means to structure and integrate knowledge in order to evaluate and predict environmental impacts.

This chapter provides an overview of how impacts are identified and predicted. How do we determine what aspect of the natural and social environment will be affected when a mine is excavated? How does the practitioner determine the range of potential impacts, assess whether they are significant, and predict the consequences? There are no standard answers to these questions, but there are established methods that provide a foundation for scoping and predicting the potential impacts of a project.

Of course, the community and the general public play important roles in this process, and these are discussed by others in this book. In the first part of this chapter, we deal with impact identification, which involves applying scoping to critical issues and determining impact significance, baseline ecosystem evaluation techniques, and how to communicate information about environmental impacts. In the second part of the chapter, we discuss the prediction of impacts in relation to the complexity of the environment, ecological risk assessment, and modelling.

Impact Identification

The initial challenge of a proposed project—identifying the potential environmental impacts—is often the most difficult task for the EIA practitioner. It is a question of scope and detail. Scoping deals with the breadth of the inquiry; detail has to do with the level of data collection and analysis. And this is all tempered by the project's budget. Impact identification consists of a series of steps that eventually lead to an understanding of the environmental consequences of a proposed action. The generic steps, as noted above, consist of scoping, determining impact significance, determining baseline ecosystem evaluation, and communicating/identifying potential environmental impacts. Each of these steps will be briefly examined.

Scoping

Scoping is generically defined as a process intended to 'identify the attributes of the environment for which there is concern (public and scientific) and a plan is provided that enables the EIA to be focused on these attributes' (Kennedy and Ross 1992, 476). In other words, the scope of the EIA determines the most important issues to be addressed and the general terms of reference. The initial goal of scoping is to be comprehensive. The question of how scoping should be done and who should be included in the scoping process provides the context for scoping: will it be narrow and restricted? or will it be a more open and inclusive process? (See Mulvhill and Baker [2001] for a more complete discussion of scoping in northern Canada.) In the final report on an international study of the effectiveness of EIA (Sadler 1996), scoping was identified as one of the four priorities that could be strengthened in practice.

Once the broad range of concerns is identified, the challenge is to reduce the issues that are not of significance to the environmental impact assessment. Often scoping has been used as a baseline study where as much data as possible are collected for the project. However, this can be a costly and time-consuming activity that might not produce the results needed for impact assessment identification. As Barrow (1997, 108) suggests, 'What scoping must do is focus on identifying significant impacts—something of a "catch 22" situation, in that the scoping has to decide what the impacts may be even though they may not be identifiable without a full impact assessment.' Thus, skilful scoping identifies the pertinent issues relevant to the environmental assessment. The tasks or activities relevant to scoping consist of the following:

- *Determining the regulatory requirements.* What are the pertinent guidelines and laws that govern an environmental assessment?
- *Developing a list of relevant stakeholders to help define the scope of the environmental assessment and the issues that people care about.* This can be accomplished through a variety of methods and entails holding open houses or inviting the participation of technical advisory committees, community working groups, and so on. A communication plan should be formulated to integrate the public input into how impacts will be addressed and mitigation formulated.
- *Assembling available scientific data to determine information gaps and potential research needs.*

- *Evaluating the concerns to determine the key issues of significance.* This requires eliminating issues, short-listing biophysical and socio-economic factors, and focusing on perceived significant impacts. However, scoping should be an ongoing exercise throughout the project so that potential issues or impacts can be kept from 'falling through the cracks'.

Developing the list of significant impacts is an important part of the scoping exercise (Beanlands 1988). Wood (1995) notes: 'In the last analysis these decisions often have to be made by individuals with the appropriate levels of knowledge and expertise who are able to say from past experience: what significant effects are likely to arise; how are they likely to impact on the environment; and what steps might be taken to deal with them' (131–2). Defining the boundaries of the assessment and identifying significant impacts are the primary functions of scoping.

Scoping is an exercise designed to reduce voluminous environmental impact statements that may be irrelevant to the pertinent issues and are costly to produce. Scoping helps to define the context of the EIA, identify the issues relevant to the project, and help determine selected alternatives. Moreover, scoping throughout the life of a project has a number of advantages, including the following:

- It improves the quality of information and the range of issues identified.
- It saves time and money and reduces duplication of efforts.
- It reduces potential conflict, improves coordination, and encourages dialogue early in the EA process.
- It encourages public and stakeholder input at the early stage of project assessment and allows for different values to be integrated into the project design.

Evaluation of Impact Significance

Once the range of potential impacts is determined by scoping, the next critical step is to determine the relative significance of the impacts. This is an important step because it further focuses the environmental assessment on the more important impacts of the project. Sadler (1996, 118) suggests that 'evaluating the significance of environmental effects is perhaps the most critical component of impact analysis.' It should be noted that the concept of *significance* is not restricted to this stage in environmental assessment, but continues throughout the process. Lawrence (2003) provides an assessment of significance in the EA process and notes that 'significance includes interpretation principles, thresholds, criteria and methods that should be applied in each EA activity' (12). Because evaluating significance is about making judgments about the importance of impacts, it should be a continuous process throughout the EA.

In the attempt to determine significance, impacts may be initially categorized as (1) beneficial or detrimental; (2) irreversible or reversible; (3) reparable by management practices or irreparable; (4) having a specific duration and frequency (short or long term); (5) having a certain magnitude; (6) project based or operational; (7) having a specific geographical extent (local, regional, national, or global); (8) accidental or planned; (9) primary, secondary, or tertiary; (10) temporary or continuous; and (11) single or cumulative (after Canter 1996, 21). Of course, at the operational level the challenge is to identify the range of impacts and flag the adverse environmental effects. The Canadian Environmental Assessment (CEA) Agency's 'Reference Guide' states that the 'concept of significance cannot be separated from the concepts of "adverse" and "likely"'. Accordingly, a three-step framework sets the relationship:

Step 1: Deciding whether the environmental effects are adverse. Table 3.1 provides an overview of some of the major factors to consider when evaluating adverse effects.

Step 2: Deciding whether the adverse environmental effects are significant. The criteria that help define significance include magnitude, geographical extent, duration and frequency, and degree of reversibility or irreversibility. In addition, the ecological context is recognized as significant, especially where it has already been adversely affected and where there is little resilience in the face of disturbance.

Step 3: Deciding whether the significant adverse environmental effects are likely. This is determined by two criteria: (a) probability of occurrence (risk); and (b) scientific uncertainty.

Significance can be determined by a variety of criteria. First, institutional regulations and guidelines can provide levels of significance. Legislation for environmental assessment may (or may not) provide guidelines for significance. Within Canada, the CEA Agency offers a significance test consisting of criteria that determine if effects are adverse, such as on biota health, on rare and endangered species, on species diversity, on habitat, or on human health. However, Sadler (1996) observes that in the Canadian system,

> public concern and social values are not factors in determining the significance of adverse impacts. The determination must be 'objective, based on scientific and credible technical and other relevant information' and 'reasonable so as to withstand court challenge'. By comparison, under NEPA, public opinion and the level of controversy associated with a proposed action help to identify and determine significance. (118)

Thus, jurisdictions provide differing guidelines for what is considered significant. At the regulatory level, this is also true. Regulations provide thresholds for standards, such as acceptable levels of suspended solids, and for other environmental indices used in describing the affected environment. However, it is important to note that guidelines and physical standards have only been prepared for a small and limited number of chemicals and hazardous materials and often do not allow for synergies when chemicals are combined.

Table 3.1   Factors in determining adverse environmental effects

Changes in the Environment	Effects on People Resulting from Environmental Changes
Negative effects on the health of biota, including plants, animals, and fish	Negative effects on human health, well-being, or quality of life
Threat to rare or endangered species	Increase in unemployment or shrinkage in the economy
Reductions in species diversity or disruption of food webs	Reduction of the quality or quantity of recreational opportunities or amenities
Loss of or damage to habitats, including habitat fragmentation	Detrimental change in the current use of lands and resources for traditional purposes by Aboriginal persons
Discharges or release of persistent and/or toxic chemicals, microbiological agents, nutrients (e.g., nitrogen, phosphorus), radiation, or thermal energy (e.g., cooling waste water)	Negative effects on historical, archaeo-logical, paleontological, or architectural resources
Population declines, particularly in top visual amenities (e.g., views)	Decreased aesthetic appeal or changes in predator, large, or long-lived species
Loss of or damage to commercial species	
The removal of resource materials (e.g., peat, coal) from the environment	Foreclosure of future resource use or production
Transformation of natural landscapes	
Obstruction of migration or passage of wildlife	
Negative effects on the quality and/or quantity of the biophysical environment (e.g., surface water, groundwater, soil, land, and air)	

Source: CEA Agency 1994, 6.

Second, significance can be identified using other structured methods, such as 'threshold of concern'. A threshold is usually applied to determine the difference between significant and insignificant effects, and it often acts as a trigger for the EA process. Different thresholds can be applied, depending on the issues being ad-dressed and the context of the EA. Haug et al. (1984) set out criteria that determine the priority of concerns and the probability that an impact will cross the threshold of concern. The threshold is determined by a maximum or minimum value which, if exceeded, causes that impact to take on a new value. The following brief sum-mary is a list of Haug et al.'s priorities:

1. Legal thresholds (highest priority): These are determined by legal or regulatory limits
2. Functional thresholds (very high priority): These are established for resource use or unavoidable adverse impacts that may disrupt the functioning of an ecosystem or destroy significant resources
3. Normative thresholds (high priority): These thresholds are established by social norms at the community or regional level that frequently deal with environmental, economic, or social concerns
4. Controversial thresholds (moderate priority): These can be impacts that have a high profile because they are a source of conflict between stakeholder groups but would not otherwise have a priority
5. Thresholds that are preferences of individuals, groups, or organizations but do not warrant higher priorities for other reasons (low priority)

Determining the significance of an impact presents a variety of challenges at both the scientific and social levels. Evaluation of thresholds based strictly on scientific data is inadequate in many cases because resources and ecosystems are linked with human values and cultural meaning. For example, First Nations people in Canada argue that there needs to be Aboriginal-based criteria for determining the significance of environmental effects (see the report prepared by Winds and Voices Environmental Services, Inc. [2003], *Determining Significance of Environmental Effects: An Aboriginal Perspective*). The business of determining an impact's significance entails both understanding the social and ecological context of the impact and proposing the appropriate mitigation measures. Quantified measures of significance have to be tempered with explanations and summaries if one is to adequately describe the context of significance and explain the significance to a general audience (Wood 1995).

Baseline Ecosystem Evaluation

Good science should inform decision makers of the potential impacts, mitigation options, and alternatives of a project. A variety of frameworks have been proposed to establish a scientific approach to defining how baseline biophysical data should be integrated into impact assessment planning. Whitney and Maclaren (1985) propose a comprehensive framework that integrates biophysical and social components into an EIA. They set up an ideal model and develop the steps and interconnections between processes for an integrated approach to project assessment.

Beanlands and Duinker (1983) propose a normative approach based on ecosystem science. They lay out specific criteria for developing an ecological perspective that include the following:

*   Derive lessons from experience. Take advantage of studies that have been done before, and use available background data (don't reinvent the wheel!).
*   Design a conceptual framework that organizes the study using a hierarchical or trophic structure of community organization. Linkages need to be

established between the project and the structural and functional components in an ecosystem.

- Scope down to the critical ecological components of the study. Define the valued ecosystem components using boundaries and interactions that can be measured. Beanlands and Duinker (1983) suggest that '[e]cological scoping can be used to determine which interaction routes offer the best opportunities for studies leading to a prediction or approximation of the changes in the valued ecosystem components' (72).
- Develop a study strategy that allows an empirical approach to deal with the prediction of impacts. Examples may include studies based on succession, bioaccumulation, or eutrophication. The study design needs to be guided by specific questions or hypotheses that define the baseline studies and set the context for the environmental assessment.

Baseline studies are often conducted for specific projects and are set up to address the perceived significant impacts of those projects. In most cases, they are isolated from other environmental planning processes and are designed for a single point in time and for only one major project decision (Beanlands 1988).

Communicating the Facts about Environmental Impacts: EIA Methods

Most textbooks on EIA have a section on the comprehensive EIA methods that are used to educate others about the potential environmental impacts of project development (see Barrow 1997; Canter 1996; Wood 1995). The methods that are frequently discussed consist of checklists, matrices, overlay mapping, networks, and simulation modelling. Each of these methods will be briefly overviewed with applied examples from Canadian environmental impact statements.

EIA methods are tools that are used to communicate complex environmental interactions and impacts. Their primary function is to provide a synthesis of information, and they are frequently used to communicate and evaluate alternatives. In order to apply these tools, one must first have in place baseline data that describe the existing environment, provide a means to describe the changes that may occur in the existing environment as a result of the project, and predict the likelihood of the proposed changes. Once this is established, an EIA method can perform four tasks:

1. Impact identification
2. Relative impact measurement
3. Impact interpretation
4. Impact communication

Different methods are used to perform different aspects of these tasks, depending on the strengths and weaknesses of the method and where it is being applied. Canter (1996, 58) suggests that the EIA method be selected according to the following criteria: (1) it should be appropriate to the necessary task; (2) it should be sufficiently free from assessor bias; and (3) it should be economical in terms of costs, investigation time, personnel, and equipment/facilities.

Checklists

The checklist is one of the earliest methods to be applied in EIS, and it is generally the simplest application. Checklists are most often used to indicate broad areas of concern and likely impacts; they are good for scoping and structuring the initial stages of an assessment. Most people are familiar with checklists (your grocery list), and this makes the checklist an effective way to relate relative impacts and include both quantitative and qualitative evaluations. Three varieties of checklists are used in EIA:

1. **Simple and descriptive checklists.** These are simple lists that make no attempt to evaluate qualitative or quantitative data.
2. **Weight-scaled checklists.** Criteria for evaluation are incorporated into the listing, usually in the form of ranking or ratings. The weightings indicate the relative significance of each impact and may indicate critical values such as 'threshold of concern' for each factor.
3. **Multi-attribute utility checklists.** These are an extension of weight-scaled checklists, for with them a measure of each impact is distilled into a single utility function. This form of distillation of data is not often used because of the complications of distilling impacts into single utility functions and the inaccuracies associated with amalgamating impacts into single functions.

Table 3.2 provides an example of a weight-scaled checklist that is often used in an EIS. The checklist is a summary of the socio-economic impacts of the Keenley-side power plant in Castlegar, British Columbia. The checklist uses ordinal scaling (relative magnitude) to rank impacts of the power plant. The ranking is done on a series of simple scales (small, moderate, large; uncertain, likely, certain; none, low, high) to provide a broad overview of the impacts.

Matrices

The matrix is a grid that is designed to link environmental factors with project activities. Generally, one set of factors (project related) is located on the horizontal axis and a second set of factors (affected environment) is positioned on the vertical axis. The intersecting cell of each axis defines the interaction, using symbols, numeric scores, or algebraic functions.

The most common example of the matrix is the Leopold Matrix, which defines impact significance and magnitude within the same cell (splitting it in half) by using a 10-point scale to rank each component. The Leopold Matrix is a standardized format consisting of (1) a horizontal axis with 10 project categories, further divided into 100 possible project actions; and (2) a vertical axis with 88 environmental factors. The intersection of the axes provides for an interaction of 8,800 cells to define potential impacts of a project (with each cell being divided into significance and magnitude).

Table 3.2 Weight-scaled checklist: Summary of socio-economic impacts of the Keenleyside power plant

	Magnitude	Study Area Dispersion	Occurrence Probability	Manage-ability	Potential for Disruption
Employment					
Direct employment	large	universal	certain	n/a	low
Indirect employment	small	universal	certain	n/a	none
Induced employment	moderate	universal	certain	n/a	none
Population					
Primary area	small	scattered	certain	easy	low
Extended area	small	scattered	certain	easy	low
Income					
Direct	large	universal	certain	n/a	none
Indirect	small	universal	certain	n/a	none
Induced	moderate	universal	certain	n/a	none
Housing Demand					
	moderate	scattered	certain	easy	low
Community Infrastructure					
Water supply – most areas	small	n/a	certain	easy	none
Water supply – some areas	small	focused	uncertain	difficult	medium
Sewage treatment	small	n/a	certain	easy	none
Garbage disposal	small	n/a	certain	easy	none
Roads and traffic	large	focused	likely	intermediate	high
Airport operation	small	n/a	certain	easy	none
Airport parking	moderate	n/a	certain	intermediate	low

Continued

Table 3.2 Continued

	Magnitude	Study Area Dispersion	Occurrence Probability	Manage-ability	Potential for Disruption
Community Services					
Education – public school	small	n/a	likely	easy	low
Education – post-secondary	small	n/a	likely	easy	none
Hospitals	small	n/a	likely	easy	low
Public health	small	scattered	likely	easy	low
Ambulance service	small	universal	likely	easy	none
Recreation – municipal	moderate	focused	likely	intermediate	high
Recreation – prov. parks	moderate	focused	likely	difficult	high
Recreation – forest sites	moderate	focused	likely	intermediate	high
Policy demand	small	universal	likely	easy	low or high
Fire protection demand	small	universal	likely	easy	low
Court services	small	n/a	likely	easy	none
Day care demand	moderate	n/a	likely	intermediate	medium
Social services	small	n/a	likely	intermediate	low
Transit demand	small	n/a	uncertain	easy	low
Transit opportunity	large	focused	uncertain	intermediate	low
Regional Transportation	small	scattered	likely	n/a	low
Regional Tourism	small	focused	likely	n/a	low
Native Interests	small	focused	uncertain	n/a	low
Fish and Wildlife	small	focused	likely	intermediate	low

Source: Canadian Resourcecon Ltd.

The primary advantage of using the matrix as a method to evaluate impacts is that it can provide a simple communication of the interaction between the project activities and the natural environment. Matrices are adaptable to different projects and formats, and can be accompanied by textual information. In this manner, a matrix can provide a good abstract for a lengthy text. The challenge is to keep the interactions relatively simple; for example, the Leopold Matrix tends to be very complex and difficult to understand. A well-designed matrix provides a good method for initially plotting the impacts of the components of a project with respect to the environment.

Table 3.3 illustrates an application of the matrix for the White Rose Oilfield. The matrix summarizes the effects of the impact of offshore oilfield development on local fisheries. Project activity is listed according to different activities with respect to impacts. Evaluation criteria for assessing impacts with respect to, say, magnitude, geographic extent, duration, reversibility, and the socio-economic context are ranked according to scales designed for each criterion (the key is located at the bottom of the table).

Networks

Networks have been developed to link primary impacts with secondary and tertiary impacts. The objective of a network diagram is to investigate higher-order linkages through directional diagrams, using either stepped matrices or systems diagrams. A common application of networks is the Sorensen Network, which attempts to identify cause and effect relationships.

Systems diagrams are frequently used to track impacts in specified components of ecosystems, using trophic levels to trace the impacts of a perturbation in the system. Systems diagrams provide a good summary of ecosystem interactions and attempt to get beyond simple primary impacts. Figure 3.1 provides an example of a systems diagram.

Simulation Modelling

Modelling is a logical extension of networks and attempts to model interactions and varied orders of impacts. Simulation models share three basic characteristics: (1) they are simplified representations of the systems under investigation; (2) they make explicit assumptions regarding the behaviour of the system; and (3) they are open to misinterpretation if used out of context. Simulation modelling is frequently used in environmental assessment, and there are numerous types of models that simulate natural processes, such as air dispersion, groundwater flow, oil slick movement, and erosion. The section below on the nature of the environment provides a more comprehensive discussion of modelling in the context of impact prediction.

The advantages of modelling include the following (after Canter 1996):

- The user is forced to clarify assumptions and causal mechanisms.
- Any form of relationship can be handled—linear or nonlinear.
- Modelling can include uncertainties of various types.

Table 3.3 An application of the matrix for the White Rose Oilfield

Project Activity	Positive (P) or Adverse (A) Environmental Effect	Mitigation	Evaluation Criteria for Assessing Environmental Effects				
			Magnitude	Geo-graphic Extent	Duration Frequency Extent	Revers-ibility	Socio-Economic Context
Malfunctions/Accidents/Unplanned Events							
Major oil spills*	A	Prevention; containment; monitoring; recovery; compensation	2–3	4–6	2–6/1	R	1
Past/Present/Future Projects (seismic testing, exploration drilling, marine transportation, Hibernia, Terra Nova, White Rose)							
Loss of access to fishing grounds	A	Discussion with fishing industry; common traffic routes	1	3	5/6	R	2
Damage to fishing vessels or gear	A	No-fishing zone; notification to mariners; reduction or elimination of debris; compensation	1	5	5/1	R	1
Biophysical im-pacts on fisheries	A		1	3	5/6	R	1
Information, com-munication and emergency response	P	n/a	1	6	5/6	R	1

Key:

Magnitude	Geographic Extent	Frequency	Ecological/Socio-cultural and economic context
1 = low	1 = < 1 km2	1 = < 11 events/year	1 = Relatively pristine area
2 = medium	2 = 1–10 km2	2 = 11–50 events/year	not affected by human activity
3 = high	3 = 11–100 km2	3 = 51–100 events/year	2 = Evidence of adverse effects
	4 = 101–1000 km2	4 = 101–200 events/year	n/a = not applicable
	5 = 1001–10,000 km2	5 = > 200 events/year	
	6 = > 10,000 km2	6 = continuous	
	Duration	Reversibility	
	1 = < 1 month	R = Reversible	
	2 = 1–12 months	I = Irreversible	
	3 = 13–36 months		
	4 = 37–72 months		
	5 = > 72 months		

* Note: Effects of major oil spills on fishing gear and on loss of access to fishing grounds can be remedied relatively quickly (often within 2 years of a spill). However, loss of market value for fisheries species depends on media coverage and public perception of fish limit. Because of this, impacts of major spills can extend over a larger area than the immediate geographic area affected by the spill and can extend long after the oil spill has been removed and/or has dissipated.

Source: *White Rose Oilfield Comprehensive Study Report* (Husky Oil Ltd 2000).

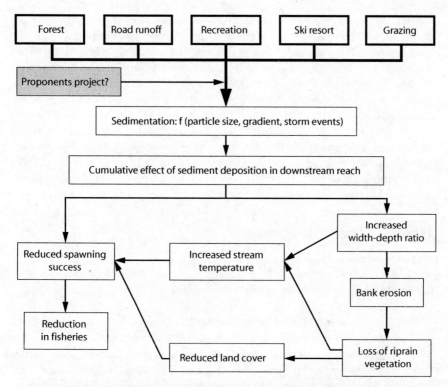

Figure 3.1 Systems diagram demonstrating the sources of sedimentation. The lines from the sources represent a weighting based on percentage of the total particles added to the stream.

- It can compare alternatives.
- It can use detailed information concerning ecosystem processes.
- It is good for communicating impacts.

Of course, the downside of using models is that they can be expensive to generate and may require a considerable amount of expertise to operate. In addition, the results can be very complex and difficult to interpret or communicate. The use of modelling is increasing in environmental assessment with the improved access to computers and more sophisticated software development to simulate natural systems.

Prediction

Prediction or forecasting aims to project the future impacts of a project in order to determine their significance, assess potential mitigation measures, and select alternatives within/to the project. The prediction of future changes in environmental quality requires assessment of two integrated components: the state of the natural system and the state of the built environment. Thus, the practitioner must recognize that the built environment is separate from, but interactive with, the physical/natural environment. For example, within an ecosystem, concentrations of nitrogen levels in the atmosphere are affected by amounts released from vegetation, soils, and ocean processes. This is known as the background concentration. In addition to this, however, are anthropocentric sources of nitrogen through fertilizers and fossil fuels. An ideal impact prediction must be able to model the current natural and built environments, and account for the dynamic relationships within and between each sector. Once this has been accomplished, the task for a good prediction is to determine how the proposed project will change the existing environment directly and indirectly. The project can affect synergies within both the natural and the built environments, making impact prediction complex.

Nature of the Environment

The environment is a complex system that differs significantly from such well-defined systems as infrastructure and machinery. Five principles need to be taken into consideration for impact prediction:

1. The environment is three-dimensional; air, water, land, and the subsurface are all a part of an integrated, complex system. While environmental impact assessments in Canada and in many other countries acknowledge the complexity of the environment, assessments tend to treat components of the environment as discrete units and keep relative impact predictions separate. Thus, while the individual impacts of each component may be low, there often fails to be a general assessment of the network as a whole to model more complex relationships and impacts.

2. All the components of the environment interact directly and indirectly, leading to the process of synergistic and emergent outcomes. Environmental processes are dynamic. Impact prediction is limited by information on how controlling factors in the environment change, and as a result, models often only predict short-term interactions.

3. The complexity of the environment is such that it is not generally possible to model it with the precision or accuracy common to science or engineering practices. Although the environment is complex, it is nevertheless possible to model and make predictions about outcomes. However, there can be both great variability and standard deviations about the mean. A good EIA should make use of known variables in prediction. This process can be validated through monitoring, which would entail obtaining information about the relevant components of the environment, the distribution of impact over time and space, and the occurrence and magnitude of the impact (Dipper et al. 1998; Glasson et al. 1999).

4. The anthropogenic environment has produced multiple sources of pollutants that have altered or impacted nearly all parts of the globe. Thus, the pollutants are spread through multiple pathways. How a project affects and

interacts with other parts of the built environment should not be seen as fixed, but rather in terms of dynamic relationships. For example, land-use transportation modelling is a defined field that has shown that changes in either the cost of driving or the routes selected by drivers can lead to dynamic effects on land-use demands and air quality (Johansson and Mattsson 1995; Rapaport 2002).

5.   In general, it is possible to control, measure, and predict the outcomes of what has been engineered by humans. From an EIA standpoint, this means that there is a high degree of credibility for verification, calibration, and validation. In other words, prediction in environmental assessment can alter the anthropogenic environment in different ways to affect project size, shape, technology used, location, and other elements in order to mitigate impacts.

For the most part, EIA is concerned with short-term changes that treat environmental components as static; hence it may convey the simplistic idea that impacts have little variation. This is limiting when one considers the high variability of many elements of the environment, including temperature, rainfall, wind direction, and groundwater level. A common assumption, which is not always valid, is that biological responses should be modelled using average conditions, whereas field data are more often determined by extreme events (Hirst and Storvik 2003; McCartney et al. 2003). Predictions in a complex environment can only be validated through monitoring (Burt 1994; Summers and Tonnessen 1998). Monitoring provides impact assessment with a 'reality check' that can lead to improvements in the model and provide plausible estimates of change in and recovery of the environment (Parr et al. 2003).

As previously discussed, modelling offers one method to evaluate the complex relationships within the environment. There has been considerable advancement in environmental modelling in the last decade, with journals focusing upon different aspects of this single topic (such as environmental modelling software). Environmental quantitative models allow for numerical outputs and have the advantage of being able to manipulate and change variables. For example, input flows of pollutants from a proposed project can be modelled in relation to variations in environmental conditions, such as their relation to average rainfall.

Generally, there are two types of models: deterministic and stochastic. A *deterministic model* has no random components, whereas a *stochastic model* includes at least one random factor. Stochastic models attempt to reproduce realistic variations in model parameters or variables, and tend to yield different results each time they are run. This provides a variety of outputs that can help determine extreme events and estimate the probability of event occurrence. In areas with extreme climate variation and low numbers of observed data, it is possible to use simulation models to generate fairly accurate data sets. Examples of applied stochastic models include Skiles and Richardson's (1998) application to weather generation in Alaska, Anisimov et al.'s freeze-thaw depth prediction (2002), and Downing and Reed's (1996) stochastic-driven, object-oriented model to predict oil spill impacts on animal migration. All these examples require constant updating with sampling in order to validate the models. It is thus clear that ecological prediction requires constant reliability testing throughout the project and that EIA needs to address a learning and adaptive component (Jones and Greig 1985). The lesson that stochastic models provide as a prediction tool is that a number of possible forecasting outcomes are possible and there is a risk that impact significance and magnitude could vary.

Impact and Risk Assessment

The relationship between EIA and risk assessment (RA) has been explored in the literature linking the potential for integrating methods and co-learning. For example, Andrews (1988) provides an overview of how those who practise EIA and RA can learn from each other. He notes that traditional risk assessment uses quantifiable measurements for hazards and emphasizes probability to evaluate risk. EIA is often criticized for using rudimentary predictions whose probability—without a rigid measurement system—would thus be described as likely or unlikely. RA is considered more valid because of the repeatability of experimentation and its reliance on quantitative data. Risk management offers an ongoing systematic framework to identify, assess, estimate, control, prevent, reduce, and communicate risks.

The field of risk management has evolved into three distinct areas: human health risk assessment, ecological risk assessment, and the less-known building structures–technology risk assessment. Examples of human health risk assessment include human exposure to contaminants or daily work-related risks. In a related area in EA, the environmental assessment panel of the Terra Nova project was pleased with the proponent's plan to implement a 40-hour work week, since excess overtime can lead to an increased risk of accidents causing injury or death (CEA Agency 1997). The Great Whale environmental assessment panel assessed the human health risk associated with electromagnetic fields through exposure to transmission lines by using analogue models and published evidence (Levallois and Gauvin 1994). With respect to the application of building structures–technology risk assessment in EA, a risk assessment for structural damage to the oil rig platform was conducted on the Terra Nova project, taking into consideration the variables of extreme winds, icebergs, and ocean currents. Ecological risk assessment has made considerable headway in EA with the *Guidelines for Ecological Risk Assessment* published by the United States Environmental Protection Agency (EPA) in 1996 (EPA 1996). Its main application in Canada, as in the United States, has been in the assessment of contaminated sites. In the following section, we give specific attention to ecological risk assessment.

Ecological Risk Assessment

Ecological risk assessment has been closely linked to eco-toxicology studies and is expected by many authors to provide a base for environmental decision making equivalent to human health risk assessment (Barnthouse et al. 1982; Khadam and Kaluarachchi 2003). It was derived from practices in human health risk assessment, environmental hazard assessment, and environmental impact assessment (National Research Council 1983). However, ecological risk assessment differs from human health risk assessment primarily because of its complexity and scale, where large numbers of species and a diversity of routes and linkages have to be considered. Thus, ecological risk assessment tends to rely on epidemiological studies rather than on modelling. It is common for modelled ecological risks to be manifestly incorrect either because the predicted effects are not occurring or because unpredicted effects are observed (Sutter et al. 2000, 15).

The main approach to ecological risk assessment recommended in Canada is a tiered approach, consisting of three levels. Level one is characterized as descriptive risk assessment based on the collection and analysis of simple, qualitative data and/or comparative methods using published literature findings. For example, the Upper Salmon hydroelectric project applied a level-one analysis using a qualitative food web model based on previously published information (Beanlands and Duinker 1983). Level two is a semi-quantitative approach that relies on models and data collected to analyse priority issues. A level-two assessment that was conducted for hydrocarbon drilling in the Davis Strait predicted impacts based on combined field information, use of reported evidence of the effects of oil slicks on wildlife, and oil slick trajectory modelling (Beanlands and Duinker 1983). The third tier uses site-specific data and predictive models to provide quantitative information and ecosystem responses to determine chronic effects, interaction between chemicals, and ecosystem impacts. However, Baker (1989) warns that the implementation and costs of a tiered analysis can be costly; not all studies warrant a level-three analysis.

A recent example of predictive risk assessment can be found in a comprehensive study of the White Rose Oilfield at the Grand Banks of Newfoundland (Husky Oil Ltd 2000). In the study, combinations of risk assessment are used. With respect to the risks associated with the potential spilling of oil, the study makes a series of predictions of spill types and assesses the probability of spills occurring. It uses world historical statistics to determine oil spill risks and notes 'that a major spill (greater than 1,000 barrels) would have a 0.5 percent chance of occurring over the project life (annual 1 in 2,600), while very large platform spills (greater than 10,000 barrels) have a 0.2 percent chance of occurring (an annual probability of 1 in 7,100)' (Husky Oil Ltd 2000). Knowledge of the impact of oil spills on wildlife is limited to secondary data and previously published information. The study claims that large spills are unlikely to occur, and thus there is little threat to the environment. The study does not include scenario outcomes of impacts if spills were to occur, and there was no higher-level risk assessment completed.

Pastorok (2002) notes that most ecological risk assessments rely on simplistic approaches and fail to incorporate basic ecological information and modelling capabilities. The simple ecological risk assessment models rely on comparison of limited-exposure estimates for each chemical of interest, taking into consideration a specific threshold or individual organism endpoint, such as survival, growth, or reproduction potential. There tends to be no extrapolation or detailed analysis that takes into consideration interactions that would extend the organism endpoints upwards to estimate effects on populations, ecosystems, or landscape endpoints (Landis 2000; Snell and Serra 2000).

Several steps are required in conducting an ecological risk assessment. Figure 3.2 represents a simple and basic framework derived from Jorgensen et al. (2000) and Pastorok (2002). The first step consists of *risk screening* and can be considered a tier-one assessment of a project's known risks, hazards, and impacts as indicated by previously published information. Then, once a project's *alternatives* have been defined and the project location is set, it is possible to determine probabilities for certain impacts to occur (e.g., oil spill probabilities and impacted surrounding environment). A set of *management objectives* that outline the expectations for the risk assessment should be determined and clearly defined. One must consider

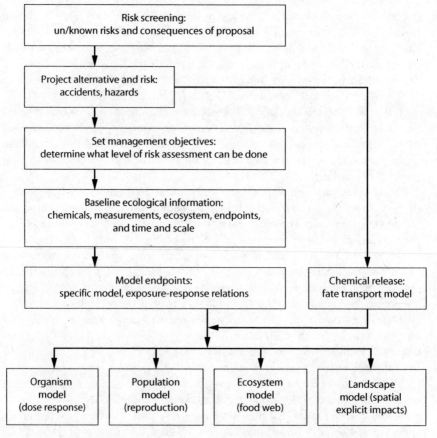

Figure 3.2   A simple framework for doing an ecological risk assessment–based model. Sources: Jorgensen et al. 2000; Pastorok 2002.

costs, data, and model availability. The management objectives might define the relative tiers of risk assessment for certain areas. For example, a tier-two ecological risk assessment might be reserved for less significant or highly uncertain impacts, while a tier-three assessment requiring extensive baseline surveys and modelling may be completed for highly sensitive organisms or environments where even small chemical releases can have significant impacts. From this, *baseline studies* can be initiated and models run to determine the effects of chemical release into the environment and the effects upon organisms, populations, ecosystems, and the landscape.

Once again, it must be emphasized that monitoring in any form improves the model predictions and verification. With respect to ecological impact prediction, Morris (1998) points out a number of limitations: (1) ecological assessments tend to be costly and time-consuming; (2) the models need to be adjusted and modified to the project or site context; (3) if accurate data is not available and monitoring is not strictly adhered to, there is little point in using such models; and (4), as pointed out previously in this chapter, some of the relationships and interactions within ecosystems are so sophisticated that models can fail to give accurate predictions.

Summary

Each consecutive step in the EIA process that identifies and predicts impacts has an iterative nature that relies on the process as a whole. For example, scoping is not a static step that only occurs at the beginning of an EIA. Rather, it is a continuous exercise that changes as a project develops, and it is inherently dependent on processes that define impact significance and prediction, because these steps may change the scope of the project. In the same manner, the interpretation of impact significance in different forms is often applied at all levels of EIA activity. Successful impact prediction based on modelling is dependent on close monitoring and on establishing a feedback loop into the model, requiring a learning and adaptive component to the process. The process of identifying and predicting impacts requires revisiting previous steps and establishing an adaptive context to the EIA, as each step improves the level of information and available knowledge.

How impacts are communicated is an important part both of the process and of the final EIA document. The traditional EIA methods outlined in this chapter are well established in practice—many EIAs employ checklists and matrices in the text. Notably, however, *transparency* lies at the heart of any method that communicates the significance of impacts. It is essential that the values that determine impact significance and magnitude are clearly articulated and that judgments are transparent. This is especially important when issues are cross-cultural and there are differing views on the determination of the significance of environmental effects. In addition, as modelling techniques and risk assessments become more complex, the inherent assumptions in the models need to be clearly stated. The public, at all levels, should play a key role in helping to identify the range and significance of environmental impacts.

References

Andrews, R. 1988. Environmental impact assessment and risk assessment: Learning from each other. In P. Wathern (Ed.), *Environmental impact assessment: Theory and practice*, 85–97. London: Unwin Hyman.

Anisimov, O.A., N.I. Shiklomanov, and F.E. Nelson. 2002. Variability of seasonal thaw depth in permafrost regions: A stochastic modeling approach. *Ecological Modelling* 153:217–27.

Baker, J.P. 1989. Assessment strategies and approaches. In W. Warren-Hicks, B.R. Parkhurst, and S.S. Baker (Eds), *Ecological assessment of hazardous waste sites, Report* 600/3–89/03. US Environmental Protection Agency.

Barnthouse, L.W.G., D.L. DeAngelis, R.H. Gardner, R.V. O'Neill, G.W. Sutter II, and D.S. Vaughan. 1982. *Methodology for environmental risk analysis.* ORNL/TM–8167. Oak Ridge, TN: Oak Ridge National Laboratory.

Barrow, C.J. 1997. *Environmental and social impact assessment: An introduction.* London: Arnold.

Beanlands, G. 1988. Scoping methods and baseline studies in EIA. In Peter Wathern (Ed.), *Environmental impact assessment: Theory and practice*, 3346. London: Routledge.

Beanlands, G., and P. Duinker. 1983. *An ecological framework for environmental impact assessment in Canada.* Hull, QC: Institute for Resource and Environmental Studies and Federal Environmental Assessment Office.

Burt, T.P. 1994. Monitoring change in hydrological systems. *Progress in Physical Geography* 18:475–96.

Canter, L. 1996. *Environmental impact assessment.* New York: McGraw-Hill.

CEA Agency (Canadian Environmental Assessment Agency). 1994. Reference guide: Determining whether a project is likely to cause significant adverse environmental effects. Available at http://www.ceaa..gc.ca/013/0001/0008/guide3_e.htm.

———. 1997. Report of the Terra Nova Project Environmental Assessment Panel, p. 25. http://www.ceaa.gc.ca.

Dipper, B., C. Jones, and C. Wood. 1998. Monitoring and post-auditing in environmental impact assessment: A review. *Journal of Environmental Planning and Management* 41 (6): 731–47.

Downing, K., and M. Reed. 1996. Object-oriented migration modelling for biological impact assessment. *Ecological Modelling* 93:203–19.

EPA (Environmental Protection Agency). 1996. *Guidelines for ecological risk assessment.* Report 61 FR 47552. Washington, DC: US Environmental Protection Agency.

Glasson, J., R. Therivel, and A. Chadwick. 1999. *Introduction to environmental impact assessment.* London: Spon Press.

Haug, P.T., R.W. Burwell, A. Stein, and B. Bandurski. 1984. Determining the significance of environmental issues under the National Environmental Policy Act. *Journal of Environmental Management* 18 (1): 15–24.

Hirst, D., and G. Storvik. 2003. Estimating critical load exceedance by combination in the EMAP model with data from measurement stations. *Science of the Total Environment* 310:163–70.

Husky Oil Ltd. 2000. *White Rose Oilfield Comprehensive Study Report.* Newfoundland: Husky Oil Operations Ltd.

Johansson, B., and L.-G. Mattsson (Eds). 1995. From theory and policy analysis to the implementation of road pricing: The Stockholm region in the 1990s. In *Road pricing: Theory, empirical assessment and policy*, 181–205. Dordrecht: Kluwer Academic Publisher Group.

Jones, M.L., and L.A. Greig. 1985. Adaptive environmental assessment and management: A new approach to environmental impact assessment. In V.W. Maclean and J.B. Whitney (Eds), *New directions in environmental impact assessment in Canada*, 21–42. Toronto: Methuen.

Jorgensen, S.E., L. Barnthouse, D.L. DeAngelis, L. Emlen, and K. van Leeuwen. 2000. *Improvements in the application of models in ecological risk assessment: Conclusion of the Expert Review Panel.* Washington, DC: American Chemistry Council.

Kennedy, A.J., and W.A. Ross. 1992. An approach to integrate impact scoping with environmental impact assessment. *Environmental Management* 16 (4): 475–84.

Khadam, I.M., and J. Kaluarachchi. 2003. Multiple criteria analysis with probabilistic risk assessment for management of contaminated ground water. *Environmental Impact Review* 23:683–721.

Landis, W.G. 2000. The pressuring need for population-level risk assessment. Society of Environmental Toxicology and Chemistry. *SETAC Globe* 1 (2): 44–5.

Lawrence, D. 2003. Significance in environmental assessment. Research and development monograph series 2000. Canadian Environmental Assessment Agency, Ottawa. http://www. ceaa.gc.ca/015/0002/0011/index_e.htm.

Levallois, P., and D. Gauvin. 1994. Risk associated with electromagnetic fields generated by electrical transmission and distribution lines. Background Paper No. 9. Great Whale Environmental Assessment, Great Whale Public Review Support Office, Montreal.

McCartney, A., R. Harriman, A. Watt, D. Moore, E. Taylor, R. Collen, and E. Keay. 2003. Long-term trends in pH, aluminum and dissolved organic carbon in Scottish fresh water; implications for brown trout (*Salmo trutta*) survival. *Science of the Total Environment* 310:133–41.

Morris, P. 1998. Ecology-overview. In P. Morris and R. Therivel (Eds), *Methods of environmental impact assessment*, 197–226. London: UCL Press.

Mulvhill, P., and D. Baker. 2001. Ambitious and restrictive scoping: Case studies from northern Canada. *Environmental Impact Assessment Review* 21:363–84.

National Research Council. 1983. *Risk assessment in the federal government: Managing the process.* Washington DC: National Academy Press.

Parr, T., A. Sier, R. Battarbee, A. Mackay, and J. Burgess. 2003. Detecting environmental change: Science and society-perspectives on long-term research and monitoring in the 21st century. *Science of the Total Environment* 310:1–8.

Pastorok, R. 2002. Introduction. In R.A. Pastorok, S.M. Bartell, S. Ferson, and L.R. Ginzbuerg, *Ecological modeling in risk assessment.* Boca Raton, FL: CRC Press.

Rapaport, E. 2002. Can regional land use transportation planning and environmental management affect forest soil acidity? *Journal of Environmental Planning and Management* 45 (6): 797–811.

Sadler, B. 1996. International study of the effectiveness of environmental assessment: Final report. Ottawa: Canadian Environmental Assessment Agency, Minister of Supply and Services, Canada.

Skiles, J.W., and C.W. Richardson. 1998. A stochastic weather generation model for Alaska. *Ecological Modelling* 110:211–32.

Snell, T.W., and M. Serra. 2000. Using probability of extinction to evaluate ecological significance of toxicant effects. *Environment, Toxicology, and Chemistry* 19 (9): 2357–63.

Summers, J.K., and K. Tonnessen. 1998. Linking monitoring and effect research: EMAP's intensive sites network program. *Environment Monitoring and Assessment* 51 (3): 367–80.

Sutter II, G.W., R.A Efoymson, B.E. Sample, and D.S. Jones. 2000. *Ecological risk assessment for contaminated sites.* Boca Raton, FL: CRC Press.

Winds and Voices Environmental Services, Inc. 2003. Determining significance of environmental effects: An aboriginal perspective. Research and Development Monograph Series 2000, Canadian Environmental Assessment Agency, Ottawa. http://www.ceaa.gc.ca/015/0002/0003/index_e.htm.

Whitney, J., and V. Maclaren. 1985. A framework for the assessment of EIA methodologies in environmental impact assessment. In J. Whitney and V. Maclaren (Eds), *Environmental impact assessment: The Canadian experience*, 1–31. Toronto: Institute for Environmental Studies, University of Toronto.

Wood, C. 1995. *Environmental impact assessment: A comparative review.* London: Longman Scientific and Technical.

Chapter 4

Public Participation in Canadian Environmental Assessment: Enduring Challenges and Future Directions

A. John Sinclair and Alan Diduck

Introduction

As observed by Devlin et al. (2005), Petts (1999, 2003), Wood (2003), and others, public participation has long been recognized as a cornerstone of environmental assessment (EA). In fact, for some, the basic legitimacy of an EA process is questionable if the process does not provide for meaningful participation (Gibson 1993; Roberts 1998). Consistent with this, most Canadian EA legislation establishes participation as an essential element of the process. For example, the preamble of the Canadian Environmental Assessment Act 1995 (CEAA) states: 'The Government of Canada is committed to facilitating public participation in the environmental assessment of projects . . . and providing access to the information on which those environmental assessments are based.'

In addition, all EA legislation, whether at the provincial or federal level, includes at least some practical measures for participation. For these reasons, it is essential to have an understanding of key theoretical and practical aspects of participation. This chapter provides an introduction to these subjects. The first section summarizes the benefits of participation, the second provides an overview of basic Canadian legislative provisions, and the third reviews enduring concerns. The fourth part examines promising future directions, with the final section discussing the implications for sustainability objectives. Throughout the chapter, we refer to illustrative Canadian EA cases.

We use the word *participation* to mean the active involvement of the public in the EA process through various means, ranging from open houses to panel reviews, as outlined later in the chapter. Reference is also made to *meaningful public participation*. This term is used in referring to participatory processes that incorporate all of the essential components of participation, from information sharing to education, including the active and critical exchange of ideas among proponents, regulators, and participants. Some authors, such as Roberts (1998), refer to this as public consultation. We prefer to use 'meaningful public participation' because consultation is often used to describe a single event, such as an open house, that is the sum total of the participatory efforts in a particular situation. In our view, an open house alone would not constitute meaningful public participation.

The Benefits of Participation

The benefits of public participation in EA have been described in both theoretical and practical terms. A key theoretical argument is that participation actualizes fundamental principles of democracy and strengthens the democratic fabric of society (Parenteau 1988; Sinclair and Diduck 1995; Shepard and Bowler 1997; Forester 2006). This argument situates EA as a key channel through which the public can choose to participate directly in the decisions that affect them, thereby contributing to a more participatory democracy. A related point is that EA provides a vehicle for individual and community empowerment (Fitzpatrick and Sinclair 2003). This recognizes that meaningful participation in decision making enables individuals and organizations to adapt to environmental change and also to generate change through the expression of human agency. It has also been suggested that participation is conducive to broad-based individual and social learning that could enable the transition to sustainability (Webler et al. 1995; Palerm 2000; Sinclair and Diduck 2001; Diduck and Mitchell 2003; Fitzpatrick and Sinclair 2003; Sinclair et al. 2008; Sims and Sinclair 2008). Such a view clarifies the link between learning at multiple levels of social organization and the achievement of sustainability goals.

In practical terms, the benefits of public participation are numerous and touch on ecology, law, politics, conflict resolution, planning, and decision making (e.g., Susskind and Cruikshank 1987; McMullin and Nielson 1991; Smith 1993; Meredith 1995; Webler et al. 1995; Shepard and Bowler 1997; Petts 1999; Usher 2000). Reflecting this interdisciplinarity, the literature suggests that public participation in EA

- provides access to local and traditional knowledge from diverse sources;
- enhances the legitimacy of proposed projects;
- helps define problems and identify solutions;
- permits a comprehensive consideration of factors upon which decisions can be based;
- ensures that projects meet the needs of the public in terms of both purpose and design;
- brings alternative ethical perspectives into the decision-making process;
- broadens the range of potential solutions considered;
- furnishes access to new financial, human, and in-kind resources;
- prevents 'regulatory capture' of EA agencies by project proponents;

- encourages more balanced decision making;
- increases accountability for decisions made;
- facilitates challenges to illegal or invalid decisions before they are implemented;
- illuminates goals and objectives, which is necessary for working through value or normative conflict;
- furnishes venues for clarifying different understandings of a resource problem or situation, which is key to resolving cognitive conflict;
- helps avoid costly and time-consuming litigation; and
- reduces the level of controversy associated with a problem or issue.

It is obvious, even from the foregoing brief summary, that public participation in EA can provide diverse and important benefits for planning and decision making. However, whether such benefits are realized in any particular case depends, to a large extent, on the legislation and policy applicable in the case. Therefore, it is important to have a basic understanding of Canadian EA legislation (Table 4.1) pertaining to public participation. The ensuing section provides an overview, focusing on the core provisions of current EA public participation regimes, namely provisions dealing with adequate notice, access to information, participant assistance, public comment, and public hearings. The discussion centres on federal legislation (the CEAA), but it includes references to various provincial and territorial processes.

Key Public Participation Provisions in Canadian Environmental Assessment

Adequate Notice

Once a development proposal is submitted to government authorities and the EA process is triggered, adequate notice is fundamental to fair and meaningful public participation. Notice should be provided in such a way that it comes to the attention of interested persons well before decisions are made. Notice normally involves some form of advertisement through local print and, in some cases, broadcast media. It provides the public with notification of the proposal and informs them where they can get further information and to whom they can submit their comments.

All Canadian jurisdictions have notice provisions in their EA statutes, regulations, or supporting policies. However, there is little consistency in how notice is provided for important milestones in the EA process, such as the filing of the proposal and the completion of the environmental impact statement (EIS) (see chapter 1 for a summary of other key steps in the EA process).

Under the CEAA, notice is mandatory for all types of assessment: screenings, comprehensive studies, mediations, and hearings. Notice of an assessment must be presented on the Canadian Environmental Assessment Agency (CEA Agency) website within 14 days of the start of the assessment. The CEA Agency is responsible for

Table 4.1   A list of primary Canadian EA legislation, plus government websites providing access to the legislation

Jurisdiction	Primary Legislation and Internet Web Address (accessed March 2009)
Alberta	*Environmental Protection and Enhancement Act*, R.S.A. 2000, c. E-12 http://www.qp.gov.ab.ca/documents/Acts/E12.cfm?frm_isbn= 9780779729241
British Columbia	*Environmental Assessment Act*, S.B.C. 2002, c. 43 http://www.bclaws.ca/Recon/document/freeside/--%20E%20--/ Environmental%20Assessment%20Act%20%20SBC%202002%20% 20c.%2043/00_02043_01.xml
Canada	*Canadian Environmental Assessment Act*, S.C. 1992, c. 37, C-15.2 http://www.canlii.org/en/ca/laws/stat/sc-1992-c-37/latest/ sc-1992-c-37.html
Inuvialuit Settlement Region	Inuvialuit Final Agreement, as implemented by the *Western Arctic (Inuvialuit) Claims Settlement Act*, S.C. 1984, c. 24, W-6.7 http://www.canlii.org/en/ca/laws/stat/sc-1984-c-24/latest
Manitoba	*The Environment Act*, C.C.S.M. c. E-125 http://web2.gov.mb.ca/laws/statutes/ccsm/e125e.php
New Brunswick	*Clean Environment Act*, R.S.N.B. 1973, c. C-6 http://www.gnb.ca/0062/PDF-acts/c-06.pdf
Newfoundland and Labrador	*Environmental Protection Act*, S.N.L, 2002, c. E-14.2 http://assembly.nl.ca/Legislation/sr/statutes/e14-2.htm
Northwest Territories	*Mackenzie Valley Resource Management Act*, S.C. 1998, c. 25, M-0.2 http://www.canlii.org/en/ca/laws/stat/sc-1998-c-25/latest/
Nova Scotia	*Environment Act*, S.N.S. 1994-95, c. 1 http://www.gov.ns.ca/legislature/legc/statutes/envromnt.htm
Nunavut	*Nunavut Land Claims Agreement*, Article 12, Part 5 http://www.canlii.org/en/ca/laws/stat/sc-1993-c-29/latest/
Ontario	*Environmental Assessment Act*, R.S.O. 1990, c. E-18 http://www.e-laws.gov.on.ca/html/statutes/english/elaws_ statutes_90e18_e.htm
Prince Edward Island	*Environmental Protection Act*, R.S.P.E.I. 1988, c. E-9 http://www.gov.pe.ca/law/statutes/pdf/e-09.pdf
Quebec	*Environment Quality Act*, R.S.Q, c. Q-2 http://www.canlii.org/qc/laws/sta/q-2/20080515/whole.html
Saskatchewan	*Environmental Assessment Act*, S.S. 1979-80, c. E-10.1 http://www.qp.gov.sk.ca/documents/English/Statutes/Statutes/E10-1.pdf
Yukon	*Yukon Environmental and Socio-economic Assessment Act*, S.C. 2003, c. 7, Y-2.2 http://www.canlii.org/en/ca/laws/stat/sc-2003-c-7/latest/

ensuring proper notice is provided (CEAA, section 55.1). In Alberta, the proponent is required to publish notice of an application in at least one paper with daily or weekly circulation in the locality of the proposed undertaking, and to announce that people who are directly affected can submit their concerns (Alberta Environment 2004). The provincial EA director can waive this requirement when the activity proposed is considered routine, when notice has already been given, or when there is an emergency. The notice provisions in Ontario's EA Act are strong; they include, for example, requirements on the content of the notice and the provision for notice to be given in the first stage of the process, that is, during development of the terms of reference for the project (Environmental Assessment Act, R.S.O. 1990, c. E-18, section 6). Given the lack of a consistent approach to requirements for notice, participants and proponents must give careful consideration to the formal notice requirements in each jurisdiction.

Access to Information

Ready access to information provided by proponents and to any comments offered by participants and regulators is essential to meaningful participation. There must be opportunities for the timely exchange of information among all parties. A registry or repository system is the basic means of public access to information used in most Canadian jurisdictions. Such registries are usually held in libraries and other public buildings to facilitate orderly access to the information available. Most jurisdictions provide Internet access (or partial access) to their public registries (Table 4.2).

Provisions for access to information in the CEAA are generally quite strong. Specifically, sections 55–55.6 of the Act set out in detail what documents must be made available and establish the Canadian Environmental Assessment Registry and CEA Agency website. As a starting point, any document that would be available through access to information legislation has to be made available proactively. Furthermore, all relevant information must be made available in a manner that ensures convenient public access. Provincially, there are diverse approaches to access to information. The Alberta legislation (section 56) has a general requirement for a document registry, but provides little guidance on what needs to be included or how it is to be made accessible (see also the Manitoba Act, section 17). Generally, provinces either leave these issues to regulation (through specific 'rules' promulgated under the authority of legislation) or have broad statutory provisions respecting access to information.

Participant Assistance

Numerous authors have established the importance of participant (or intervenor) funding (e.g., Gibson 1993; Wood 2003), and this view has long been echoed by Canadian environmental non-governmental organizations (ENGOs). For instance, the Canadian Environmental Network (1988) argued that '[e]ffective participation by the public requires funding. The disproportion of resources between pro-

Table 4.2 Public registry websites for those jurisdictions providing full or partial Internet access

Jurisdiction	Internet Web Address (accessed March 2009)
Alberta	http://environment.alberta.ca/1283.html
British Columbia	http://a100.gov.bc.ca/appsdata/epic/html/deploy/epic_home.html
Canada	http://www.acee-ceaa.gc.ca/050/index_e.cfm
Manitoba	http://www.gov.mb.ca/conservation/eal/registries/index.html
New Brunswick	http://www.gnb.ca/0009/0377/0002/index-e.asp
Newfoundland and Labrador	http://www.env.gov.nl.ca/env/Env/EA%202001/pages/index.htm#Projects
Northwest Territories	http://www.mveirb.nt.ca/registry/index.php
Nova Scotia	http://www.gov.ns.ca/nse/ea/projects.asp
Ontario	http://www.ebr.gov.on.ca/ERS-WEB-External/
Prince Edward Island	No on-line public registry but EA info is available at http://www.gov.pe.ca/infopei/index.php3?number=40190&lang=E
Quebec	http://www.bape.gouv.qc.ca/sections/mandats/
Saskatchewan	http://www.environment.gov.sk.ca/Default.aspx?DN=dd506e76-4819-4493-a22b-6411133ca469
Yukon	http://www.yesab.ca/assessments/public_registry.html

ponents and the public necessitates the establishment of an independent funding body to provide adequate amounts of funding to allow full and meaningful participation, at all steps, to committed members of the public.'

Participant assistance, when it is provided, is usually reserved for large-scale EAs. In such cases, funding can create a more substantive dialogue by allowing participants to gather independent technical expertise related to specific issues in the assessment (see, for example, Lynn and Wathern 1991; Lynn 2000; Hayward et al. 2007). Public participants can use financial resources to prepare and participate in scoping meetings, review draft assessment guidelines, review the proponent's EIS, and prepare and participate in public hearings.

Few Canadian EA processes provide mechanisms for participant assistance. At one time, the province of Ontario was a leader in this field and enacted a participant assistance statute (the Intervenor Funding Project Act). However, the legislation had a time limit and lapsed in 1996. In practice, only the governments of Canada and Manitoba provide assistance at present.

Under the CEAA, assistance is available for both public hearings (discussed further below) and comprehensive studies. See the CEA Agency website (http://www.

acee-ceaa.gc.ca/010/0001/0002/index_e.htm (accessed 23 March 2009) for a good overview of the federal Participant Funding Program. In Manitoba, the participant funding regulation (Participant Assistance Regulation, Manitoba Regulation 125/91) was used only once from 1990 to 2001 (in the case of the proposed Conawapa hydroelectric generating station), but in recent years the government has been far more willing to support public interest intervenors. A guide to the province's Participant Assistance Program can be downloaded at http://www.cecmanitoba.ca/File/CEC%20Admin/assistance-guide%20Feb.%2008.pdf (accessed March 2009). In July 2003, $870,000 (Cdn) was announced for the Wuskwatim EA, which dealt with a proposed 200 MW generating station on the Burntwood River in northern Manitoba. The funding award, one of the largest in Canadian EA history, was allocated in the following manner:

- $190,000 for the Consumers Association of Canada/Manitoba Society of Seniors
- $160,000 for Pimicikamak Cree Nation
- $145,000 for Time to Respect Earth's Ecosystems/Resource Conservation Manitoba
- $115,000 for the Canadian Nature Federation
- $80,000 for the Manitoba Metis Federation
- $60,000 for Opaskawayak Cree Nation
- $60,000 for the Community Association of South Indian Lake
- $20,450 for Mosakahiken Cree Nation
- $20,450 for Puckatawagan Fisherman's Association
- $20,000 for Trapline #18
- $5,488 for York Factory First Nation

Public Comment

Meaningful participation processes include reasonable opportunities for the public to comment on the project proposal, respond to the government's position on the proposal, and react to input from other participants. Most often, proponents provide opportunities for public comment during open houses at which the proponents typically supply information on their proposals and invite input. Government regulators normally allow for public comment on EA documents by inviting people to provide written comments. These and other opportunities for public comment typically occur irrespective of whether there is a public hearing (see below) for the proposal under consideration.

Provisions in the CEAA for public comment on screenings are weak, generally speaking. The decision on whether (and if so how) to permit comment is at the discretion of the responsible authority (RA), which may be the proponent, the regulator, or both (for a discussion of responsible authorities, see chapter 14). For comprehensive studies (defined in chapter 14), public comment is mandatory, but only regarding project scope and on the final comprehensive study report—not on the ongoing assessment or development of the report. It is important to note that only a handful of assessments out of thousands conducted under the CEAA have gone through a panel review, and less than 50 have undergone a comprehensive study. For more than 99 per cent of the assessments, therefore, the CEAA includes no legislative requirement for public comment.

Some provinces, such as Ontario (e.g., sections 6(3.6), 6.4, and 7.2 of the Act), Newfoundland (e.g., sections 3(3), 6(1), and 7(3) of the Environmental Assessment Regulations, 2003, O.C. 2003-220), and Manitoba (e.g., sections 7(2) and 10(4) of the Act), make general provision for public comment without providing details on the timing or quality of the opportunities. Consequently, although notice must be given, there is no assurance of the types of opportunities that will be available for members of the public to act on the notice. Most provinces have little in the way of specific requirements for when and how public comments are to be sought and considered, outside the public hearing process.

Public Hearings

To support decision making and add transparency to the EA process, environment ministers in all Canadian jurisdictions have the authority to call public hearings (or, in some cases, public meetings). Hearings are sometimes equated with impartial decision making because hearing panels typically operate at arm's length from governments and proponents. However, in no Canadian jurisdictions do hearing panels have ultimate decision-making authority. The role of panels is restricted largely to providing advice to government decision makers. In spite of this, interest groups and other members of the public often favour hearings for their procedural certainty and transparency. In addition, public participants who take part formally in a hearing (often referred to as intervenors) receive access to key documents of the other parties to the hearing. They also receive, in a timely fashion, the formal written reasons for the ultimate decision in the case. With all that said, it is noteworthy that, contrary to popular opinion, very few assessments go to public hearings—less than 2 per cent of all EAs nationally. Further, Hazell (1999) noted that 'there is much cynicism about CEAA public review panels such as the BHP diamond mine [in the Northwest Territories] review among others.' In the overall EA context, therefore, panel reviews are probably less important than is suggested by the media attention they typically attract.

The CEAA does include provisions for the conduct of public hearings. Section 34 requires review panels to make information available and hold hearings in a manner that offers the public an opportunity to participate. Under section 35(3), hearings are generally open to the public, although the Act does provide for exceptions. Otherwise, the CEAA leaves the conduct of public hearings to the panel.

Among the provinces, the Ontario (Part III), Quebec (Division II.1), and Newfoundland and Labrador (sections 62–66) Acts provide considerable detail regarding public hearings. In other jurisdictions, hearing protocols are adopted from general public inquiry legislation (e.g., Saskatchewan—section 14), established in detail in regulations (e.g., Nova Scotia Environmental Assessment Board Regulations, O.I.C. 95-221, N.S. Reg. 27/95) or developed by hearing panels or government EA officials (e.g., the Manitoba Clean Environment Commission's hearing guidelines, http://www.cecmanitoba.ca/index.cfm?pageID=54 [accessed March 2009]). It should also be noted that other provincial decision-making processes, which often occur before, or concurrently with, a project EA, can include hearings as well. Energy utility regulators, such as the Alberta Utilities Commission and the Manitoba Public Utilities Board, are good examples of this.

The CEAA and several provincial acts also provide for various forms of alternate dispute resolution, such as mediation and arbitration. The CEAA permits the use of mediation as a partial or complete alternative to a panel review. Nova Scotia also provides, under section 14 of its Act, for the use of alternate dispute resolution in place of more conventional EA decision making. In both jurisdictions, however, uncertainty over the proper application of the legislation, especially with respect to the identification of interested parties and what happens when the alternate dispute resolution fails, has discouraged the use of these mechanisms. In the case of the federal act, these provisions have not been used even once since the Act came into force. Ontario (section 8), Manitoba (section 3), and other provinces also have general references to mediation in their legislation.

Enduring Concerns Regarding Public Participation in Environmental Assessment

Despite the potential benefits of public participation in EA and the diversity of legislative approaches to implementation, meaningful participation has proven elusive (Blaug 1993; Meredith 1997; Prystupa et al. 1997; Petts 1999; Diduck and Sinclair 2002; Sinclair and Fitzpatrick 2003). Long-standing concerns range from substantive matters, such as lack of shared decision making, to procedural issues, such as inadequate notice (Sinclair and Doelle 2003).

Lack of Shared Decision Making

For citizen activists and some ENGOs, lack of public participation in decision making is a serious shortcoming of Canadian EA processes. A helpful tool for understanding this problem is Arnstein's (1969) classic ladder of citizen participation (also see Dorcey et al. 1994; Connor 2001). Arnstein identified eight levels of public participation and associated degrees of power sharing. The bottom rungs of the ladder, manipulation and therapy (characterized as non-participation), describe participation that amounts to no more than a public relations exercise designed to gain support for a predetermined decision. The middle rungs (forms of tokenism) include informing, which involves flows of information from managers to citizens but does not include dialogue; consultation, in which citizens are given a voice but are not necessarily heard; and placation, in which managers seek citizen advice but retain decision-making authority. The top rungs describe situations where forms of citizen power occur, such as partnerships (where trade-offs are negotiated), delegated power, and citizen control, the last of which involves the highest degree of 'decision-making clout' (Arnstein 1969, 217). Activists and ENGOs are often highly critical of processes that only consist of lower levels of involvement, and argue that public participation should entail redistribution of power over decision making. In other words, for them, public participation should involve shared decision making.

A related concern is lack of clarity over the purposes of any given public participation exercise. As suggested above, the purposes of participation can range from relatively modest goals, such as information sharing, to more ambitious objectives, such as partnerships and power sharing. If purposes are unclear, members of the public could easily expect to participate in decision making when the very purpose of the process in which they are involved is considerably less ambitious. In the end, lack of clarity and concomitant unfulfilled expectations frequently leave participants cynical and dissatisfied with their participation experiences (Shepard and Bowler 1997; Hazell 1999; Petts 1999; Stewart and Sinclair 2007). This occurred in the Halifax Harbour clean-up EA where key choices concerning siting and technology were made in a less than transparent way, leaving participants with little role in the decision making (Doelle and Sinclair 2006). Lack of clarity can sometimes be traced to muddled public participation jargon. Diverse terms are used, often interchangeably, to describe both the ends (i.e., the goals) and the means (i.e., the strategies and specific techniques) of participation programs. Such terms include involvement, consultation, engagement, information sharing, education, communication, and collaboration. Given this problem, regardless of the terms used, EA practitioners need to be clear about whether the purpose of participation is information exchange, shared decision making, or some other goal.

Lack of Participation at Normative and Strategic Levels of Planning

Another concern for activists, ENGOs, and academics is lack of public participation at early stages of project planning (Diduck and Sinclair 2002; Diduck and Mitchell 2003). Participation can occur at the normative level of planning (in which decisions are made about what *should* be done), the strategic level (in which decisions are made to determine what *can* be done), or the operational level (in which decisions are made to determine what *will* be done) (Smith 1982). The earlier that participation occurs in the EA process, the more influence the public is likely to have on important issues such as project need, purposes, and alternatives. A good early example of normative involvement is the Mackenzie Valley Pipeline Inquiry of the 1970s; it was groundbreaking in the extent to which it engaged potentially affected communities early in the planning process (Berger 1977). As that case showed, through early participation, basic choices can be considered (including whether the project should proceed at all) before political momentum builds for a project and substantial amounts of time and money are invested.

In Canada, participation is not required legally until well into the planning process and sometimes just at the operational stages of planning. Participation at earlier junctures is left to the discretion of the proponent, and early involvement is sporadic at best. This has proven to be a barrier to widespread participation, as exemplified by the Maple Leaf hog-processing EA in Manitoba (in which the proponent was ultimately licensed to slaughter up to 54,000 hogs per week under a single daily work-shift scenario). In that case, many members of the public as well as ENGO activists wanted to discuss large normative questions (e.g., Should we have an industrial hog-processing plant in our community?) rather than detailed operational issues (e.g., How thick should the liner be for the anaerobic lagoon in the waste treatment facility?) (Diduck and Sinclair 2002). For these people, the fact that they could not discuss the normative questions was an indication that the ultimate decision in the case was a foregone conclusion and that there was therefore little point in participating actively in the EA.

The recent Red Chris case—*MiningWatch Canada v. Minister of Fisheries and Oceans et al.* (F.C.: T-954-06); (F.C.A.: A-478-07, A-479-07)—underscores the importance of decisions made by RAs about how to engage the public in early scoping decisions. In this case the proponent needed approvals from Fisheries and Oceans and Natural Resources Canada for a proposed copper and gold mining operation in British Columbia. Since metal mines processing more than 3,000 tonnes of ore per day are on the federal comprehensive study list, the project was destined for this type of assessment until the RAs scoped the project very narrowly and changed it from being a comprehensive study to a screening. Section 21 of the CEAA indicates that

> [w]here a project is described in the comprehensive study list, the responsible authority shall ensure public consultation with respect to the proposed scope of the project for the purposes of the environmental assessment, the factors proposed to be considered in its assessment, the proposed scope of those factors and the ability of the comprehensive study to address issues relating to the project.

In contrast to a screening, comprehensive studies require that the public be consulted on the scope of the project, have the opportunity to comment on the comprehensive study report, and be given access to a funding program to help facilitate their participation. MiningWatch argued that the public had not been actively engaged in the process that led to the decision to narrow the scope of the project, and had been excluded from the actual screening assessment undertaken. Justice Luc Martineau agreed. This decision was 'considered a major victory as it upheld a law passed by Parliament which required a greater role for public consultation in environmental assessment' (MiningWatch 2008). The Crown appealed the Federal Court's decision, and on 13 June 2008 the Federal Court of Appeal ruled in favour of the Crown, citing the Crown's discretion regarding decisions related to whether an assessment is to be a comprehensive study or a screening. In early 2009, the Supreme Court of Canada agreed to hear MiningWatch's appeal of the Federal Court of Appeal's decision.

In other cases, facilitating public participation early in the planning process has been a challenge, and asking normative questions before specific projects and sites have been identified has created a barrier to broad participation. When the Regional Municipality of Ottawa-Carleton started planning a waste management system, it organized public workshops to help design the project. Despite adequate promotion and apparent widespread public interest, only a handful of people participated. However, once the municipality identified the preferred components of the system (e.g., a landfill and specific potential sites), the rooms that had been booked for public meetings were not big enough to hold the participants who wanted to hear what was planned and have their voices heard. A number of these participants also indicated that there should have been opportunities for them to be involved earlier in the process. Herein lies an important lesson for public participation practitioners: reasonable provision must be made to allow those who join the process late to be brought up to speed and have their concerns heard.

To complicate matters further, even if early participation is permitted, restrictive policies and practices can sometimes undermine the very purposes of such involvement. For example, restrictions that limit the number and type of alternatives that can be considered for any given project (an important strategic phase of project planning) scuttle an important benefit of participation, namely bringing in diverse and innovative ideas. Further, such restrictions confound the ability of participants to influence the selection of alternatives. Some of these elements can be seen in the Cheviot coal mine EA from Alberta (which involved a proposed open-pit mine less than 2 km from Jasper National Park, a UNESCO World Heritage Site). In that case, there was limited public participation in the consideration of alternatives for the project, as well as a general lack of attention to important reasonable alternatives. Members of the public and ENGOs argued that the underground mining alternative, as opposed to open-pit mining, had not been considered fully by the proponent and regulators. They challenged the validity of the assessment on this and other grounds in court and won, with the judge requiring the proponent and regulator to do the assessment again but this time with all reasonable alternatives included (see *Alberta Wilderness Association v. Cardinal River Coals Ltd,* [1999] 3 F.C. 425). (Recent decisions of the Federal Court of Canada can be found on the Internet at http://decisions.fct-cf.gc.ca/fct/index.html [accessed March 2009].) In most EAs involving projects of large magnitude, important issues to consider are the range of alternatives considered by proponents to meet the need or problem being addressed by a project and the public's participation in the choice of an alternative.

The questions surrounding why normative and strategic planning sometimes engenders and sometimes inhibits broad public participation—and why there are so often difficulties associated with implementing early participation—are challenging and require additional research. Further, to some degree, these questions are related to ongoing policy initiatives that seek to involve the public in wide-area planning, which is typically broader, more abstract, and more overtly value laden than most EAs. Two examples of such initiatives are discussed later in the chapter, in the section on future directions.

Information and Communication Deficiencies

Yet another enduring concern pertaining to public participation in EA is information and communication deficiencies, including inadequate notice, inaccessibility of project documents, and lack of dialogical (or interactive) participation opportunities.

With respect to notice, as discussed earlier, there is little consistency in legal provisions in Canadian EA processes. Further, some processes, such as CEAA screenings, have traditionally had discretionary notice provisions. With respect to screenings, which are about 99 per cent of all federal assessments, this has meant that assessments are often concluded before the public is given an opportunity to review and comment on the information available. In the case of Bounty Bay Seafoods, which dealt with the development of a major aquaculture project in St Ann's Harbour, Nova Scotia, the Sierra Club of Canada challenged the approvals that were granted by the federal Department of Fisheries and Oceans (*Sierra Club of Canada v. Canada [Attorney General]* [2003], Federal Court of Canada, Trial Division, T–765–02). The Sierra Club argued that its right to participate fully in the EA (a screening done under the CEAA) was denied because access to key EA documents was overly restrictive. The environmental impact statement, for example, was available only for viewing but not for copying (to protect proprietary commercial information, according to the proponent), and it was only available publicly in five locations, all of which were in or near Baddeck, Nova Scotia (May 2003). In addition, the final screening report was not placed in the public domain until four days before the project approval was granted. In the end, the court quashed the approval permit and ordered a new EA to be done.

Inadequate notice has also been a legal issue at the provincial level. In *Caddy Lake Cottagers Association v. Florence-Nora Access Road Inc.* (1998), 129 Man. R. (2d) 71 (Man. C.A.), for example, the court considered whether the provincial EA office was required to publish project notices in newspapers with circulation outside the directly affected local area. While the applicants, who argued for wider public notice, lost the case and the judge agreed with a narrow definition of *public*, the case brought into focus the issue of notice provincially and has since precipitated a policy change on the part of the provincial government (Sinclair 2002).

Despite recent improvements in public registry systems and the growing influence of access to information legislation, problems still exist with the inaccessibility (both physical and cognitive) of project documents. With respect to physical inaccessibility, some public registries in Canada present ongoing problems, including limited hours of operation, lack of harmonization among registry locations, poor condition of documents, incomplete indices, and expensive copying charges (Burcombe 1998; Kidd 1998; Sinclair and Diduck 2001). The Sunpine EA, which involved a logging road and two bridges designed to provide access to forest on the eastern slope of the Rocky Mountains in Alberta, shows the controversy that can arise because of physical inaccessibility (also see the *Bounty Bay Seafoods* case discussed earlier). In this case, the public registry was in Sarnia, Ontario, while the affected publics were over 2,000 km away in southwestern Alberta. This was challenged in court, and the judge ruled that the registry must be made more accessible (*Friends of West Country Association v. Ministry of Fisheries and Oceans et al.,* [1998] Federal Court of Canada, Trial Division, T–1893–96).

With respect to cognitive inaccessibility, the overly technical language and general lack of readability of EISs and other EA documents remain a concern for public participation practitioners because these problems tend to impede broad and active participation (Gallagher and Jacobson 1993; Sullivan et al. 1996; Petts 2003; Diduck and Sinclair 2002). Related to the cognitive inaccessibility of EA documents are the complexity and difficulty of EA legislation itself. In most Canadian jurisdictions, EA legislation is highly technical and jargon filled, which has led to the creation of guidance (or public information) documents. Both the public and proponents rely on guidance material to aid in their understanding of the EA process and ensure their effective participation. However, guidance material is often lacking, as is education about the EA process, including case-specific training opportunities (Sinclair and Diduck 1995; Croal and Grady 1998; Clark 1999; Petts 1999). Recognizing the need for improved guidance, the CEA Agency released new guidance, including the publications 'Public Participation Guide' and 'How to Determine If the Act Applies'. All of the agency's guidance material is available at http://www.acee-ceaa.gc.ca/012/newguidance_e.htm (accessed March 2009).

Although proponents frequently use open houses (and similar consultation methods) and government officials occasionally convene hearings, dialogic participation techniques are rarely used in Canadian EA. These types of methods (e.g., advisory committees, task forces, community boards, mediation, and non-adversarial negotiation) emphasize ongoing dialogue and communication among project proponents, EA officials, and civic organizations, and serve important mutual learning, relationship building, and conflict resolution functions. Table 4.3 summarizes the full range of public participation techniques available for use in EA, including various dialogic practices. Canadian EA has tended to rely on the public information and public input methods, while making little use of the more dialogic (i.e., group problem solving) techniques.

Insufficient Resources for Participants

As noted earlier, participant funding mechanisms in Canadian EA legislation are varied in their clarity and design, and sporadic in their application. Consequently, ENGOs and other civic organizations frequently lack sufficient resources to make effective interventions. Important voices and issues, reflecting critical yet valid points of view, are therefore often silenced in formal EA processes. This is an enduring concern for EA practitioners and policy makers. The problem is compounded when proponents have a high degree of control over public participation programs, which has been the case in many Canadian jurisdictions (outside of hearing situations) (Sinclair and Diduck 2001); that is, in many EA processes, the proponent has often had considerable control over how the public is consulted and how the information obtained is assessed and utilized. Those promoting the project are thus left to collect, implement, and report to EA decision makers the issues and concerns raised by the public. This can result in an unbalanced representation of the results of the public consultations (in favour of the proposed project) and contributes to negative perceptions of public participation processes.

Table 4.3   Public participation techniques available for use in EA

Passive public information techniques

Advertisements	Feature stories	Information repositories
News conferences	Newspaper inserts	Press releases
Print materials	Technical reports	Television
Websites		

Active public information techniques

Briefings	Central contact person	Community fairs
Expert panels	Field offices	Field trips
Information hotline	Open houses	Technical assistance
Simulation games		

Small-group public input techniques

Informal meetings	In-person surveys	Interviews
Small-format meetings		

Large-group public input techniques

Public hearings	Response sheets
Mail, telephone, and Internet surveys	

Small-group problem-solving techniques

Advisory committees	Citizen juries	Community facilitation
Consensus building	Mediation and negotiation	Panels
Role playing	Task forces	

Large-group problem-solving techniques

Workshops	Interactive polling	Sharing circles
Websites and chat rooms		
Future search conference		

Sources: International Association for Public Participation 2001; Diduck 2004; Rowe and Frewer 2005.

Calls for enhanced resources must, however, be weighed against the sentiment of some project proponents that public participation is too expensive (Petts 1999). Governments have been hesitant to take on more costs—or to pass costs on to proponents—that would add to the overall expense of project planning and development.

Accelerated Decision Processes

All EA laws in Canada restrict the amount of time given to the public to review and respond to EA documentation, typically 30 days, and they also restrict the time that regulators have to make a decision. While these timelines are meant to be minimum standards, they tend to be interpreted strictly by regulators regardless of the complexity of the EA case at hand. This has occurred because a key issue of concern to project proponents is the amount of time it takes for a decision to be rendered. Proponents often contend that public review periods unnecessarily extend the time it takes to make decisions, thereby creating process inefficiencies and economic uncertainties. As a result, decision makers have established timelines in an attempt to balance the needs of proponents and those of the public. The result, however, is accelerated decision processes that often do not allow adequate time for the public, and some would argue the regulator, to properly review and respond to the EA information provided and for a two-way dialogue, aimed at issue resolution, to occur (Petts 2003; Sinclair and Diduck 2001).

Weak Public Participation in Follow-up

As discussed in chapter 1, follow-up is a critical component of EA because it gauges the way a project actually affects the environment. It provides the basis for designing further mitigation efforts if they are necessary, assessing the accuracy of the predictions made in the original EIS documents, and learning from EA experiences. The CEAA provides for consideration of follow-up for all types of federal assessment, but the implementation of follow-up programs is required only for comprehensive studies and panel reviews. If follow-up measures are required, there is an obligation to advise the public of the results of the measures (section 38(2)(d) and (e)). Most provinces do not make any reference to follow-up, and none provide for public participation in the follow-up process itself. There is also no provision in any Canadian legislation, including the CEAA, requiring notice of and public comment on the results of monitoring or follow-up measures. The fact that the follow-up processes do not require public participation has likely meant that valuable information gathered during follow-up has not been readily available to the public and that valuable local knowledge has been excluded from the assessment. Despite the general lack of follow-up activities in Canada, there are notable exceptions, such as the low-level flying EA from Labrador and the BHP diamond mine EA from the Northwest Territories, both of which are discussed in more detail below.

Promising Future Directions

The foregoing concerns suggest that EA practitioners and policy makers face formidable challenges in creating more meaningful roles for the public in EA planning and decision making. However, consideration of these concerns helps them identify important future directions. These are discussed below, unified by the proposition that public participation in EA needs to be reconceptualized in more collaborative, community-based terms. This would involve fresh views of both the ways people are involved and the very purposes of their participation.

Early and Ongoing Participation

An important future direction for public participation in EA lies in resolving the tensions surrounding normative and strategic planning issues. As noted earlier, some cases suggest that lack of opportunities for early involvement can be a barrier to broad participation, and other cases provide powerful examples of normative involvement. Still others suggest that opportunities for participation can occur too early in the planning process, namely when they arise prior to the development of a concrete proposal to which members of the public can react. In such cases, the very earliness of the opportunity presents a barrier to broad, active, and meaningful participation.

The importance of finding ways to encourage early participation was reiterated in a survey of Canadian EA practitioners and ENGO officials (Sinclair and Doelle 2003). In that study, the respondents expressed clearly that the way public participation is envisioned by most EA statutes in Canada limits early involvement and fragments later participation opportunities. They pointed out that opportunities for participation in EA have tended to occur at discrete points in the process, such as during scoping, preparation of the EIS, and public hearings. Further, the respondents promoted moving to a more holistic process—that is, to one that encourages meaningful early and ongoing participation. In their view, the process should enable participation from the outset of the project or program planning development phase, when decisions are being made about the objectives and goals regarding what ought or ought not to be done. Later provisions should then require opportunities for participation through the operational stages of project development, implementation, follow-up, and eventual decommissioning. Such changes would not only address criticisms concerning the timeliness of involvement, they would create opportunities for ongoing exchange among stakeholders and could have positive implications for power imbalances. They would also be in line with policies in British Columbia, Ontario, and Manitoba that encourage proponents to consult with the public long before the formal EA process is initiated.

In recent years, regional (or wide-area) planning has been an important testing ground for new ideas on the subject of early and ongoing involvement. In British Columbia, the Commission on Resources and Environment and the Land and Resource Management Plans of the 1990s established broad-based round tables that brought local people together in an attempt to resolve regional land-use planning issues. Each table offered innovative experiments in multi-stakeholder planning, consensus building, and shared decision making (Owen 1998; Penrose et al. 1998; Williams et al. 1998; Jackson 2001). The Government of Manitoba recently started a broad-area planning initiative for the east side of Lake Winnipeg, a large geographic area bounded by Lake Winnipeg on the west, the Ontario-Manitoba border to the east, Sagkeeng First Nation to the south, and Oxford House to the north. The objective was to 'bring together local communities, First Nations, industry and environmental organizations to develop a vision for land and resource use in the area that respects both the boreal forest and the needs of local communities' (Manitoba Conservation 2003). A round table and advisory committees were established to meet with local communities and organizations in the development of a vision and plan for the region that would precede any project EAs for activities such as forestry operations, parks development, and road construction. As part of this ongoing planning effort, the government has supported traditional area land-use planning activities with thirteen affected First Nations. Coordinators have been hired to assist with these efforts, and the East Side Policy Coordination Committee will ensure that broad-area planning is maintained as these local plans are developed.

Given these experiences and similar policy initiatives in Canada, it is our view that EA policy makers and practitioners should continue to seek ways to integrate regional planning (with its increasingly concomitant early and ongoing public participation) and formal EA processes. If wide-area planning precedes assessment, EA legislation should then require that the plan be considered during the assessment (unless the EA itself was done in a sufficiently open and participative way). In cases where broader plans do not exist, normative issues would continue to be raised by members of the public during the EA of individual projects. To facilitate the development of broader regional plans, government departments involved in EA processes should be required to alert decision makers when an undertaking is proposed in a region where no land-use plan or policy exists. It would then be up to the decision makers to decide whether EA consultation activities should include broader issues than those raised in the context of the narrow scope of a proposed undertaking or whether a broad planning exercise needs to be carried out before any decision is rendered on the individual undertaking.

Mutual Learning

Building on the idea of ongoing participation, another key future direction for public involvement in EA relates to taking greater advantage of the learning opportunities resulting from participation. As noted earlier, participation in EA creates important opportunities for learning related to sustainability at various organizational levels. First, it affords opportunities for individual practitioners and decision makers to identify social values and learn from local or traditional knowledge. Second, it provides opportunities for members of the public to acquire scientific and technical knowledge, learn about their community and the interests of fellow citizens, and engage in collective political action. In addition, participation in post hoc and cumulative effects assessments creates opportunities for ENGOs, communities, and other forms of social organization to learn from past development decisions.

Promising forums for mutual learning are created by mechanisms designed to provide ongoing participation, such as community advisory committees and co-management boards. Organizations of this type are being used increasingly in EA, including in research and monitoring functions. Two prominent examples can be found in the low-level flying EA from Labrador and the BHP Ekati diamond mine EA from the Northwest Territories.

The first example involved a federal assessment of an agreement between Canada (the Department of National Defence) and some of its NATO allies (the United Kingdom, Germany, and the Netherlands) to permit low-level military training flights over the Quebec-Labrador peninsula. A key outcome of this controversial EA was the creation of the Labrador Institute for Environmental Monitoring and Research, a co-management organization involving local and Aboriginal partners. Since so little was actually known about the impacts of low-level flying on people and the environment, the institute was established to conduct comprehensive, multidisciplinary scientific research on the flying program. For more information on the institute and its current activities, visit http://www.iemr.org/ (accessed March 2009).

In the second example, a federal review panel approved the BHP Ekati diamond mine but with 29 conditions, several dealing with environmental monitoring and mitigation. In 1996, when the federal government conditionally approved the project, it required an agreement to be drawn up among Canada, the Northwest Territories, and the project proponent (BHP Diamonds Inc.) to ensure that the conditions of the review panel were met. The agreement called for the establishment of the Independent Environmental Monitoring Agency, which included representation from local and Aboriginal communities. The tasks of the agency included reviewing monitoring and management plans and results, encouraging the use of traditional knowledge, bringing concerns of Aboriginal peoples and the general public to the proponent and government, and keeping Aboriginal peoples and the public informed about agency activities and findings. The agency's most recent annual report can be reviewed at http://www.monitoringagency.net/ (accessed March 2009).

Although the reviews are mixed regarding the efficacy and results of the organizations described above (Macleod Institute 2000; Ross 2004), these types of institutional arrangements for ongoing public participation are highly conducive to mutual learning. They not only create opportunities for communication and deliberation, but also provide forums for trust building, risk taking, problem solving, and conflict resolution. All of these are key dynamics for integrating knowledge systems and modifying basic attitudes, values, and behaviour (Diduck et al. 2005). Our position, therefore, is that proponents, EA practitioners, and decision makers should continue to expand the use of mechanisms for ongoing participation. Doing so will help maximize the learning benefits that can be derived from participation in EA.

Alternative Dispute Resolution

The use of mediation and other forms of *alternative dispute resolution* (ADR) has been almost non-existent in Canadian EA. There is increasing recognition, however, that various ADR tools could be applied usefully at most points of the process when there is conflict among the parties. Such adjustments in process would signal the need for approval agencies and proponents to give serious consideration to more collaborative techniques of participation. In this regard, the default forms of participation, such as open houses and town hall meetings, would be viewed as mere on-ramps to more participatory involvement techniques.

Indications that such a change in thinking is happening can be found on a few fronts. For years at the federal level, the CEAA included the provision that if a mediation attempt failed, the EA would move directly to a panel review. Many viewed this as a deterrent to the use of mediation and the main reason why no mediations were attempted in the first five years of the CEAA's existence. When the Act was amended in 2005, this requirement was dropped in an attempt to encourage the use of mediation.

At the provincial level, Quebec is by far the most active in using ADR tools. Under certain circumstances, the Bureau d'audiences publiques sur l'environnement (BAPE), an independent body responsible for environmental assessment hearings in the province, can engage in environmental mediation. Under the EA process in Quebec, any individual may request that a public hearing be carried out by the BAPE. In reviewing the input received, the BAPE can request that the minister of environment allow it to proceed by way of mediation rather than public hearing. Since 1990, there have been 39 projects that proceeded in this fashion, and there are now formal rules for BAPE mediation processes.

Manitoba's Environment Act includes provisions for mediation, as noted earlier. These have been used very rarely, but through a combination of political will and administrative support on the part of the Clean Environment Commission, some mediations have occurred in recent years. One of these, the Rothsay Rendering EA, was quite successful in that the community and company were able to agree on steps the company needed to take to reduce odours and the company was able to secure funding to install the technology required. The commission is taking a cautious approach to using mediation and is trying to find ways to level the playing field for all participants at the mediation table. It has had at least one case where the public participants requested mediation rather than a hearing.

In our view, these activities in Quebec and Manitoba and the legislative change at the federal level are positive reforms that should guide change in other jurisdictions.

Community-Based Assessment

New approaches to working with residents of small communities to assess the impacts of local projects are being tested in various parts of the developing world. What has become termed *community-based environmental assessment* (CBEA) has merged aspects of conventional EA with aspects of participatory local appraisal (Chambers 1994) to form an innovative way to assess smaller, community-based projects that utilize natural resources for basic livelihood needs (Spaling 2003; CIDA 2005; Neefjes 2000; Pallen 1996). Typical projects include boreholes, gravity water systems, small reservoirs, agro-forestry, fish ponds, construction of latrines, clinics, schools, and small bridges. Since these projects interact directly with bio-

physical systems, many already under stress, there is potential for resource degra-
dation through over-extraction, land clearing, soil erosion, contamination, and
other forms of exploitation. Application of EA to these projects is emerging as
a way to facilitate management of local resources and ensure continued project
benefits (Spaling 2003).

CBEA is a highly participatory approach in which local communities are directly
involved in conducting EAs. In fact, local residents do much of the assessment
themselves with the help of facilitators and environmental practitioners. Recent
studies in Kenya and Costa Rica show the benefits of CBEA from the community's
perspective, including individual and social learning outcomes, greater social cohesion, and better decision making (Montes 2008; Sinclair et al. 2009). While the
context for CBEA has generally been the developing world, many of the reasons
for using the approach are applicable to small projects in Canada—reasons such
as the desire to ensure that projects are assessed (and assessed cost effectively and
efficiently) and to build local capacity for participation in EA. Furthermore, in the
right Canadian setting—one involving a localized project with broad support in
a small community (perhaps in a First Nation)—it is reasonable to expect to see
some of the same types of benefits witnessed in other parts of the world.

Legal Requirements

The benefits of having a legislated and mandatory basis for EA are well established
in the Canadian literature (Jeffery 1991; Gibson 1993; Hazell 1999). Such a foundation reduces administrative discretion, enhances procedural certainty, clarifies
authority, creates rights and responsibilities, and presents opportunities for judicial remedies. In spite of this, as noted throughout this chapter, high levels of
discretion and critical legislative gaps characterize public participation regimes in
Canada. Government EA officials have argued that discretion is necessary to allow
for flexibility in tailoring public participation to the multitude of situations created by different EA cases. The opposite view is that with too many discretionary
provisions, the law provides little in the way of direction for achieving meaningful
public participation. Sinclair and Doelle (2003) attribute many of the weaknesses
found in EA public participation activities to missing, discretionary, or ineffectual
legal provisions. In essence, too much discretion was given to regulators in deciding whether or not there should be public participation in any given case, when
it should occur, and what it would entail, and this has led to weak and ineffectual
participatory events.

Given this, we advocate a rethinking of the legal underpinnings of public participation in Canadian EA legislation and suggest that changes are necessary to
create a clear and mandatory foundation for public participation. Legislated reforms are necessary to address the concerns and enable the future directions outlined above. As argued elsewhere (Sinclair and Doelle 2003; Doelle and Sinclair
2006), many of the changes needed are relatively modest and are consistent with
the views of diverse EA reformers both within and outside of government. They
also pertain to familiar aspects of EA public participation that have been debated

since the inception of legislated EA. At the same time, in totality a coherent package of reforms aimed at resolving the concerns and enabling future directions could make a significant contribution to reconceptualizing participation in EA, seeing it transformed from a series of proponent-driven opportunities for input at discrete points in a process to meaningful public-driven participation occurring throughout the project/policy cycle from development to decommissioning.

Conclusion

Public participation is essential to effective and fair environmental assessment in Canada and elsewhere in the world. Meaningful participation provides diverse social, political, and environmental benefits. It actualizes democratic principles, enables individual and community empowerment, facilitates individual and collective learning, informs planning and decision making, and affords numerous other practical benefits. However, whether such advantages are realized in practice depends on various factors, including pertinent legislative and policy frameworks. Key aspects of EA public participation regimes in Canada include adequate notice, access to information, participant assistance, public comment, and public hearings.

Owing to Canada's relatively long EA history and the resulting knowledge gained by EA practitioners, this country is often looked to as a leader in public participation. However, this does not mean that public participation practice in Canada is flawless. There are long-standing concerns about weaknesses in the way important aspects of participation are implemented, including lack of shared decision making, lack of early (i.e., normative and strategic) involvement, information and communication deficiencies (e.g., inadequate notice and inaccessible information), and insufficient intervenor funding.

These concerns highlight the challenges facing practitioners, policy makers, and others who care to strengthen the role of public participation in EA. They also illuminate promising future directions, such as the promotion of early and ongoing involvement, the enhancement of opportunities for mutual learning, the increased use of alternative dispute resolution, and the creation of a clear, mandatory legislative base for public participation. With respect to this last point, only relatively modest legislative reforms would be necessary to achieve these goals. At the same time, a coherent, focused package of reforms would help reconceptualize public participation in EA in a community-based rather than a managerial (or top-down) paradigm. This would facilitate the formation of partnerships among industry, government, and civic organizations, thereby promoting economic development that is community oriented, equitable, and sustainable (both environmentally and socially) over the long term.

Acknowledgements

We are grateful to Richard Roberts, Patricia Fitzpatrick, and Peter Duinker for reviewing early drafts of this chapter. Shortcomings in the final manuscript remain our sole responsibility.

References

Alberta Environment. 2004. *Alberta's environmental assessment process*. Pub No. I/990, September. Edmonton: Alberta Environment. http://environment.gov.ab.ca/info/library/6964.pdf. Accessed 20 June 2008.

Arnstein, S. 1969. A ladder of citizen participation. *Journal of the American Institute of Planners* 35 (4): 216–24.

Berger, T.R. 1977. *Northern frontier, northern homeland: The Report of the Mackenzie Valley Pipeline Inquiry*. Ottawa: Minister of Supply and Services.

Blaug, E.A. 1993. Use of environmental assessment by federal agencies in NEPA implementation. *Environmental Professional* 15 (1): 57–65.

Burcombe, J. 1998. *Federal Environmental Assessment Index and public registries*. Brief to President of CEAA on behalf of Mouvement au Courant, Montreal.

Canadian Environmental Network, Environmental Planning and Assessment Caucus. 1988. *A federal environmental assessment process: The core elements*. Ottawa: Canadian Environmental Network, Environmental Planning and Assessment Caucus.

Chambers, R. 1994. The origins and practice of participatory rural appraisal. *World Development* 22 (7): 953–69.

CIDA (Canadian International Development Agency). 2005. *Environment handbook for community development initiatives: Second edition of the handbook on environmental assessment of non-governmental organizations and institutions programs and projects*. Gatineau: CIDA. Available at http://www.acdi-cida.gc.ca/CIDAWEB/acdicida.nsf/En/JUD-47134825-NVT.

Clark, B.D. 1999. Capacity building. In J. Petts (Ed.), *Handbook of environmental impact assessment*, 145–77. Oxford: Blackwell Science.

Connor, D.M. 2001. *Constructive citizen participation: A resource book*. Victoria, BC: Connor Development Services Ltd.

Croal, P., and K. Grady. 1998. Environmental assessment in Canada. *EIA Newsletter* 17, University of Manchester.

Devlin, J.F., N.T. Yap, and R. Wier. 2005. Public participation in environmental assessment: Case studies on EA legislation and practice. *Canadian Journal of Development Studies* 26 (3): 487–500

Diduck, A.P. 2004. Incorporating participatory approaches and social learning. In B. Mitchell (Ed.), *Resource and environmental management in Canada: Addressing conflict and uncertainty*, 497–527. Toronto: Oxford University Press.

Diduck, A.P., N. Bankes, D. Clark, and D. Armitage. 2005. Unpacking social learning in social-ecological systems: Case studies of polar bear and narwhal management in Northern Canada. In F. Berkes, R. Huebert, H. Fast, M. Manseau, and A. Diduck (Eds), *Integrated management, complexity and diversity of use: Responding and adapting to change*. Calgary: University of Calgary Press.

Diduck, A.P., and B. Mitchell. 2003. Learning, public involvement and environmental assessment: A Canadian case study. *Journal of Environmental Assessment Policy and Management* 5 (3): 339–64.

Diduck, A.P., and A.J. Sinclair. 2002. Public participation in environmental assessment: The case of the nonparticipant. *Environmental Management* 29 (4): 578–88.

Doelle, M., and A.J. Sinclair. 2006. Time for a new approach to environmental assessments: Promoting cooperation and consensus for sustainability. *Environmental Impact Assessment Review* 26 (2): 185–205.

Dorcey, A., L. Doney, and H. Ruggeberg. 1994. *Public involvement in government decision-making: Choosing the right model*. Victoria, BC: Round Table on the Environment and the Economy.

Fitzpatrick, P., and A.J. Sinclair. 2003. Learning through public involvement in environmental assessment hearings. *Journal of Environmental Management* 67 (2): 161–74.

Forester, J. 2006. Participatory governance as deliberative empowerment: The cultural politics of discursive space. *American Review of Public Administration* 36 (1): 19–40.

Gallagher, T.J., and W.S. Jacobson. 1993. The typography of environmental impact statements: Criteria, evaluation, and public participation. *Environmental Management* 17 (1): 99–109.

Gibson, R.B. 1993. Environmental assessment design: Lessons from the Canadian experience. *Environmental Professional* 15:12–24.

Hayward, G., A.P. Diduck, and B. Mitchell. 2007. Social learning outcomes in the Red River Floodway environmental assessment. *Environmental Practice* 9 (4): 239–50.

Hazell, S. 1999. *Canada v. the environment: Federal Environmental Assessment 1984–1998.* Toronto: Canadian Environmental Defence Fund.

International Association for Public Participation. 2001. IAP2 Public Participation Toolbox. http://www.iap2.org/practitionertools/index.html. Accessed 6 May 2002.

Jackson, L.S. 2001. Contemporary public involvement: Toward a strategic approach. *Local Environment* 6 (2): 135–47.

Jeffery, M.I. 1991. The new Canadian Environmental Assessment Act—Bill C-78: A disappointing response to promised reform. *McGill Law Journal* 36:1071–88.

Kidd, S. 1998. My adventures at the public registry. *Eco-Journal* 11 (3): 5–6.

Lynn, F.M. 2000. Community-scientist collaboration in environmental research. *American Behavioral Scientist* 44(4): 648–62.

Lynn, S., and P. Wathern. 1991. Intervenor funding in environmental assessment processes in Canada. *Project Appraisal* 6 (3): 169–73.

Macleod Institute. 2000. *Independent Environmental Monitoring Agency evaluation report.* Calgary: Macleod Institute.

McMullin, S.L., and L.A. Nielson. 1991. Resolution of natural resource allocation conflicts through effective public participation. *Policy Studies Journal* 19 (3/4): 553–9.

Manitoba Conservation. 2003. *East side of Lake Winnipeg: Broad area planning initiative.* Winnipeg: Government of Manitoba.

May, E. 2003. Sierra Club of Canada wins one for public participation. EA Reporter (Environmental Planning and Assessment Caucus, Canadian Environment Network), no. 6 (Spring): 13–14.

Meredith, T.C. 1995. Assessing environmental impacts in Canada. In B. Mitchell (Ed.), *Resource and environmental management in Canada*, 360–83. Toronto: Oxford University Press.

———. 1997. Information limitations in participatory impact assessment. In A.J. Sinclair (Ed.), *Canadian environmental assessment in transition*, 125–54. Waterloo, ON: University of Waterloo.

MiningWatch. 2008. *Federal government denies need to consult public on major mining projects: Ruling will test strength of revised environmental assessment law.* Retrieved from http://www. miningwatch.ca/index.php?/bcmetals/red_chris_fedl_court_of_appeal.

Montes, J. 2008. Community environmental assessment in rural Kenya—Decision making for a sustainable future. Unpublished master's thesis, Natural Resources Institute, University of Manitoba.

Neefjes, K. 2000. *Environments and livelihoods: Strategies for sustainability.* London: Oxfam.

Owen, S. 1998. Land use planning in the nineties: CORE lessons. *Environments* 25 (2/3): 14–26.

Palerm, J.R. 2000. An empirical-theoretical analysis framework for public participation in environmental impact assessment. *Journal of Environmental Planning and Management* 43 (5): 581–600.

Pallen, D. 1996. *Environmental assessment manual for community development projects.* Ottawa: Asia Branch, Canadian International Development Agency.

Parenteau, R. 1988. *Public participation in environmental decision-making.* Ottawa: Minister of Supply and Services.

Penrose, R.W., J.C. Day, and M. Roseland. 1998. Shared decision making in public land planning: An evaluation of the Cariboo-Chilcotin CORE process. *Environments* 25 (2/3): 27–47.

Petts, J. 1999. Public participation and environmental impact assessment. In J. Petts (Ed.), *Handbook of environmental impact assessment*, 145–77. Oxford: Blackwell Science.

———. 2003. Barriers to deliberative participation in EIA: Learning from waste policies, plans and projects. *Journal of Environmental Assessment Policy and Management* 5:269–93.

Prystupa, M., D. Hine, C. Summers, and J. Lewko. 1997. The representativeness of the Elliot Lake uranium mine tailings areas EARP public hearings. In A.J. Sinclair (Ed.), *Canadian environmental assessment in transition*, 51–76. Waterloo, ON: University of Waterloo.

Roberts, R. 1998. Public involvement in environmental impact assessment: Moving to a 'Newthink'. *Interact* 4 (1): 39–62.

Ross, W.A. 2004. The independent environmental watchdog: A Canadian experiment in EIA follow-up. In A. Morrison-Saunders and J. Arts (Eds), *Assessing impact: Handbook of EIA and SEA follow-up*. London: Earthscan.

Rowe, G., and L.J. Frewer. 2005. A typology of public engagement mechanisms. *Science Technology and Values* 30 (2): 251–90.

Shepard, A., and C. Bowler. 1997. Beyond the requirements: Improving public participation. *Journal of Environmental Planning and Management* 40 (6): 725–38.

Sims, L. and A.J. Sinclair. 2008. Learning through participatory resource management programmes: Case studies from Costa Rica. *Adult Education Quarterly* 58 (2): 151–68.

Sinclair, A.J. 2002. Public involvement in sustainable development policy initiatives: Manitoba approaches. *Policy Studies Journal* 30 (40): 423–44.

Sinclair, A.J., and A.P. Diduck. 1995. Public education: An undervalued component of the environmental assessment public involvement process. *Environmental Impact Assessment Review* 15 (3): 219–40.

———. 2001. Public involvement in EA in Canada: A transformative learning perspective. *Environmental Impact Assessment Review* 21 (2): 113–36.

Sinclair, A.J., A.P. Diduck, and P. Fitzpatrick. 2008. Conceptualizing learning for sustainability through environmental assessment: Critical reflections on 15 years of research. *Environmental Impact Assessment Review* 28 (7): 415–28.

Sinclair, A.J., and M. Doelle. 2003. Using law as a tool to ensure meaningful public participation in environmental assessment. *Journal of Environmental Law and Practice* 12 (1): 27–54.

Sinclair, A.J., and P. Fitzpatrick. 2003. Provisions for more meaningful public participation still elusive in new Canadian EA bill. *Impact Assessment and Project Appraisal* 20 (3): 161–76.

Sinclair, A.J., L. Sims, and H. Spaling. 2009. Community-based approaches to strategic environmental assessment: Lessons from Costa Rica. *Environmental Impact Assessment Review*, 29(3): 147–156.

Smith, L.G. 1982. Mechanisms for public participation at a normative planning level in Canada. *Canadian Public Policy* 8 (4): 561–72.

———. 1993. *Impact assessment and sustainable resource management*. New York: J. Wiley.

Spaling, H. 2003. Innovations in environmental assessment of community-based projects in Africa. *Canadian Geographer* 47 (2): 151–68.

Stewart, J.M., and A.J. Sinclair. 2007. Meaningful public participation in environmental assessment: Perspectives from Canadian participants, proponents and government. *Journal of Environmental Assessment and Policy Management* 9 (2): 1–23.

Sullivan, W.C., F.E. Kuo, and M. Prabhu. 1996. Assessing the impact of environmental impact statements on citizens. *Environmental Impact Assessment Review* 16 (3): 171–82.

Susskind, L., and J.L. Cruikshank. 1987. *Breaking the impasse: Consensual approaches to resolving public disputes*. New York: Basic Books.

Usher, P.J. 2000. Traditional ecological knowledge in environmental assessment and management. *Arctic* 53 (2): 183–93.

Webler, T., H. Kastenholz, and O. Renn. 1995. Public participation in impact assessment: A social learning perspective. *Environmental Impact Assessment Review* 15 (5): 443–63.

Williams, P.W., J.C. Day, and T. Gunton. 1998. Land and water planning in BC in the 1990s: Lessons on more inclusive approaches. *Environments* 25 (2/3): 1–7.

Wood, C.W. 2003. *Environmental impact assessment: A comparative review*. 2nd ed. London: Prentice Hall.

Chapter 5

The Hearing Process in Environmental Impact Assessment: As Concept and as Practised in Ontario

Len Gertler

One of the features of our times is the conspicuous emergence of environmental issues on the public policy agenda of industrial societies. Two centuries of cumulative effects on air, water, land, climate, and biotic species came to a head in 1972 at the United Nations Stockholm Summit on the Human Environment, the first-ever international gathering on the health of planet earth. The ensuing 'Declaration' was a historic statement, pinpointing an incipient crisis:

> We see around us growing evidence of man-made harm in many regions and dangerous levels of pollution in water, air, earth and living beings; major and undesirable disturbances to the ecological balance of the biosphere, destruction and depletion of irreplaceable resources; and gross deficiencies, harmful to the physical, mental and social health of our man-made environment, particularly in the living and working environment.

and the urgent *need for action:*

> A point has been reached in history when we must shape our actions throughout the world towards more prudent care for their environmental consequences. . . . To defend and improve the human environment for present and future generations has become an imperative goal for mankind. (United Nations Environment Programme, 15 June 2003)

Concurrent with the heightening of environmental awareness and the commitment signalled by Stockholm, governments in Canada began to establish the first environmental agencies to create and administer new policies and legislation—for example, federally, in Ottawa, in 1970, with the Government Organization Act, and provincially, in Ontario, in 1971, with the Environmental Protection Act.

Environmental Assessment: The Ontario Model

The role of the hearing process in environmental decision making will be considered in this chapter, mainly with reference to the experience in the province of Ontario. This is because I have had direct experience, from 1990 to 2001, as an environmental adjudicator in that province and because of the centrality of the Ontario story, both in development and its repercussions and in the evolution of public policy on the environment. One of the first initiatives of the new Environment Ministry in Ontario was the release for discussion in September 1973 of its *Green Paper on Environmental Assessment*. The paper led to the adoption of the Environmental Assessment Act (EAA) in 1975, and when the Act came into force on 20 October 1976, it was the first legislated environmental assessment (EA) process in Canada (Valiante 1998, 217, 218). And, notably, it provided for public hearings.

The main features of the EA system in Ontario are summarized in Table 5.1, which presents two versions:

- *Baseline*, as contained in the Revised Statutes of Ontario, 1990—the original Act with the addition of minor amendments up to an Office Consolidation of January 1994 (R.S.O. 1990; c. E.18 and c. 27)
- *Altered*, reflecting the major amendments introduced in Bill 76, changing the name of the Act to the Environmental Assessment and Consultation Improvement Act (EACIA), 1996, after the change of government in 1995

The two versions are presented comparatively in terms of the purpose of the Act, the definitions of *environment* and *undertaking*, the content of the EA, types of EA, public hearings, other public involvement, and decisions. What the table reveals is a system designed, in its basic elements, to have the potential for broad scope and meaningful public involvement. This potential, however, is limited by some serious constraints, inherent in the baseline version and intensified in the post-1995 altered version.

The major qualifiers of both versions are

- ministerial discretion to exempt public sector projects from the EA process;
- the exclusion of private sector undertakings from the EA process unless explicitly designated by the minister; and
- the minister's discretion, very broadly defined, to refuse the request of 'any person' for a public hearing on a proponent's application under the environmental assessment legislation.

Under the altered Ontario EA model, the EA process may be further constrained by

- ministerial authority to determine the scope of an EA through the decision on the approval of the EA terms of reference (TOR);

Table 5.1   Environmental assessment: The Ontario model

FEATURES	VERSIONS	
	BASELINE, 1975 to 1995 R.S.O. 1990, amended to January 1994	ALTERED, 1995 to 2003 R.S.O. 1990, amended to 15 June 2001
Purpose	The betterment of the people of the whole or any part of Ontario by providing for the protection, conservation, and wise management in Ontario of the environment. s. 2	Same, s. 2
'Environment'	Air, land, or water; plant, animal, and human life; social, economic, and cultural conditions; buildings and other physical structures and their effects; and the relationships between the foregoing. s. 1	Same, s. 1(1)
'Undertaking'	A proposal, plan, or program in the public sector (provincial and municipal) or the private sector (note 'Types of EA' below)	Same s. 1(1)
Content of EA	For the undertaking: *description*—purpose; rationale; alternative methods; alternatives to; of the affected environment; of the expected effects; of the action necessary to prevent, change, mitigate or remedy the effects; and *evaluation* of the advantages and disadvantages to the environment of the options: the undertaking, alternative methods, and alternatives to. s. 5(3)	Same s. 6.1(2), but minister may set scope of EA terms of reference s. 6(4)—with addition of 'a description of any consultation about the undertaking by the proponent and the results of the consultation.' Note: Separate acceptance of EA, not required.
Types of EA	*Individual* by proponent for public sector undertakings, unless *exempted* by the minister; and for private sector undertakings *designated* by the minister; acceptance of EA document and approval of undertaking required before proceeding. ss. 29, 39. *Class*, not in Act, but by 1990s was common practice for small-scale, recurring projects with minor potential impacts; governed by provincial guidelines.	*Individual* same, s. 3.2(1) *Class*, new provision for a 'class of undertakings'; approval of the class exempts individual undertakings in the class from EA process; but no criterion limiting the use of the class mechanism to small, recurring projects with minor impacts. Part II.1

Continued

Table 5.1 Continued

FEATURES	VERSIONS	
	BASELINE, 1975 to 1995 **R.S.O. 1990, amended to** **January 1994**	**ALTERED, 1995 to 2003** **R.S.O. 1990, amended to** **15 June 2001**
Public Hearings	Any person, by written notice to the minister, may require a hearing by the board re the EA, the review of the EA and the undertaking. s. 7(2) (b). But the minister has 'absolute discretion' to refuse a hearing on grounds that the request is 'frivolous or vexatious' or 'unnecessary or may cause undue delay'. s. 13(d); referral to board by minister where he/she deems it 'advisable', s. 12(2); complementary —Intervenor Funding Project Act (April 1989), funds for public interest intervenors in hearings (a pilot project).	Same, s. 7.2(3) and s. 9.3(1) (2) New powers—minister may refer only a part of the EA application to the tribunal; 'give such directions or impose such conditions on the referral as the Minister considers appropriate'; and set a deadline for the tribunal decision. ss. 9(1) (5); 9.2(1) (2); same powers of ministerial referral, 9.1(1); intervenor funding legislation allowed to lapse, 1996.
Other Public Involvement	Notification of public re submission of EA document and completion of government review; and right to examine and comment and have written submissions considered in decisions. ss. 14(2), 8	Baseline features retained; also, proponent shall consult with interested persons on preparation of EA terms of reference and docs. for both types of EA. ss. 5.1, 13.1
Decisions	Minister decides on acceptability of EA and approval of the undertaking; or the board, if referred by minister. Approval of Cabinet required; if decision by the board, Cabinet may review and vary within 28 days. ss. 9, 14, 12	Same, s. 9(1) (2) re minister and tribunal and s. 11.2 re review of tribunal decision by Cabinet

- the wide range of the new class of EA provisions, which could have the effect of extending exemptions from the requirements of the individual EA process to additional undertakings;
- the power of the minister to further control the public hearing process by
  – assigning only a part of an EA application to a tribunal,
  – giving directions and imposing conditions on an EA referral to the Environmental Review Tribunal (ERT) for a hearing, and
  – setting a decision deadline on a referral to the tribunal; and the demise, in 1996, of the Intervenor Funding Project Act.

A 2003 decision of the Ontario Divisional Court (released 17 June) has created a strong presumption that there are principled limits to the exercise of the minister's additional discretionary powers under the altered version of the Ontario EA model. The undertaking was an application from Canadian Waste Services Inc. for an expansion of the existing Richmond Landfill Site near Napanee from a permitted limit of 125,000 tonnes/year to a capacity of 750,000 tonnes/year of 'non-hazardous waste' from an all-Ontario service area for 25 years. The case resulted from an application for judicial review filed by those in opposition, the Mohawks of the Bay of Quinte and some local residents. The court ruled that the Ontario minister of the environment, under the power to 'approve the proposed terms of reference' of an environmental assessment (section 6 (4)), sanctioned an EA that excluded several key features of the EA prescribed by the Act (section 6.1 (2)), namely,

- whether there was a demonstrable need for the expansion (i.e., the rationale for the undertaking);
- whether there were preferable 'alternatives to' (e.g., the 3 Rs—reduce, reuse, recycle) the expansion; and
- whether there were more environmentally suitable sites.

Accordingly, the court supported the applicants' submission that terms of reference with a putative narrow scope contravened the Environmental Assessment and Consultation Improvement Act and were, as such, inadmissible. The disposition of the court was that 'the decision of 16 September 1999 approving the ToR is quashed' (Divisional Court, Ontario, 17 June 2003).

For those readers who may be unfamiliar with the scope of, the various participants in, and the mechanisms of the public hearing process, particularly as practised by the Environmental Review Tribunal, Ontario, Tables 5.2 and 5.3 explain the hearing process and the steps leading to a tribunal decision.

Unfortunately, in a decision of 25 August 2004, the Court of Appeal for Ontario, responding to an appeal, set aside the 17 June 2003 decision of the Divisional Court.

Table 5.2 Environmental hearings, Ontario: Overview and preliminaries

(A) The Hearing Process	(B) Public Information Sessions
• Governed by the procedures provided by: – The enabling legislation (e.g., Environmental Assessment Act) – The Statutory Powers Procedure Act – The Tribunal's Rules of Practice – The Tribunal members conducting the hearing	• Conducted by staff/members • Information about the hearing process • Information about the applicant

Continued

Table 5.2   Continued

• Usual steps in the hearing process:

– Tribunal publishes Notice of Hearing

– Public information session

– Preliminary hearing

– The hearing

(C) Preliminary Hearings	**(D) Exchange of Information**
• Identify/designate persons who wish to participate	• Procedural directions
– Presenters	– Hearing documents
– Participants	– Exchange of all relevant documents
– Parties	– Meetings of consultants, professional and technical Staff
• Determine and simplify concerns/issues	– Consultants are required to meet and exchange information
• Determine whether any agreed facts	– Joint submission of consultants, et al.
• Set deadlines for exchange of information	
• Schedule dates and location for the hearing	
• Develop procedural directions	
• Hear preliminary motions	

Source: Environmental Review Tribunal, Ontario.

Table 5.3   Environmental hearings, Ontario: Participation to decision

(E) Participating in a Hearing	**(F) Witnesses/Sharing Information**
• As a Presenter	• Guidelines for Technical/Opinion Evidence
– presents views during day or evening session	– fair and full disclosure
– may provide supplemental written statement	– plain language
– only needs to attend to make presentation	– resolve issues raised by parties

Continued

Table 5.3   Continued

• As a Participant	• **A witness statement** should include
– receives a copy of exchanged documents	– an indication of the witness's interest in the application
– attends site visits	– an indication of whether the evidence is factual or opinion
– makes submissions	
– cannot call or cross-examine witnesses,	– a complete statement of the evidence
– seek review	• Practice Directions
• As a Party	– hearing documents
– all of the above plus be, call, or cross-examine witnesses	– exchange of all relevant information
	– consultants meetings
– make submissions, bring motions	– consultants are required to meet and exchange information
– seek costs (if legislation permits), seek review	
– often represented by counsel	
(G) Science and Adjudication	**(H) The Tribunal's Decision**
• Different Kinds of Evidence	• Approve, approve with conditions, or refuse
– oral	
– documents	• Relevant considerations?
– transcripts	• Review of the decision
• Complementary Evidence	– the minister (with Cabinet approval)
• Conflicting Evidence	– the court (judicial review on points of law)
• Public Values and Scientific Findings	• Common-law reconsideration

Source: Environmental Review Tribunal, Ontario.

Environmental Hearings: The General Case

While keeping the Ontario EA model in mind, in considering the role of public hearings in environmental decision making in the province of Ontario, we need to cast a wider net. To that end, *assessment* must be considered in its broadest meaning—namely, making evaluations or judgments. Thus, hearings concerned with EA in that sense will include those that may be held under the entire range of relevant legislation. This will encompass, apart from the Environmental Assessment Act, the Environmental Protection Act (EPA), the Ontario Water Resources Act (OWRA), the Consolidated Hearings Act (CHA), and the Niagara Escarpment Planning and Development Act (NEPDA). These encompass the great majority of hearings, for both applications and appeals, that are the responsibility of the Environmental Review Tribunal, Ontario. In making this perhaps heroic choice, there is an obligation to set out the common attributes of such hearings that entitle them to the imprimatur of 'environmental assessment'. There are eight key features:

1. They are all governed by legislation that explicitly or implicitly invokes a public purpose.

2. The public purpose relates to maintaining the integrity of the environment, defined within a continuum from the physical and biophysical (structures and air, land, water) to the human (social, economic, cultural).

3. All of these governing statutes make provision, for certain defined purposes, for public hearings issuing in decisions or recommendations, conducted by a tribunal or hearing officers (under NEPDA).

4. While the scope of the undertaking, subject to evaluation, may be broad or narrow, the adjudicative process is based on evidence submitted primarily by parties expressing the viewpoints of proponents and the impacted persons, including intervenors affected both directly and indirectly, in addition to evidence from the responsible public agencies.

5. The outcomes of such hearings, the *decisions* or *recommendations* prescribing approval with or without conditions or rejection, will both individually or cumulatively have an impact on the quality of the environment—protecting, enhancing, or degrading.

6. From this, it follows that the hearing authority has a special responsibility, in the words of R.W. Macaulay, '[t]o conclude a hearing with a balanced record which reflects not only the interests of the parties, but most important, the public interest' (Macaulay 1989).

7. The seventh common feature, not necessarily related to the environmental connection, is that the frequency and number of hearings, under all of this legislation, have been inconsistent and erratic. Some data are available (courtesy of the office of the ERT, Ontario), beginning in1990, in terms of decisions rendered on applications under the various acts for each fiscal year, 1 April to 31 March. Under the EPA, for example, the number of hearings ranges from a high of six in each of 1993/4 and 1994/5 to two, five, one, two, two, and zero in each of the last six years; under the CHA, there were 33 hearings with decisions rendered in the six years, 1990/1 to 1995/6,

compared to 14 in the following six years. Nine of the CHA hearings in the first period involved matters under the EAA, in association with matters under other legislation, notably the Planning Act, involving joint hearings with the Ontario Municipal Board. In the same period, there were only two hearings, exclusively under the EAA. After the 1996 amendments to mid-2003, there were two referrals to the tribunal, in December 1997, and only one was brought to a decision. With regard to the OWRA, the record shows that there were very few applications during the reporting period—one in 1993/4 and one in 1997/8. Appeals under the Act were somewhat more frequent—10 in the 1990/1–1995/6 period; eight in the 1996/7–2000/1 period, and then, following the water pollution tragedy at Walkerton in the spring of 2000, a spurt in appeals to 15 in a single year, in 2001/2. These latter were mainly appeals by municipalities in response to ministry upgrade orders on municipally owned water systems. While some data are available for numbers of hearing decisions before 1990, conveying the specifics is problematic because there is not a common base year for all of the legislation owing to differences in years of origin and reporting records. The common feature is a marked downward trend (Dunn 2003; Faeita 1995).

There is one more attribute of EA hearings, as defined here, that moves the discourse from description to analysis and, ultimately, to prescription, and this is the aspect that lies at the core of my position. It is this:

8. Not only is a genuinely participatory hearing on environmental matters important for its intrinsic consequences, but it is one of the institutions that has emerged in our times that can play a meaningful role in the struggle for a truly democratic society.

The implication of this last proposition is that the opportunity for effective public hearings is the primary, critical, and transcending requirement for sound environment-related legislation. This is my own personal conviction, based partly on my more than 10 years of experience as an adjudicator—particularly the experience of having presided over hearings with both considerable and minimal participation and with reasonably well- and decidedly ill-balanced information bases, and of having agonized over how to adhere to a state-of-the-art hearing process, grounded on the evidence brought forward. But there is a more important consideration—that the rationale for participatory public hearings is grounded in the fundamental values of democracy. This assertion is given credence by the converging ideas of three highly qualified witnesses who, starting from different origins, reinforce the assertion about the linkage between EA hearings and democracy. My witnesses are James P. Boggs, recently of the Public Policy Research Institute, University of Montana (in Missoula, Montana); the late C. Brough Macpherson, former professor of political science, University of Toronto, and author of numerous works on democratic theory; and John Friedmann, former director of the Urban Planning Program, University of California in Los Angeles, and a writer, informed by extensive international experience, on the philosophy of planning. A concise interpretation of their testimony follows.

Democracy and Public Hearings

In the fall 1991 edition of the *Environmental Impact Assessment Bulletin*, Boggs published an article titled 'EA within Democratic Politics—Contradiction in Terms or Emerging Paradigm?' His central idea is that 'EA . . . democratizes knowledge use by throwing open public policy dialogue to the perspectives of different groups and interests, and even to different forms of knowledge' (Boggs 1991, 8). And in asserting this, he rejects the critiques of EA that are typically based on two assumptions:

- 'that . . . there is no rationally determinable common ground but only competing interests; and
- that power structures within democratic polities are such that in reality only elites—special interests—use knowledge . . . to guide policy, and, therefore, knowledge use subverts the democratic process' (Boggs 1991, 4, 5).

Boggs counters these assumptions very directly—holding, first, that the stated purposes of environmental legislation invoke common interests, and second, that this reflects underlying conditions and concerns. He writes: 'Events press upon us current understandings of the common good: environments free of chemical and radiation pollution, an intact ozone layer, sustainable balanced ecosystems, and so on' (Boggs 1991, 5).

His response to the view that knowledge in EA is simply an instrument of elite privilege and power is that EA, when it is properly designed, provides a unique opportunity for concerned people to participate in 'hitherto closed decision processes' and that their ability to do so 'represents a significant step toward a more open and adaptive social order' (Boggs 1991, 6, 7).

Providing a synoptic account of Macpherson's line of thought is challenging because his view of democracy has historical depth. He published his ideas in 1977, in a little classic with the intriguing title of *The Life and Times of Liberal Democracy*. The key point that he makes is that the liberalism associated with contemporary democracy has evolved through several stages: there was the *protective* stage in the early nineteenth century, when liberalism meant market freedom and the protection of the owning classes from the interference of government; then there was the *developmental* stage, which has evolved since John Stuart Mill asserted in the mid-nineteenth century the centrality in a democracy of the opportunity for full development of human potentials; *equilibrium* democracy, the prevailing model, comes next, featuring 'competition between elites', which seeks a degree of stability by putting limits on popular participation; and then, finally, there is the ideal, struggling-to-be-born *participatory democracy,* based on the notion that 'a more equitable and humane society requires a more participatory political system

... substantial citizen participation in government decision-making' (Macpherson 1977, 22, 93, 94).

In his historical overview, Macpherson convincingly concludes that 'the central ethical principle of liberalism [is] the freedom of the individual to realize his or her human capacities' and that this is associated with a change in the prevailing value system, away from seeing people primarily as consumers and acquisitors to 'exerters and enjoyers of their own capacities'. Hence, for example, we see the emergence in democratic societies of broadly accessible public education (Macpherson 1977, 2, 99). In looking at prospects for moving from where we are to the next stage of evolution, he makes a very interesting observation about environmental issues. First, he notes the rising awareness of the costs of air, water, and earth pollution—'costs largely in terms of the quality of life.' And then he poses a question: 'Is it too much to suggest that this awareness of quality is a first step away from being satisfied with quantity, and so a first step away from seeing ourselves as infinite consumers, towards valuing our ability to exert our energies and capabilities in a decent environment' (102). Macpherson's conjecture is that the humanist ideal of developmental democracy has a greater possibility of being realized in a regime in which the ordinary person can have some influence on decision making, and that greater participation, in turn, is a precondition for moving from an equilibrium to a participatory democracy (102–5).

The pertinent concepts in Friedmann's world view, as set out in his book *Retracking America, a Theory of Transactive Planning*, are a guidance system, mutual learning, and transactive planning. A *guidance system* refers to the pattern of institutional arrangements (political, legal, administrative, economic, planning, etc.) that guide the processes of change in society. *Mutual learning* is an approach to problem solving that features the linking of processed—that is, scientific and technical—knowledge with the personal knowledge of the affected and interested public. *Transactive planning* is a process in which mutual learning is integrated with action, with 'an organized capacity and willingness to act'. 'Transactive planning', Friedmann states, 'is a style that humanizes the acquisition and uses of scientific and technical knowledge' (Friedmann 1973, 190, 191, 245).

Friedmann argues that sound planning, along with the action or decision making that flows from it, depends on effective dialogue between the planner as expert and the client or public. The expert's informed analysis is joined to the public's intimate knowledge of context and of realistic alternatives. 'In this process,' he writes, 'the knowledge of both undergoes a major change. A common image of the situation evolves through dialogue; a new understanding of the possibilities for change is discovered' (Friedmann 1973, 185, 187).

The philosophical route has been taken to the subject of public hearings in environmental decision making for three good reasons: to inspire, hopefully, some fundamental thinking about such processes; to draw attention to their centrality in our society; and, perhaps, to contribute to the philosophy of environmental hearings. Some reflections along these lines, experience suggests, are necessary as a bulwark against the capricious use of environmental hearings. The record, as we have seen, is very inconsistent. We seem to change our attitude to and use of hearings much as we change our shirts and underwear. And that, considering the evidence of our witnesses, is decidedly not wise. To clarify this point, I think it might be helpful to draw out some of the implications of their views, briefly.

Boggs speaks directly to the issue; one can infer the following:

- On matters of the environment, the common good is best served if decision making has a strong knowledge base.
- Given the prevailing power structures in our kind of society, the elites have an inordinate advantage in mobilizing and delivering information to serve their special interests.
- The environmental assessment process has emerged as a promising innovation in what Boggs calls 'democratic polities', opening up environmental decision making to a wide range of groups and interests.
- This is intrinsically a step forward (indeed, it is an 'emerging paradigm') both because of the inherent nature of environmental issues, which affect our common habitat, and because it is a leavening in the body politic, making it more 'open and adaptive'.

Mapherson has the virtue of giving us perspective—of helping us see our own times in the flow of history, particularly in the evolution of the democratic ideal. Specifically, we need to understand the connection between substance (the values that we cherish) and process (how we pursue those values). The rise of environmental awareness in our time is a harbinger of change, an affirmation of quality of life as a priority in the face of the prevailing pecuniary value system. Translating awareness into policy and decision making requires meaningful opportunities for the participation of the concerned publics. And this, in turn, will foster a social order (Macpherson's participatory democracy) that will sustain those opportunities. It would not be stretching these ideas too much to say that EA hearings are opportunities that could meet Macpherson's prescription.

Friedmann's work comes into focus if we take a small mental leap and consider the EA process, including hearings, as part of—as a sub-institution of—the guidance system. Then his recipe of mutual learning and transaction very closely approximates the adjudicator's ideal of attaining the balanced record of evidence that would yield a decision that is fair and that responds to the public interest embedded in the governing legislation.

Hearing Imperatives

The testimony of Boggs, Macpherson, and Friedmann provides a broad but compelling rationale for public hearings on the environmental dimensions of development; it also leads to three guiding requirements:

- The initiative imperative
- The equity imperative
- The efficiency imperative

The Initiative Imperative

The *initiative imperative* refers to the circumstances or conditions leading to a statutory requirement for a hearing. One of the overwhelming impressions arising from my tenure at the Environmental Assessment Board and then the Environmental Review Tribunal (May 1990 to May 2001) is just how uncertain and elusive are the criteria for initiating hearings. There was for a time, in the late 1980s and the first half of the 1990s, one bedrock requirement, under section 30 of the Environmental Protection Act (EPA), R.S.O. 1990, c. E–19, that provided for mandatory hearings for waste disposal sites capable of serving 1,500 or more persons. In a series of hearings, none taking more than a few days, each with a diverse array of parties and sometimes supported by intervenor funding, the board undertook some essential work to protect the environment—mainly by taking meticulous care through technical monitoring and by reporting the conditions of the decision. These efforts included hearings to reach decisions on the establishment or expansion of landfills in Kenora; Sudbury; Niagara Falls; Fort Erie; Sarnia; London/Green Lane; Guelph; Orillia; Brockville; the Townships of Charlottenburg, Asphodel, Alice, and Fraser; South Gower; and so on. In addition, it is important to note two landfill decisions governed by the Consolidated Hearings Act (CHA), namely those made with respect to the Regional Municipality of Halton (24 February 1989) and Steetley-South Quarry (17 March 1995), in different ways both landmark decisions. All of these cited hearings were about individual projects. Having occurred, however, right across the province within a relatively few years, they constituted in their environmental implications a virtual program. But, as Alan Levy has explained in his recent CELA study, under the Environmental Assessment and Consultation Improvement Act, 1996, the amended assessment legislation, a regulation was promulgated, O.Reg. 206/97, that has the effect of eliminating mandatory hearings for landfills of a certain size (Levy 2002, 236–7).

What happened? Is the need for public scrutiny any less in 2003 than in 1993? Are our ecosystems and groundwater sources now sufficiently safe and secure that we can treat applications for certificates of approval as routine matters to be regulated by handbook alone? Not very likely. If you go back to the text of those decisions, you will find abundant evidence of sensitive factors that emerged in the hearing process. Some critics might legitimately say, 'Why bother about landfills, an outmoded method of disposal? Bring on the 3 Rs (reduce, recycle, reuse).' Yes, indeed bring them on, and the sooner the better, but in the meantime there remains a need for the larger and more complex landfills to be subject to public scrutiny.

Similarly, EA hearings, as indicated at the beginning of this chapter, have become practically an endangered species. The use of discretionary powers carried forward from the original Act—EAA, R.S.O. 1990, c. E–18, s.12(2), s. 13(d)—and as amended in 1996—ss. 9.1(1) (5); 9.2(1) (2) (3) and 9.3(1) (2)—has resulted in the thorough domestication of this so sensitive and critical process. 'Intervention' is just not part of the vocabulary of regulation.

These concerns are not raised to suggest that the requirement for public hearings and indeed for individual assessments should not be governed by some limiting principle. The Ministry of the Environment, using the powers of Part II.I of the amended Act, has established what appears to be a rigorous process for addressing those undertakings that are amenable to a standardized evaluation approach—namely (to use some now classical words), 'projects which occur frequently, have a predictable range of effects, and are likely to have minor impacts on the environment' (Levy 2002, 230). Section 14(2), on the prescribed contents of a proposed class EA, requires a lot of useful information that could lead to the defining of a niche for a class EA, but the limiting principle is just not there. By not providing any basis for limiting the use of the class mechanism in that way, the prevailing practice illustrates the hazard of fuzzy criteria—fuzzy criteria for requiring individual rather than class EAs in addition to, as noted above, fuzzy criteria for the necessity of hearings on EAs, whether they be individual or class (Levy 2002, 228–30).

The kind of feature that we need enshrined in all of our legislation involving environmental decisions is illustrated by the test for *leave to appeal* in the Environmental Bill of Rights (EBR). Section 41 states that leave to appeal a decision through an appeal hearing shall not be granted unless (a) there is good reason to believe that no reasonable person, having regard for the relevant law and for any government policies developed to guide decisions of that kind, could have made the decision; and (b) the decision in respect of which an appeal is sought could result in significant harm to the environment.

This, of course, is not literally advocated but is presented merely as an example of a principled environmental criterion. This is the kind of explicit touchstone that we need, particularly if the parts of the test are looked at holistically, as intimated in my Environmental Bill of Rights decision in *Hannah v. MOE* (16 September 1998), as follows:

> While the EBR does not explicitly deal with the relationship between these two dimensions, there is a strong presumption—inherent in the Preamble and Part I of the Act—that the two aspects of the test are related. The reasonableness of the Director's decision depends on whether it 'could result in significant harm to the environment'. And any decision which could result in significant harm to the environment would be an unreasonable decision.

It is important to recall that the opening features of the EBR provide the explicit context of environmental concerns that qualify the provisions of that statute. For example, in the preamble,

• 'The People of Ontario have a right to a healthful environment.'

In the definition of *environment*,

• '"environment" means the air, land, water, plant life, animal life and ecological systems of Ontario.'

And in the purposes of the Act, three aspirations are conditioned by 'the means provided in this Act':

- 'to protect, conserve and, where reasonable, restore the integrity of the environment
- to provide sustainability of the environment
- to protect the right to a healthful environment' (Environmental Bill of Rights, S.O. 1993, c. 28).

Unless we invoke some clear environmental guidelines of that nature, together with respected protocols for determining the applicability of the test—to have or not have a hearing—then the province of Ontario is not very serious about protecting and enhancing the environment, and down the road, the price will be paid.

On this question of environmental criteria in statutes, the Niagara Escarpment Planning and Development Act provides an edifying case. The point of importance is that notwithstanding some of the limitations of the Act, such as section 25(4) on the delegate's (e.g., commission's) power of decision and section 25(14) on the power of the minister, the so-called privative clause, and notwithstanding that these barriers must be addressed to bolster the integrity of the Niagara Escarpment Plan, the substantial normative character of the Act, which provides explicitly in its 'purpose' for the preservation of the 'natural environment' of the Niagara Escarpment, has been a powerful motivator for protecting that unique landscape. It should be noted that the vast majority of the 332 appeal hearings in the nine years from 1993/4 to 2001/2, which have been conducted by tribunal members designated as hearing officers, have served with a few exceptions as a substantial force for conservation while conscientiously respecting the public's right to be heard (ERT, Ontario, April 2003). The recommendations and the outcomes are on the record.

The Equity Imperative

If the initiative imperative directly follows from the democratic context of environmental assessment, then the *equity imperative* is about democratic participation in public hearings on the environment. This is the injunction that a sound decision depends on the interests involved in the undertaking, on both the proponent and intervenor sides, having a more or less equal opportunity to provide salient facts and opinions.

Some time ago, I presided at a hearing, under the EPA, involving the processing of polychlorinated biphenyls (PCB) wastes. On the proponent side, there were three witnesses with professional qualifications, namely in dispersion modelling for air emissions, environmental health, and quality control and groundwater impacts. The Environment Ministry, assuming its statutory responsibilities, brought forward four qualified witnesses who testified positively about the application.

These individuals were an expert in air quality and hazardous waste, the regional hydrogeologist, the regional coordinator of hazardous waste, and the senior environmental officer of the district. On the other side, there was one intervenor/witness, a conscientious local resident, one little man, concerned about impacts but lacking any special knowledge about the matter before us. A ratio of 7 to 1. There was not much opportunity for mutual learning there. Did the two members of the panel achieve Macaulay's 'balanced record' of evidence as a basis of a decision? I have never believed in the passive or sponge model of the adjudicator—'just sit there and soak it in.' We took a strongly probing posture, searching for information miscarriages or gaps, an effort that, hopefully, was reflected in the quality of the decision—notwithstanding that neither panel member was a technical specialist.

The implicit concern is whether hearings that are destined to produce skewed findings fulfil the purposes of the governing legislation. Do they serve the public interest? To allay the justifiable fear that the answer to both questions may be 'No!', something must be done to make the playing field less grossly uneven. One thing that has been tried in this province is intervenor funding. Another device, one that has not had much of a workout (except by the Ontario Energy Board) is the strategic use of an information fund by the tribunal itself for the purpose of calling upon technical witnesses to address perceived shortcomings in the evolving record. Both options have some pluses and some minuses. On balance, intervenor funding is probably the best option, provided that it is used wisely, without the perception (or reality) of abuses. That is where the efficiency imperative, the third necessity, comes in. But before going on to that, I would like to unburden a discomfort I experienced during my years as an adjudicator. This was an unease about the source of the intervenor's money—the proponent across the table. Underlying this is the purely visceral feeling that the monetary relationship between proponent and intervenor, using an award under the Intervenor Funding Project Act, somehow compromised, however slightly, the position of the beneficiaries.

This issue can be expressed more positively. If, as suggested in the philosophical preamble, the public hearings on environmental issues are, in a sense, sacred rituals of a democratic society—on the premise that decision processes are opened up without prejudice—then ensuring the integrity of such events is in the public interest. And intervenor funding should be arranged accordingly—from the public purse.

This type of provision under the Canadian Environmental Assessment Act has not, it seems, always unfolded effectively, mainly because of some of the features of the Participant Funding Program; for example, authorized expenses for 'expert advice' may include legal fees, but 'applicants are expected to represent themselves' (Canadian Environmental Assessment Agency Operational Policy Working Group, December 2000; updated to 31 December 2002). This does not quite approximate a level playing field with proponents. Notwithstanding, the principle of public rather than proponent funding is compelling. It would be interesting to see the outcome of a concerted creative effort to produce an innovative design for Ontario, one based on public funding.

The Efficiency Imperative

The third imperative—the *efficiency imperative*—follows from the preceding two. One cannot, in good conscience, insist that public hearings be held whenever they are deemed essential for protection of the environment, and that intervenors be ensured, by appropriate funding, of the opportunity to make their case, without also committing to hearings that are efficient in time and cost, both absolutely and relatively, that is, compared to other ways of addressing the matter at hand. At the same time, the benefits side of the equation should not, of course, be overlooked.

I have great confidence that the tribunal system will be able to fulfil this requirement because I saw, experienced, and, to a degree, facilitated over a number of years in the 1990s the emergence of a robust culture of efficiency. This trend to reform began in January 1991, when the Environmental Assessment Board, led by Grace Patterson, held a series of round table discussions, with a very diverse dramatis personae, on the board's hearing process. Detractors will hasten to cite certain 'notorious' examples—Timber Management, Ontario Waste Management Corporation, and Ontario Hydro Demand and Supply. These were decidedly the exception to the rule, and the duration of such hearings must be seen in the context of their challenging subjects. Now that we have seen what has happened to the energy sector since the abrupt withdrawal of Ontario Hydro in January 1993, thoughtful reflection in 2003 might lead us to regret the abortive end of, as the record shows, a highly comprehensive, conscientious, and environmentally sensitive exercise in strategic planning. Be that as it may, the principal point is about the evolution of the environmental hearing process towards increasing efficiency.

Considerably before retiring from the Environmental Review Tribunal in May 2001, I had the pleasure of helping to create a new hearings model—a hybrid of the adjudicative and investigative approaches, which, backed up by the tribunal rules and procedural directions for specific hearings, was vigorously pursued. Without going into details, I will provide here some key words and phrases of that model:

- case management
- exchange; consultation; mediation; settlement
- preliminary hearing; meetings of expert witnesses; scoping by parties
- joint submissions on agreed facts; opinions; and issues, resolved and unresolved
- pre-hearing settlement conference; narrowing of issues
- guidelines for the presentation of evidence
- case outlines and a programmed hearing schedule

It is important to appreciate the place of the preliminary hearing and mediation in the new hearing model. The first, the initial formal, scheduled event in the proceedings, involves the critical functions of confirming the parties, participants, and presenters in the hearing; searching for consensus on the priority issues; and, through the tribunal's procedural directions, shaping the agenda of the hearing proper. In doing all this, the preliminary hearing can set the tone and spirit of the entire process.

The philosophy of the board commencing early in the 1990s was to temper the adversarial encounter of parties in favour of cooperative decision making (Gertler 1992). This was not to obscure fundamental differences, but to give priority to environmental stewardship, to search for the presumed common interest in a sustainable habitat. Hence facilitation, mediation, negotiation (sometimes called 'alternative dispute resolution' [ADR]) were built into the proceedings long before mediation found an explicit place in EA legislation as part of the 1996 amendments (EACIA 1990, ss. 6, 8 (4) (7) (9)). Thus, the Rules for Practice for the Environmental Appeal Board and Environmental Assessment Board (published in November 1998, about two years before the formal unification of the two boards, in December 2000, within the Environmental Review Tribunal) reflected the procedures of the time, as follows:

> *Purpose of Mediation*
> 47. Mediation, which is part of the proceeding, but not a part of the Hearing, may be held for the purpose of attempting to reach a settlement of the issues, or at least their simplification. It is conducted, in confidence, by a member of the Board, by Board staff or a person appointed by the Board. If no full settlement is reached on the issues, the Hearing will take place with no reference to the information disclosed during mediation, except with consent of the parties.

The connection between mediation and the efficiency imperative was strongly implied, for example, in the board's *Reason for Decision and Decision*, on the Town of Kenora landfill expansion, in the following way:

> The Board recognizes that the hearing would have lasted far longer than three days had it not been for the considerable effort made and success achieved by the parties in their attempt to narrow the issues and resolve areas of dispute. The Board was pleased to learn that Kenora had voluntarily agreed to advance funding . . . prior to the preliminary hearing and funding hearing in order to speed up the process and permit these parties to retain consultants. The parties and their counsel are to be commended for their cooperation in making the environmental review process more efficient. (Environmental Assessment Board 1991)

Behind the efficiency imperative is nothing less than a sense of social responsibility, a determination to be prudent in the use of public resources. A tribunal can, indeed must, set the stage, but experience suggests that it really works when other participants in the process—environmentalists, lawyers, planners, other professionals—along with intervenors, share a common commitment.

Concluding Observations

Reflecting on the theme of this chapter, I find that three caveats come to mind: first, that efficiency should never be sought at the expense of effectiveness—the transcending obligation to fulfil the public interest as defined in or implied by the governing legislation; second, that in giving emphasis to the commonalities of all the environmental statutes that may govern hearings, one does not intend to minimize the value of the classical EA process, which involves needs, alternatives, and the prescription of effective conditions for addressing a wide spectrum of impacts, from the physical to the cultural, and is framed broadly enough to be a vehicle for environmental enhancement; and third, that while the elements of a comprehensive and participant environmental assessment process are present in both the baseline and altered versions of the Ontario EA model, the EIA thrust can, unfortunately, be fundamentally frustrated by the exercise of ministerial discretion. Indeed, the track record to date suggests that the opportunity for public hearings involving environmental assessments of undertakings has been foreclosed.

The Ontario system, notwithstanding its potential, is not, as it is currently practised, a demonstration of the words of the three wise witnesses:

- 'EA democratizes knowledge use.'
- 'A more equitable and humane society requires a more participatory political system.'
- 'Dialogue and mutual learning.'

At the same time, the hard-won insights of experience, in the form of the three imperatives, show the way towards fulfilling the compelling promise of public hearings in environmental decision making.

References

Boggs, J.P. 1991. EIA within democratic politics, contradiction in terms or emerging paradigm. *Environmental Impact Assessment Bulletin* 20:1–11.

Canadian Environmental Assessment Agency Operational Policy Working Group. 2000, updated 31 December 2002. *Participant funding program, guide for assessments by review panels*. Ottawa: Canadian Environmental Assessment Agency.

Court of Appeal, Ontario. 2004. *Decision on Appeal of Divisional Court Judgement on Ministerial Discretion in EAA re Richmond Landfill, Canadian Waste Services Inc.* 25 August.

Divisional Court, Ontario. 2003. *Reasons for Judgement, between Ben Sutcliffe and Helen Kimmerly and the Mohawks of the Bay of Quinte; and Minister of the Environment (Ontario) and Canadian Waste Services Inc.* 17 June.

Dunn, S. 2003. *Environmental Review Tribunal*. Data on numbers of hearings by act, email attachment. Toronto, 10 April and 11 July.

Environmental Appeal Board. 1998. *Hannah v. Ontario (Ministry of Environment), Reasons for Decision and Decision*. Toronto, 16 September.

Environmental Appeal Board and Environmental Assessment Board. 1998. *Rules for Practice for the Environmental Appeal Board and Environmental Assessment Board*. Toronto, November.

Environmental Assessment Board. 1991. *Kenora Landfill, Reasons for Decision and Decision*. Toronto, 8 November.

———. 1992. *City of Orillia, Expansion of Kitchener Street Landfill Site* (EP–90–03), *Reasons for Decision and Decision*. Toronto, 24 February.

———. 1997. *Gary Steacy Dismantling Limited* (EP–97–03), *Reasons for Decision and Decision*. Toronto, 4 December.

Faieta, M. 1995. *Summary of Board Decisions Involving the Environmental Assessment Act*. Environmental Assessment Board, internal document. Toronto, 8 December.

Friedmann, J. 1973. *Retracking America, a theory of transactive planning.* Garden City, NY: Anchor Press/Doubleday.

Gertler, L.O. 1992. Cooperative decision making. Address to a course in Toronto of the Technical University of Nova Scotia. Toronto, 14 May.

Gibson, R.B. 1992. The New Canadian Environmental Assessment Act: Possible responses to its main deficiencies. *Journal of Environmental Law and Practice* 2:223–55.

Levy, A.D. 2002. A review of environmental assessment in Ontario. *Journal of Environmental Law and Practice* 11:173–283.

Macaulay, R.W. 1989. *Directions: Review of Ontario regulatory agencies.* Prepared for Management Board of Cabinet. Toronto: Queen's Printer.

Macpherson, C.B. 1977. *The life and times of liberal democracy.* Oxford, London, New York: Oxford University Press.

United Nations Environment Programme. 2003. *Stockholm 1972—Declaration of the United Nations Conference on the Human Environment.* Nairobi, Kenya: UNEP, 15 June.

Valiante, M. 1998. Evaluating Ontario's environmental assessment reforms. *Journal of Environmental Law and Practice* 8:216–64.

Chapter 6

Strategic Environmental Assessment

Bram F. Noble and Jill Harriman Gunn

Introduction

Since the introduction of the US National Environmental Policy Act (NEPA) in 1969 and the Canadian Environmental Assessment and Review Process (EARP) in 1973, environmental impact assessment (EIA) has undergone a number of evolutionary changes. Of particular importance is the growing interest in the environmental implications of decisions and actions above the project level, at the strategic tier of policies, plans, and programs (PPPs). Procedurally, EIA is concerned about the most likely impacts resulting from a proposed project development and about finding ways to mitigate those impacts before they become a reality; those conducting EIAs, however, do not ask whether the proposed project is the most appropriate form of development or whether the development is consistent with broader environmental goals or desired future conditions. If we are concerned about the sources of environmental impacts and the real drivers of environmental change, then a more strategic form of environmental assessment is needed.

Strategic environmental assessment (SEA), the environmental assessment of PPP initiatives and their alternatives, including future development scenarios, is gaining widespread recognition as a tool for integrating environmental considerations at the earliest possible stage of decision making. SEA is about integrating environmental concerns into higher-order PPP development and decision-making processes; it is about applying environmental assessment above the project level. This chapter provides an overview of SEA in the Canadian context, focusing on the principles of SEA and on SEA methodology. This is followed by a review of SEA systems in Canada and a discussion of the current state of practice.

Definition and Characteristics

Strategic environmental assessment is a higher-order process by which PPPs and their alternatives are developed and assessed based on a much broader set of objectives and constraints than is project-level EIA. The need for SEA evolved on at least three fronts: first, in recognition of the need to promote the development of more environmentally sensitive PPPs; second, in recognition of the need to focus and streamline project-level assessments, making them more relevant to policies and programs by ensuring that development actions are set within a broader environmental framework; and third, in recognition of concern about the capacity of project-level EIA to consider development alternatives, non-project impacts, and cumulative effects that occur beyond the individual project. In other words, SEA was seen as providing the planning-type framework and decision-making environment necessary within which environmental effects may be addressed in a much broader, comprehensive, and objectives-led context (Fischer 2002; Therivel 2004; Harriman and Noble 2008).

Definition

Strategic environmental assessment has gained considerable momentum in recent years. Its support is illustrated by the volume of international literature and special workshops and conferences dedicated to the subject and by the increasing numbers of national SEA systems and regulatory frameworks. While SEA is generally acknowledged to involve the early consideration of environmental issues in PPP decision making, there is no single best definition for SEA. The concept itself has been defined in various ways, including as

> . . . a systematic process for evaluating the environmental consequences of proposed policies, plans or programmes initiatives in order to ensure they are fully included and appropriately addressed at the earliest appropriate stage of decision making on par with economic and social considerations. (Sadler and Verheem 1996)

> . . . the proactive assessment of alternatives to proposed or existing PPPs, in the context of a broader vision, set of goals, or objectives to assess the likely outcomes of various means to select the best alternative(s) to reach desired ends. (Noble 2000, 215)

> . . . a systematic, on-going process for evaluating, at the earliest possible stage of publicly accountable decision-making, the environmental quality, and consequences, of alternative visions and development intentions incorporated in policy, planning, or program initiatives, ensuring full integration of relevant biophysical, economic, social and political considerations. (Partidário and Clark 2000, 3)

> . . . a decision support tool, designed to integrate environmental and social issues into higher-order PPP decision making processes, bringing together different aspects of problems, different perspectives, and providing possible solutions in an accessible form to the decision maker. (Sheate et al. 2003)

Achieving a common definition of SEA, however, is far less important than understanding the principles upon which SEA is based.

SEA Principles

It is generally acknowledged that SEA differs from current forms of project-level EIA, but the distinction between SEA and other types of environmental assessment and appraisal is not always clear. As Bina (2007, 586) observes, 'scholars and practitioners appear divided on such fundamental matters as the concept of and approach to SEA' and SEA's foundations remain unclear. On the one hand, Clark (2000), for example, suggests that 'SEA has different features to other types of impact assessment' and 'recognizing this difference may be a crucial condition for understanding SEA and allow process and practice improvement.' On the other hand, it has been suggested that the features of SEA are simply those that would also apply to 'good-practice' EIA. However, we argue here that SEA is clearly set within a different context (that of policies, plans, and programs) and asks different types of questions than project EIA. The international environmental assessment literature identifies several principles that characterize SEA (Table 6.1); these are discussed in detail by Noble (2000) and are summarized below.

Principle #1: Strategically Focused

There has been an ongoing debate over the 'strategic' in SEA, and the term itself is too often indiscriminately used in an attempt to add more importance or significance to a variety of topics, including EIA (see Bina 2007). 'Strategic' in SEA cannot be simply explained in terms of its application to initiatives above the project level; (Partidário 2000); rather it should be seen in terms of the relationship between SEA and the broader planning process (Bina 2007), including the types of questions being asked in SEA (Noble 2000). The word *strategic* is derived from the Greek word *strategos*, meaning that which has to do with determining the basic objectives and finding the means to achieve them. A strategy is the process of defining goals or visions in terms of the desirable (and feasible) principles to be established, proposing alternative possibilities for achieving these principles, and selecting the most desirable approach. Strategic EA, then, is a process or means that identifies strategic initiatives, evaluates alternatives, and formulates a strategy for moving forward. The key component is strategy—the determination of objectives and means and the identification of courses of action to achieve the desired, rather than the most likely, ends.

Principle #2: Futures Oriented

SEA is inherently futures oriented. It is focused on identifying desired outcomes and on what is required to achieve those outcomes, and is designed to determine the consequences of alternative means. Future goals and objectives are defined, and alternative means of achieving those goals and objectives are evaluated. For example, an EIA of a proposed hydroelectric project would aim to predict the most likely effects of hydro development and to make the necessary adjustments to mitigate potentially negative impacts. The option to pursue hydroelectric power would be determined before the assessment took place. Variations might be assessed, such as technical design, but the strategic action itself (hydroelectric development) is predetermined. In contrast, an SEA would assess the feasibility of, say,

Table 6.1   Characteristics of 'strategic' environmental assessment (SEA)

1. Strategically focused
   - places emphasis on strategy
   - determines objectives and the means to achieve them
   - identifies strategic initiatives

2. Future oriented
   - is forward looking
   - backcasts desirable ends and alternative means

3. Focused on alternatives
   - assesses alternative policy, plan, and program options
   - considers alternatives to meet a need or formulate a policy, plan, or program
   - considers alternatives to a proposed or existing policy, plan, or program

4. Objectives led
   - examines particular goals and objectives to be accomplished
   - alternatives assessment set within the context of a broader vision

5. Proactive
   - attempts to avoid, eliminate, and minimize potentially negative actions
   - attempts to enhance and create potentially positive PPP actions
   - creates and examines alternatives to identify the best practicable environmental option

6. Integrated
   - addresses the interrelationships of biophysical, social, and economic systems
   - incorporates multiple objectives, criteria, and sources of knowledge
   - is, ideally, an integral part of policy, plan, and program formulation

7. Broad focus
   - not project-specific
   - often more broad-brush than project-level assessments
   - scope broadens as assessment moves from program, to plan, to policy-level assessment

8. Tiered
   - is set within the context of previous and subsequent decision outcomes and objectives
   - sets the stage for subsequent assessment and decision-making processes, including EIAs

Source: Based on Noble 2000.

meeting the electricity demand and would propose and assess the implications of a range of alternative means to supply that electricity, including the option of a reduction in electricity demand.

Principle #3: Focused on Alternatives
SEA does not contain a 'no action' alternative, but rather assumes that something will be done to address the need or objective, including the 'status quo' alterna-

tive. To select the status quo is to decide consciously that current conditions and trends will lead to the most desirable future, and that one should implement the necessary measures to ensure that status quo conditions continue. The alternatives are ideally set within the context of a broader environmental vision, such as sustainable development or balanced environmental and economic growth, and once goals are identified and objectives are set, the environmental effects of those alternatives are assessed against particular goals, objectives, or changes from the baseline condition. There is no one specific type of alternative that must be incorporated, as SEA is adaptive to the nature and objectives of the particular PPP process and question(s) at hand. In this regard, alternatives considered in SEA, as well as the subsequent strategic outputs, can be grouped as follows (Table 6.2):

1. *Alternatives to address the need or problem or to identify desirable futures.* There may be a particular demand for action or a problem that needs to be addressed. Alternative PPP options are presented and evaluated, and the preferred alternative approach is selected. Although PPPs are likely to be one outcome of the selected alternative, they need not be a part of the initial need, problem, or purpose of the strategic assessment. Alternatively, the focus can be on identifying the environmental implications associated with accomplishing alternative futures.

2. *Alternatives to a proposed PPP.* For example, when a PPP is proposed, strategic alternatives are developed and assessed. The assessment evaluates the proposed PPP and suggests alternatives within the context of the broader vision and specific targets. The most desirable (or feasible) alternative(s) is selected, which may be the original PPP or variations thereof. In addition, the assessment may result in the selection of a new vision of environmental targets.

3. *Alternatives to an existing PPP.* For example, an existing PPP may not be meeting its intended objectives and targets, or new objectives and targets may have emerged. In either case, strategic alternatives may be suggested and assessed, and a more desirable PPP, or an alternative form of the existing one, may be selected as a more effective means of meeting specified objectives and targets.

Principle #4: Objectives Led

SEA involves identifying the particular sustainability goals and objectives to be accomplished and assessing the various alternative options by which these can be achieved, based on setting specified targets and criteria. In other words, SEA is an objectives-led process. Objectives represent the desired outcomes as a result of the SEA decision; they may address a particular problem or need, or lead to the development of alternative goals and objectives. *Targets* are certain milestones that we aim to accomplish; these may range from a specified timetable to certain budgetary requirements. *Criteria* are the specific standards that must be met, such as a carrying capacity or a set limit of environmental change, and are usually set in the context of the broader environmental vision.

Table 6.2   Classification of SEA alternatives and strategic outputs

Alternatives (SEA Inputs)	Direction (SEA Outputs)
Type I Alternatives to address a need or problem or to formulate PPPs based on alternative futures	**Type I** Desired course of action, PPP development, or alternative future
Type II Alternatives to a proposed PPP	**Type II** The proposed PPP, the proposed PPP with modifications, or the identification of an alternative PPP
Type III Alternatives to an existing PPP that is not meeting its objectives or is in conflict with other PPPs	**Type III** Identification of a new PPP, changes to the existing PPP, or refinement of PPP objectives

Principle #5: Proactive

Project-based EIA focuses primarily on mitigation of the adverse environmental impacts of proposed activities, whereas SEA determines the justification for activities and attempts to avoid, eliminate, or minimize potentially negative impacts and enhance and create positive ones. Numerous SEA reviews, as well as recent SEA case studies, note the importance of a proactive approach to SEA (Harriman and Noble 2008; Noble and Storey 2001; Buckley 2000; Connelly 2000; DEAT 2000). SEA acts in anticipation of future problems, needs, or challenges, and creates and examines alternatives leading to the preferred option. In other words, a proactive approach is one that identifies *desired ends* and seeks the preferred option among a variety of alternatives. In this way, SEA informs the planning process by assessing possible alternatives to arrive at a preferred course of action within the context of a broader environmental vision.

Principle #6: Integrated

An integrated approach involves combining different strands of knowledge and information to identify and analyse problems. There are three dimensions to SEA integration. First, and at the most basic level, SEA's integrated approach is one that addresses the interrelationships of biophysical, social, and economic systems and objectives (IAIA 2002). Second, issues and problems at the strategic level rarely adhere to disciplinary boundaries. The implications of an energy policy or development program, for example, extend well beyond the knowledge of economics and engineering. Thus, a key principle of SEA is the integration of multiple objectives, criteria, knowledge systems, and interests in evaluating the environmental implications of strategic alternatives. The third dimension of SEA integration is one in which SEA is conceived as an integral part of PPP development, rather than as an add-on process to validate a PPP decision (Noble and Christmas 2008; Vicente and

Partidário 2006). As an integrated assessment process, SEA influences the selection and course of PPP development as decisions unfold, rather than serving as the yardstick against which they are measured.

Principle #7: Broad Focus
SEA is focused on assessing alternative options and opportunities for PPPs, regions, and sectors. Its scope will differ depending on the tier of application, but in general the higher the order of decision making (moving from the program to the policy level), the broader the scope of SEA. An SEA of alternatives for a national energy policy, for example, will be broader in scope than an SEA of alternatives for a regional energy efficiency program. As the scope of SEA broadens, so do the methods and techniques used. Techniques that are applied at the project level may become less useful as SEA broadens from the program level to the planning and policy levels. At the policy level, SEA applications typically involve methods or combinations of methods (e.g., scenario analysis, spatial analysis) rather than phenomenon-specific techniques per se (e.g., Gaussian dispersion modelling of airborne particulates), reflecting the scope and scale of higher-order assessment and decision making.

Principle #8: Tiered
SEA is a tool that can complement and support project-level assessment by identifying the project-level issues, alternatives, and effects that could result from policy and plan initiatives. Thus, SEA is often set within the context of a tiered planning and assessment framework where, ideally, SEA and EIA are considered in sequence (Table 6.3)—SEA proactively examines a range of alternatives and selects the preferred course of action, and EIA is initiated to determine in greater detail the potential impacts of the particular development activities.

This tiered, forward-planning approach to SEA implies both a clear distinction between actions and a hierarchical and even chronological order of actions. While such an approach is advantageous and attractive from a planning perspective, in practice it rarely works like this. Higher-order decisions will set the context for actions at other levels of the decision-making process, but SEA at the policy level may develop bottom-up as a result of combinations of strategic decisions made at the planning, program, or project level. While a tiered arrangement is a fundamental feature of SEA, it need not proceed in a top-down fashion from policies to plans to programs. What is important is that SEA at one level is set within the context of the policy, plan, program, or project at the previous or subsequent level of decision making. In other words, SEA is not an isolated assessment process; rather it sets the context for, and is set within the context of, subsequent and previous decision-making and assessment processes.

SEA Methodology

Debates over which methodological approach should form the basis for SEA are far from resolved (Harriman and Noble 2008; Noble and Storey 2001; Brown and

Table 6.3   Simplified representation of tiered planning and SEA: Category of action and type of assessment

Government Level	Energy Use and Efficiency Plans (SEA)	Policies (SEA)	Plans (SEA)	Programs (SEA)	Projects (EIA)
National	National energy efficiency plan	National energy policy	Long-term energy efficiency plan	Renewable energy development program	Hydroelectric facility construction
Regional	Regional energy efficiency plan	National environ-mental policy	Regional strategic plan		
Sub-regional	Sub-regional energy efficiency plan			Sub-regional investment program	
Local	Local energy efficiency plan				Local infra-structure project

Sources: Based on Barrow 1997; Therivel 1993; Lee and Wood 1978.

Therivel 2000; Bond and Brooks 1997). On the one hand, the Canadian Environmental Assessment and Research Council (CEARC 1990) suggested that SEA methodology could be adopted in large part from approaches already applied at the project level. The United Nations Economic Commission for Europe (UNECE 1992) presented the same idea, suggesting that environmental assessment procedures for PPPs should reflect project-level EIA principles and involve a basic shift of EIA methodologies upstream. On the other hand, several authors and SEA practitioners (e.g., Noble 2009; Harriman and Noble 2008; Noble and Storey 2001; Bailey and Renton 1997; Boothroyd 1995) propose that an alternative approach to the extension of EIA methodology upstream is required. Rather than simply extend EIA frameworks to address the outcomes of development, including the effects of multiple project developments within an administrative or ecological region, the objective is to address higher-order strategic initiatives in order to assess the sources of change before irreversible decisions are taken. Grafting SEA on to PPP formulation and assessment procedures will not be achieved by attempting to translate existing project-based EIA upstream (Brown and Therivel 2000). New or adapted methodologies appropriate for the types of questions asked at the strategic levels of assessment are required.

Requirements of SEA Methodology

SEA methodology is best described as 'one concept, multiple forms' (Verheem and Tonk 2000). However, on the basis of recent international SEA experiences and according to the principles of SEA summarized in Table 6.1, several guidelines emerge to facilitate the development of effective SEA methodologies (see Noble 2009; Harriman and Noble 2008; Noble and Storey 2001).

Accommodating a Range of Options and Interests

First, SEA methodology must be capable of accommodating a broad range of alternatives, interests, and assessment criteria, and of balancing competing and often conflicting goals and objectives. In other words, SEA methodology must be able to address a multi-criteria problem. A multi-criteria problem arises when a decision-making process involves the simultaneous evaluation of assessment criteria and of competing objectives and decision alternatives. Solving multi-criteria problems requires an assessment that is aimed at rationalizing decision problems by systematically structuring all relevant aspects of assessment choices. It requires an approach that enables the practitioner to use a variety of information, including both scientific, quantifiable data and qualitative information derived from intuition, experience, values, and judgments. The purpose of SEA application is to assist decision makers in choosing a course of action. Thus, SEA methodology serves as a decision aid; it clarifies the problem by presenting the alternatives and assessing their relative effects and attractiveness.

An Integrated Approach

Second, addressing SEA problems requires a certain degree of integration. The effects of PPP decisions are almost always multidisciplinary and involve multiple levels of interest, ranging from the interests of political decision makers to those of disciplinary specialists (Jones and Greig 1985). The Canadian Council of Science and Technology Advisors (CSTA 1999), an independent body established to provide the Cabinet Committee on Economic Union with advice on federal government science and technology issues, noted the importance of an interdisciplinary and interdepartmental approach to scientific research. In the CSTA report *Science Advice for Government Effectiveness*, the council emphasizes an interdisciplinary approach, one that enables decision makers and experts to identify and address horizontal issues and to appreciate where, and in what form, their information is useful to others.

Adaptive to Different Tiers of Decision Making

Third, SEA methodology must be flexible and thus adaptable to different types of applications and to different tiers of decision making. The scope of SEA broadens as SEA moves upstream from programs to plans to policies. SEA methodology must be capable of adopting a variety of methods and techniques, depending on the level of application. When choosing assessment techniques, for example, the practitioner should be aware of the appropriateness of the technique for the task involved within the context of related PPP initiatives, available resources, baseline data, the geographic scale of the assessment, and the nature of the impact data required (Glasson et al. 1999).

A Structured Approach

Fourth, although SEA should be a flexible process and adaptable to the planning and decision context within which it operates, Retief (2007) cautions that some have perceived being flexible and adaptable in SEA as synonymous with being vague and confusing. As such, there is an emerging appeal to ensure that there be systematic and structured methodological frameworks for SEA. This is particularly the case in Canada, where there has been difficulty in advancing SEA frameworks and in ensuring common understanding of SEA methodology and process (Noble and Christmas 2008; Auditor General 2004). A structured approach to SEA allows for

- systematic identification of PPP choices and interpretation of the decision structures underlying such choices;
- explicit analysis of trade-offs between alternatives across various interests such that 'satisfying' decision outcomes are possible;
- sensitivity analysis of selected PPP options and identification of critical thresholds that affect implementation and adoption;
- reassessment under a variety of scenarios and at multiple spatial scales without having to reconstruct the entire impact assessment process; and
- assurance that the assessment output is based on an explicit set of decision rules, thereby addressing the 'fuzziness' of PPP-level impacts (Noble and Christmas 2008).

A Framework for SEA Application

As an *impact assessment* process, SEA adopts such ex-ante tasks as scoping, identifying and comparing alternatives, evaluating according to technical and publicly adopted criteria, reporting, and monitoring—and all in a consistent and systematic form, ensuring open and accountable decision making, and contributing to the improvement in quality of subsequent decisions (Partidário 2000). In an operational sense, this demands a framework within which a variety of methods and techniques can be used to address particular questions at the strategic levels of decision making. Based on the requirements discussed above, Figure 6.1 outlines a generic methodological framework for SEA application, adapted from Noble and Storey's (2001) initial seven-phase framework for SEA. The framework is intended to provide the structure that allows decision makers to evaluate alternatives and assess their impacts in order to determine the preferred strategic action, without in any way being constricted in their choice of which methods and techniques to use. The framework consists of three broad components:

- A pre-impact assessment phase focused on developing a reference framework for the assessment, scoping the environmental baseline, and identifying key baseline trends and stressors of concern

- An impact assessment phase, frequently technical in nature, that serves to identify and assess the impacts of alternative options
- A post-impact assessment phase focused on moving SEA output forward to PPP implementation and on following up on the results

While on the surface the framework is somewhat similar in structure to 'good' project-level EA, its application is different in that the focus is on PPPs and on resolving different types of questions than are dealt with in project-level EIAs.

Pre-Assessment

1. *Developing a reference framework.* The first step in any SEA is to develop a reference framework and establish the context within which the SEA will take place. The purpose is to identify the question(s) or strategic problems to be addressed, the intended objectives of the assessment, the type of SEA to be undertaken (see Table 6.2), the relevant publics, data requirements and availability, opportunities for tiering, and the parties responsible for undertaking the SEA and monitoring and following up on PPP implementation.

2. *Scoping the baseline and issues of concern.* In the early stages of SEA, the current baseline conditions must be determined, including the primary issues of concern regarding the proposed PPP or strategic initiative to be developed and/or assessed. This means undertaking a description of the existing regional environmental and socio-economic conditions and the relevant policy and planning environment. An important aspect of baseline scoping is the delineation of the *valued ecosystem components* (VECs) to be included in the assessment; these often involve such broad and strategic issues as biodiversity, but also more localized phenomena such as particular species or water quality and quantity, and are important factors in the sustainability of the environmental and human system of concern. A measurable indicator or assessment criterion should define each VEC. *Indicators* should be measurable and are often scientific in nature, providing an early warning of the

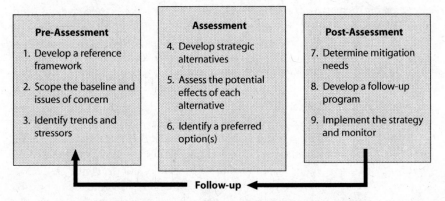

Figure 6.1   Strategic environmental assessment methodological framework

state or health of the VEC of concern. *Criteria* tend to be used as objectives or standards against which potential changes in VEC conditions or indicators are assessed and are often derived or translated from previously stated goals and objectives or, ideally, on the basis of sustainability principles and criteria (see Gibson 2001). The objective is to identify a comprehensive set of indicators and criteria that reflects all concerns relevant to the VECs and the problem at hand.

3. *Identifying trends and stressors.* This is the retrospective phase of the SEA and serves to identify and delineate key trends and the driving forces of change. Depending on the nature and application of the SEA, this phase may involve establishing relationships between environmental or VEC conditions and human disturbances. In other cases, where SEA is aimed at broader policy problems, the emphasis may be on identifying past policy initiatives and changes, and determining public and environmental responses. The objective in either case is to identify key trends or stressors that can be carried forward as a projection of the current baseline or PPP conditions in the future.

Assessment

4. *Identifying alternatives.* Once the basic issues or problems are identified, the next step is to identify potential or feasible PPP alternatives. The alternatives represent the decision options, or decision variables, among which the decision maker(s) must choose, and these options can be identified in several ways, including by consulting with the public and experts, building future scenarios, or borrowing from other, somewhat similar situations (see Bartlett 2001; CEC 1993). Depending on the nature of the SEA, the decision maker will identify alternatives to a proposed or existing PPP, alternative ways to develop a particular strategy for action, or alternative futures. Unless there is more than one potential and feasible way to proceed, there is no decision choice to be made and therefore no SEA is required. However, one alternative is always to carry forward the current baseline, or the status quo PPP, into the future.

5. *Assessing the potential effects of alternatives.* This is the prospective or futures phase of SEA. Each alternative will likely result in a different future condition or change relative to the current and future baseline; the objective is to identify how the VECs identified during baseline scoping might respond under each alternative or how the issues identified as being of concern under the baseline condition might vary. How far into the future one forecasts depends largely on the SEA objectives, the nature of the environmental variables under consideration (e.g. forest regeneration rates versus climate change), and the intended PPP reach. A horizon of 20 to 25 years is the most commonly used futures outlook, with some extending 50 to 100 years and beyond; however, even predictions concerning modest 5–10 year horizons are often highly uncertain. SEA thus does not deal with predictions of what will happen per se, but rather offers contingency statements of possible effects and changes under each alternative relative to the future baseline, assuming that certain relationships, trends, patterns, or expected outcomes hold true. The methods used to make forecasts and assess effects can vary from simple expert-based forecasting to more complex spatial analysis and simulation modelling. Scenario-based approaches are recommended in SEA when dealing with longer-term and highly uncertain future conditions.

6. *Identifying the preferred option.* Once the potential effects or possible future conditions under each alternative are identified, the impacts must then be assessed and a preferred alternative selected. Impacts are determined by assigning meaning to effects. Here, the forecasted conditions under each alternative are evaluated against specified goals, objectives, or particular criteria identified during the pre-assessment phase (e.g., sustainability criteria, acceptable levels of environmental change, economic and financial criteria, public acceptability). Where baseline data permit, the significance of impacts is assessed in terms of the sensitivity, vulnerability, or other characteristics of the affected VECs or according to the importance of meeting or exceeding specified VEC targets or thresholds. The aim is to identify, summarize, and compare the impacts associated with each alternative in order to determine the preferred strategic option or PPP direction. The final choice of strategic option depends in part on the need for and availability of mitigation options, identified during the post-assessment phase. Thus, identifying the preferred option may be an iterative process.

Post-Assessment

7. *Identifying mitigation options.* The preferred alternative represents the 'satisfying' solution and provides an overall sense of direction based on the range of possible alternatives and their associated futures and impacts. However, even the preferred alternative may have some potentially adverse environmental effects, although, ideally, it may also present an opportunity to create or enhance positive ones. Hence, the need for and types of mitigation or 'best management practices' should be identified and prescribed prior to implementation of the PPP or strategic initiative.

8. *Developing a follow-up program.* *Follow-up* refers to the variety of activities that take place after the approval and implementation of a PPP or strategic initiative, including the post-decision monitoring of conditions and indicators, PPP performance evaluation, adaptive management, and communication. Follow-up is essential in that strategic initiatives are often formulated under greater uncertainties than is the case for project-level EIAs, are potentially larger in the scope of their impacts, and are more sensitive to changing social and economic conditions. A good follow-up program ensures that both SEA and the strategic initiative are delivering their intended results, that impact management and enhancement measures are working, and that the PPP is able to recognize and adapt to emergent and external factors that may influence its success or trigger the need for reassessment.

9. *Implementing and following up.* The final stage of the SEA process is to actually implement the preferred alternative. This requires an institutional commitment on the part of the responsible parties identified during the pre-assessment phase, as well as a commitment of the necessary financial, institutional, human, and political resources. It is during the implementation stage that SEA and PPP results are monitored and information communicated so that those responsible may measure performance and prepare for the next round of PPP revision or regular review.

SEA Systems in Canada

Canada is recognized as a nation that has contributed significantly to the development of SEA systems (Table 6.4). On paper, Canada has been committed to assessing the potential environmental implications of policies since 1984, when the Environmental Assessment and Review Process (EARP) Guidelines Order defined a *proposal* as including 'any initiative, undertaking or activity for which the Government of Canada has a decision-making responsibility.' Under the EARP Guidelines Order, the reach of environmental assessment extended well beyond individual projects and encompassed broader regional, conceptual, and policy-level review processes (Sadler 2005). Early strategic forms of impact assessment, such as the Mackenzie Valley Pipeline Inquiry (1974–7), the Beaufort Sea hydrocarbon review (1982–4), and Atomic Energy of Canada Limited's nuclear fuel waste management concept (1988–94), were put into operation as area-wide reviews, public review panels, and concept-based assessments. Although none of these early assessments were formally recognized as SEA, they offered much to the future of SEA development in Canada (Noble 2009).

It was not until 1990 that SEA was formally established in Canada, by way of a federal Cabinet directive and as a separate process from project impact assessment, 'making it the first of the new generation of SEA systems that evolved in the 1990s' (Dalal-Clayton and Sadler 2005, 61). Procedural guidance for SEA was provided in *The Environmental Assessment Process for Policy and Programme Proposals* (FEARO 1993), with implementation subject to oversight by the Federal Environmental Assessment Review Office and later the Canadian Environmental Assessment Agency. In 1992 the EARP was replaced by the Canadian Environmental Assessment Act, which came into effect in 1995. While the Act was intended to make impact assessment more rigorous, it restricted EA to 'projects' or physical works, meaning that policy decisions would no longer be subject to a formal impact assessment. In many respects, the evolution and formalization of SEA as a separate process was a step backward for impact assessment in general insofar as the directive created a non-statutory system for PPP assessment that would remain separate from any legislated environmental assessment process (Noble 2009). In 1999, after a decade of increased attention to growing environmental concerns (yet less than stellar SEA performance), Canada reinforced its commitment to SEA with its release of the *1999 Cabinet Directive on the Environmental Assessment of Policy, Plan and Program Proposals*.

From 2000 onward, SEA experienced considerable growth in Canada. This new era of SEA, however, is in sharp contrast to the conceptual, public, and area-wide reviews conducted under the EARP and has been criticized as being narrowly fo-

Table 6.4 Summary of influential developments in SEA in Canada

1984	• Environmental Assessment and Review Process Guidelines Order defines 'proposal' as including any initiative, undertaking, or activity for which the federal government has a decision-making responsibility.
1990	• Bill introduced to establish the Canadian Environmental Assessment Act; policies, plans, and programs not included within the scope of the proposed Act
	• Requirement for SEA established in the 1990 Cabinet Directive on the Environmental Assessment of Policy, Plan and Program Proposals
	• National Round Table on Environment and Economy (NRTEE) and CEARC workshop on the integration of environmental considerations into government policy
1991	• Federal government reform package introduces Canada's first initiative in the development of a system of strategic environmental assessment: *Environmental Assessment in Policy and Program Planning: A Sourcebook.*
1992	• Canadian Environmental Assessment Act receives legislative approval; section 16(1) requires proponents to consider the cumulative environmental effects of their projects, and section 16(2) emphasizes the role and value of regional studies outside the Act in the consideration of cumulative effects.
	• North American Free Trade Agreement: report of the Canadian Environmental Review released just two months prior to the signing of the trade agreement
1993	• FEARO procedural guidelines released to federal departments on the EA process for policy and program proposals (FEARO 1993)
	• Natural Resources Canada releases guidelines for the integration of environmental considerations in energy policy.
	• SEA of the Net Income Stabilization Account is carried out, albeit not under the SEA name tag, in accordance with the Farmers Income Protection Act.
1995	• FEARO procedural guidelines released for assessing policy, plan, and program proposals
	• Amendments to the Auditor General Act requiring that all federal departments and agencies prepare a sustainable development strategy.
	• Federal government releases *Strategic Environmental Assessment: A Guide for Policy and Program Officers.*
1996	• Environmental assessment of the new metals and minerals policy
1997	• EUB-CEA Agency Joint Review Panel for the Cheviot mine project identifies the difficulty of attempting to address cumulative effects at the project level.

Continued

Table 6.4   Continued

	• Department of Foreign Affairs and International Trade tables Agenda 2000, outlining its commitment to conduct environmental reviews of all memoranda to Cabinet.
1998	• Parks Canada releases its Management Directive 2.4.2. Impact Assessment, stating: '[T]he impact of management plans, business plans and related planning products and policies will be assessed in conformance with the … Cabinet Directive.'
1999	• Agency releases its guidelines to implementation of the Cabinet directive on SEA.
2000	• Frameworks for regional environmental effects assessment appears in the CEA Agency's research and development priorities for 2000–1.
	• Department of Foreign Affairs and International Trade prepares an SEA manual for federal program officers.
2001	• Canada–Nova Scotia Offshore Petroleum Board publishes Canada's first SEA for the offshore oil industry.
	• Transport Canada releases a policy statement demonstrating its commitment to and framework for the SEA of transport policies, plans, and programs.
	• Department of Foreign Affairs and International Trade announces EA framework for trade negotiations.
	• Bill C-19 to amend the Act is introduced.
2004	• Commissioner of Environment and Sustainable Development to report on the state of SEA within federal government departments and agencies
	• Guideline on the Cabinet directive on SEA updated
2007	• Agency identifies SEA, in particular the integration of regional and cumulative effects assessment, as a research and development priority for 2007–8.
	• Minister of Environment's Regulatory Advisory Committee, Sub-committee on SEA, commissions report on the state of SEA models, principles, and practices in Canada.
	• Government of Saskatchewan releases the Great Sand Hills Regional Environmental Study, a regional, strategic environmental assessment with direct consideration of cumulative effects.
2008	• Canadian Council of Ministers of the Environment commissions a report to produce guidelines for SEA methodology and good practice.

Sources: Based on Noble 2003, 2009.

cused on the implications of federal government initiatives and confidential memoranda submitted to Cabinet (Noble 2009). It was not until January 2004, under an updated SEA Directive, that Canadian federal departments and agencies were required to prepare a public statement whenever a full SEA had been completed. Outside the federal process, however, SEA is still practised largely on an ad hoc basis and there remains only limited knowledge of the diverse nature and scope of SEA and the value it has added to PPP decisions (Auditor General 2004; Noble 2004, 2009).

Directive-Based Systems and Practices

There are essentially three types of provisions for SEA in international practice. The first involves a legislative or mandatory provision, such as in Western Australia, where SEA is carried out under the umbrella of a formal environmental assessment process. The second type is an administrative order, such as in Denmark; this is a quasi-mandatory provision and requires SEA by way of administration or directive. The third type is an advisory or operation policy provision, such as provided for by the World Bank; this type suggests a more discretionary approach to SEA.

In Canada, SEA is required within federal departments and agencies by a Cabinet directive that states, as a matter of policy, that an SEA be conducted when

i)   a proposal is submitted to an individual Minister or Cabinet for approval; *and*
ii)   implementation of the proposal may result in important environmental effects, either positive or negative. (CEA Agency 2004, s. 1.0)

The purpose of the directive is to 'strengthen the role of strategic environmental assessment at the strategic decision-making level by clarifying obligations of departments and agencies and linking environmental assessment to the implementation of sustainable development strategies.' The Directive outlines the guidelines and requirements for federal government departments on implementing SEAs, and notes that, by addressing environmental issues at the strategic level, departments and agencies will be better able to

- optimize the positive impacts and minimize or mitigate the potentially negative effects of a proposal;
- consider potential cumulative environmental effects;
- implement department and agency sustainable development strategies;
- save time and money by drawing attention to issues at an early stage;
- streamline project-level environmental assessments;
- promote accountability and credibility; and
- contribute to broader policy commitments and obligations. (CEA Agency 2004, s. 2.1.1)

Several federal departments and agencies have adopted some form of SEA framework in recent years. In 2001, for example, Transport Canada released a policy statement on SEA detailing the types of transport initiatives subject to SEA and the key components of the SEA process.[1] However, and notwithstanding the advancement of federal SEA systems, Canada's experience with SEA application at the federal level is characterized by mixed success.

Case Study #1: Canada–Nova Scotia Offshore Petroleum Board, Misaine Bank Area SEA

The Canada–Nova Scotia Offshore Petroleum Board (CNSOPB) is an independent joint agency of the governments of Canada and Nova Scotia responsible for the regulation of petroleum activities in the Nova Scotia offshore area. SEA under the CNSOPB is undertaken by way of compliance, in principle, with the Cabinet directive and prior to a call being issued for bids for downstream offshore exploration and project development. In 2005, an SEA was undertaken for the Misaine Bank offshore area, Nova Scotia, to provide an overview of the offshore region, identify the potential effects associated with exploration activity, and assist in determining whether exploration rights should be offered for the area (CNSOPB 2005). A scoping document was released for public comment in early 2005, and the SEA was completed that same year. The assessment consisted of a review of seismic surveys and exploratory drilling activities, followed by a baseline description and discussion of the potential individual and additive effects of, and mitigation options for, exploration activities in the area.

Only a relatively narrow set of alternatives was considered in the SEA—namely, to proceed or not to proceed with offshore licensing, options that are consistent with alternatives typically considered at the project-specific level. However, the scope of the alternatives is consistent with the intended role of SEA under the CNSOPB, as well as with the scope of SEA practices elsewhere in the offshore sector, such as in the United Kingdom offshore area and the US Gulf of Mexico region. The Misaine Bank Area SEA concluded that the area is not more sensitive to the potential effects of oil and gas than other areas of the Nova Scotia offshore area, but it did identify sensitive marine areas that should be considered downstream in subsequent project-based EIAs. Unfortunately, there is no formal mechanism to ensure enforcement of SEA output and downstream project compliance. SEA experience under the CNSOPB reflects a proactive approach to offshore exploration, but at the same time SEA is restrictive in the scope of alternatives considered. Its primary role is to streamline the EIA and approvals process, but even this remains to be demonstrated in practice.

Case Study #2: National Capital Commission Core Area Sector Plan Assessment

The National Capital Commission (NCC) is an arm's-length federal Crown corporation with the mandate to plan lands in Canada's national capital region. In 2003, the NCC commenced a process to develop a Core Area Sector Plan (CASP) for the national capital region. The CASP was the third plan in a hierarchy of plans for the region, intended to guide planning and decision making over the next 20 years. Project-related actions under the CASP would be addressed in subsequent project-based EIAs under the Canadian Environmental Assessment Act. Based on the policy direction of the NCC and in compliance with the Cabinet directive, the CASP itself was subject to SEA. The SEA was conducted parallel to the development of the CASP, with the intention that its results would be fed into the planning process (NCC 2005). The SEA was objectives led, and it adopted a structured approach to assessing the implications of future planning initiatives resulting from the CASP on the basis of specified VECs and broader planning objectives. Alternatives to the plan were not explicitly assessed; rather, emphasis was placed on identifying a range of future planning actions or initiatives most likely to result from the CASP itself and evaluating their potential impacts. Future initiatives and projects under the plan, for example, were identified and assessed for potential environmental effects and cross-referenced with other known activities in the region with foreseeable environmental conditions or trends. Mitigation and monitoring measures were recommended for potentially adverse environmental effects. Since the SEA was conducted while the CASP was being developed, it proved difficult to coordinate and integrate its results with the development of the plan (see Noble 2009) However, the CASP is reported by the NCC to have been improved by SEA application, and the results of the SEA are reported to have improved subsequent planning initiatives in the National Capital Region.

Provincial Systems and Practices

Canadian provinces do not have any formal system of SEA comparable to that of the federal Cabinet directive. The requirements for EA in each province are established under various provincial acts and regulations (see chapters 15 through 21), and thus any provisions and requirements for any form of PPP assessment vary by jurisdiction (Table 6.5). In Alberta, Manitoba, Newfoundland, Prince Edward Island, and Quebec, existing EA legislation does not require or explicitly provide for the higher-order assessment of policies, plans, or programs. In Manitoba, the Manitoba Clean Environment Commission for the review of the Wuskwatim Hydroelectric Generating Station Project (2003–4) identified the potential for SEA in future planning and development in northern Manitoba; however, at the time of writing this chapter, provisions for SEA were still under review, with no specification as to whether or when the decision to implement a system for PPP assessment would be made.

In Newfoundland, PPP assessment is directed through Cabinet submissions; however, there is no formal provision for SEA under existing legislation. Policies and plans were once included under EA regulations, but in 1995 a government white paper on proposed reforms to the EA process excluded policies and plans from Newfoundland's EA requirements (Newfoundland 1995). Legislation in British Columbia, New Brunswick, Nova Scotia, and Ontario does provide for PPP assessment to varying degrees (Table 6.5), but not for any formal SEA process. In Ontario, for example, the scope of the Environmental Assessment Act, 1990, is defined as including 'enterprises or activities or proposals, plans, or programs'. That said, assessments for undertakings above the individual project level typically

Table 6.5   Provincial EA legislation and 'provisions' for EA above the project level

Jurisdiction	EA Legislation	Strategic Scope*
Alberta	Environmental Protection and Enhancement Act	———
British Columbia	Environmental Assessment Act	Policies, enactments, plans, practices or procedures of government (sec. 49, chap. 43)
Manitoba	Environment Act	———
New Brunswick	Clean Environment Act	Programs (sec. 31.1; schedule. A(l)(u))
Newfoundland	Environmental Protection Act	———
Nova Scotia	Environment Act	Policy, plan, program (sec. 3(az))
Ontario	Environmental Assessment Act	Plans, programs (sec. 3, chap. E.18)
Prince Edward Island	Environmental Protection Act	———
Quebec	Environment Quality Act	———
Saskatchewan	Environmental Assessment Act	Forestry plans (sec. 9.1)

*Refers to provisions or requirements for EA above the project level at the level of policies, plans, or programs.

Source: Based on Noble 2004.

occur under a Class EA system, setting out a planning process to be followed for a group or class of undertakings that do not require separate approval provided the class process is followed. There are no specific provisions for 'strategic' EA. Saskatchewan does have a formal legislative requirement for higher-order assessment, although this is not explicitly referred to as SEA and requirements are limited to the forestry sector and, in particular, to 20-year forest management plans. That being said, the definition of 'projects' under the Saskatchewan EA system is interpreted as including plans.

Case Study #1: Ontario Power Authority's Integrated Power System Plan

The Ontario Power Authority (OPA) is Ontario's planning authority for electricity. It is responsible for power system planning, generation development, conservation and demand management, and electricity sector development. In May 2005, the Ontario minister of energy directed the OPA to begin the process of developing an Integrated Power System Plan (IPSP)—a comprehensive plan for Ontario's electricity system to 2027 that would outline the steps to a reliable electricity supply. Power system planning is exempt from the Ontario EA process, but it is legislated under the Electricity Act and reviewed by the Ontario Energy Board. The basis for the IPSP was a supply-mix directive that identified a preferred set of electricity supply and distribution objectives for the province based on a review of possible supply-mix alternatives prior to the IPSP. The IPSP itself, then, did not take into consideration the impacts of a range of electricity alternatives or to identify a preferred supply mix, but rather focused on the implications of a prescribed supply mix. The IPSP set out to (1) identify demand-reduction strategies and new generation technologies; (2) increase alternative and renewable energy-generating capacity; (3) replace coal-fired generation; and (4) develop new programs and targets for electricity production, planning, and delivery. The IPSP process did involve consideration of a number of more limited alternatives within the context of the preferred supply mix, including a range of scenarios for electricity conservation and demand management, and competing options for plan procurement. The results of the final phase of the IPSP— assessment and identification of procurement options—were released in January 2007.

Although the IPSP was not an SEA by name (nor was it reviewed under any system of environmental assessment), it is illustrative of an attempt at integrated SEA for sector-based plan development, albeit restrictive in the sense that the SEA process was limited to a prescribed electricity mix and to the evaluation of alternatives within the scope of a predetermined policy outcome. The IPSP did not result in the identification of a preferred alternative but rather in the rationalization of a prescribed electricity mix. In this sense, while the IPSP is comprehensive, the model of SEA depicted by the planning process is relatively restrictive in comparison to SEA applications for electricity planning elsewhere, such as the United Kingdom's privatized electricity sector (see Marshall and Fischer 2006).

Case Study #2: Saskfor MacMillan 20-Year Forest Management Plan

The Environmental Assessment Act in Saskatchewan was originally passed into law in 1980, predating the formal development of SEA on the Canadian national scene. The scope of EA under the Act, however, does apply to plans for potential 'developments' and, in particular, for 20-year forest management plans. Forestry companies wishing to apply for harvest permits in Saskatchewan are required to prepare a 20-year management plan and an environmental impact statement for review by the Saskatchewan government and the public. Five plan-level assessments have been carried out in Saskatchewan's forestry sector in recent years, including the Pasquia-Porcupine Forest Management Plan assessment. While assessments of 20-year forest management plans are not formally titled SEAs in Saskatchewan, they do resemble the principles and practices of SEAs.

In 1995, MacMillan Bloedel Limited and a subsidiary of Saskatchewan Crown Investments formed a partnership (the Saskfor MacMillan Limited Partnership— SMLP) and submitted an application to the Government of Saskatchewan to develop the Pasquia-Porcupine forest management area. This area, located along the Saskatchewan-Manitoba border within the Boreal Plain ecozone, encompasses approximately two million hectares, of which half is suitable for commercial timber production (SMLP 1997). In 1996, in compliance with the Saskatchewan Act, SMLP commenced public consultations, following up with an integrated harvest plan and an environmental assessment. The SMLP assessment involved the consideration of five forest plan alternatives, including the baseline alternative of 'no timber harvesting'. SMLP assessed the environmental and socio-economic implications of each alternative by using a combination of qualitative measures and quantitative indices and models. Assessment criteria were derived on the basis of public consultations, forest management goals and objectives as contained in various provincial and national forest resource management and harvest regulations, and SMLP's broader goals of integrated resource management, ecosystem process maintenance, forest restoration, timber removal, and the provision of local social and economic benefits. These criteria were set within the Canadian Standards Association's vision of sustainable forest management, and the potential implications of each alternative were evaluated in relation to the existing environmental and socio-economic baseline conditions. Once the preferred alternative was selected, potentially residual negative effects were identified and impact mitigation measures proposed. SEA was applied early in the process, informed the development and selection of the preferred planning option, and ensured that the plan outcome would contribute to improved industry standards, operating regulations, and operating procedures. The management plan and assessment were endorsed in 1997.

SEA Challenges and Directions

This chapter has presented a brief overview of the nature, methodology, and current state of SEA systems and practices in Canada. SEA in Canada is ongoing, albeit under a range of systems and provisions, including

- legislated requirements for EA above the project level, albeit with no SEA label;
- policy requirements for formal SEA application in compliance with the Cabinet directive;
- one-time conceptual and policy reviews carried out under the auspices of legislated EA guidelines and provisions; and
- applications demonstrating good-practice SEA methodology but one carried out under neither the SEA label nor any formal system of environmental assessment.

Given the current state of SEA in Canada, several outstanding issues deserve immediate attention if SEA is to advance in practice and effectiveness.

Knowledge and Understanding

Despite general acceptance of the notion of SEA among researchers, the formal adoption of SEA has been slow to advance. 'The complexity of the processes associated with SEA, the consequent need for additional resources, the fact that it is often indicated as having little added value in relation to project's EIA, are some of the factors that are limiting the more extensive adoption of SEA' (Partidário and Clark 2000, 4). In the Canadian context, very little seems to have changed since

Davey's (1999) investigation of SEA in Nova Scotia, suggesting that the lack of SEA application and of a formal legislative framework for SEA has been due primarily to the limited understanding of SEA concepts. SEA is typically defined as a short, concise analysis to help in the investigation of the potential environmental implications of proposed and existing PPPs and their alternatives, such that a direction based on the best practicable environmental option can be identified. However, the term *strategic* is used in a variety of ways in environmental assessment, often meaning different things in different contexts. As a result, not all assessments that carry the SEA label are actually SEAs. Equally important, many assessments that do not carry the SEA label are in fact good SEAs. Research into the different uses of SEA is required, under both formal and informal applications, so that some common framework and a set of agreed-upon principles for SEA in Canada may be established.

Methodological Pluralism

Canadian SEA is currently characterized by methodological pluralism, the boundaries of which are not well defined (Noble 2009). One of the main difficulties experienced in most countries in relation to the operationalization of SEA is the lack of appropriate methodological frameworks that specifically address SEA requirements. While a common framework is being developed at the federal level in Canada,[2] measurable progress to date in SEA methodological development is limited. SEA asks different types of questions than do project-level assessments, and therefore different assessment methodologies are required. A *best* SEA process may not exist. Partidário suggests: 'Rather than offering SEA as an environmental impact assessment procedure to decision makers, or as a new type of planning or policy making, . . . SEA should be conceptualized as a framework, with core elements that are incrementally tailor-made through policy and planning procedures and practices, whatever the decision-making system in place' (2000, 661). This flexibility in SEA methodology, however, does not mean that a structured framework is not required. The lack of a structured SEA framework may create confusion among practitioners and lead to inconsistencies in SEA application. Practitioners would benefit from further applications of the framework illustrated in this chapter and of similar frameworks in a variety of sectors and at different tiers of decision making. Research is required into international practices and experiences so that we can see what has worked, what has not worked, and why. This sort of knowledge is most likely to be gained through experience and by recognizing that there are examples of good-practice SEAs and SEA frameworks that do not carry the SEA name tag.

Tiering

SEA is intended to involve the examination of a range of alternatives above the project tier and then the identification of a preferred course of action. Project assessment is then initiated to determine in greater detail the potential impacts and implementation options of the 'best' alternative. While SEA and project assessment do not always occur in such a hierarchical fashion, if SEA is to be influential, then its results should inform decisions, including EIA, at the next tier. A major challenge in SEA, however, is the limited transmittal of strategic-level output to actions and assessments at the next tier (see Noble 2009). In practice, SEA is often a 'one-off' process, with only limited influence over subsequent assessment processes. If there is no tiered system of assessment and only limited commitment to ensuring that SEA output is carried forward, it is unlikely that the added value of SEA will be fully realized. At the federal level in Canada, as in most other Canadian jurisdictions, there does not exist a formal tiered system of PPP assessment to carry SEA results forward effectively to the project level.

Collaboration

Essential to the SEA process is the ability of government agencies and stakeholders to work together in partnership. This requires a common vision and commitment; strategies for communication, data sharing, and joint decision making; and clear delineation of roles in and responsibilities for the implementation of recommendations emerging from SEA. One challenge is that different government agencies often express different views concerning appropriate PPPs, and further, many of the recommendations emerging from SEA are beyond the scope and authority of the government agency and environment ministry in charge of the assessment process (Noble 2009, 2008). Effective SEA requires a degree of inter-agency collaboration not typical of traditional project-based EAs.

Institutional Commitment

In Canada, an SEA is required at the federal level, by matter of Cabinet Directive, (a) when a proposal is being submitted to an individual minister or Cabinet for approval *and* (b) when the implementation of the proposal may result in important environmental effects. Any PPP that does not meet both requirements is not subject to the SEA process. At the provincial level, there are few if any formal requirements for SEA. If SEA is to be used effectively, then it must include not only formal Cabinet submissions but also any instrument that gives effect to a PPP that may have potential environmental effects (either positive or negative) (Noble 2002a). Buckley (2000) suggests a non-exclusive list of government instruments to which SEA should apply, including

- formal government PPP documents and instruments;
- any government documents that describe, set out, or establish government policy or perspectives on any topic or issue;
- any bill or legislation;
- any government document that defines a government intention, budget, trade agreement, or expenditure of funds;
- any government involvement in, or accession to, any international agreement; and

- any other document or component of government activity likely to have an effect on the environment.

An additional criterion can be added to this list: any government decision-making process that might result in a PPP-related strategy or course of action. Institutional commitment through both policy and legislation is critical to the implementation of SEA frameworks and practices. As Partidário suggests, implementation depends on effective political will, and furthermore, 'there may be no procedural or technical mechanisms that can replace political accountability and effective . . . institutional frameworks' (1996, 39). Political and institutional limitations, rather than technical or procedural ones, are often the underlying cause of insufficient SEA systems and practices. Without the appropriate institutional support, even the most robust SEA frameworks will be of little significance to decision making. There is a need for comparative research on international SEA legislation and requirements and on the characteristics of success under various legislative frameworks.

Conclusion

SEA has the potential to improve environmental decision-making and assessment processes at the earliest possible stages of planning and development. Currently, SEA in Canada is occurring, albeit largely informally and mostly within the realms of government policy, planning, and decision making. However, if SEA is to be realized to its full potential, then the scope of SEA research and SEA application must be broadened beyond the realm of government policy. Specifically, there is a need to address the potential role and benefits of SEA in industry-based planning and decision-making processes. Marshall (2003), for example, argues that the introduction of SEA either as a government requirement or as a business planning tool may lead to improved planning and decision making. If SEA is to fulfil its potential in the business community, it must become relevant and responsive to the environmental governance of business. To limit SEA to government legislation and government PPPs is to fail to recognize the potential of SEA as a tool for improved and integrative decision making across all sectors and industries.

Notes

1. Under Transport Canada's framework, several types of initiatives are subject to an SEA, including (a) policies that would lead to financial support for transportation systems; (b) decisions that would affect the mode or location of transportation services; (c) policies that would affect the level of use of different transportation modes; (e) policies that involve economic deregulation; and (f) policies that would affect the pricing of transportation services. (See Transport Canada 2001.)
2. In 2008, the Canadian Council of Ministers of the Environment Environmental Assessment Task Group commissioned a study to develop a structured methodological framework and good-practice guidelines for SEA in Canada.

References

Auditor General. 2004. Assessing the environmental impact of policies, plans, and programs (chap. 4). In *Report of the Commissioner of Environment and Sustainable Development*. Ottawa: Auditor General of Canada.

Bailey, J., and S. Renton. 1997. Redesigning EIA to fit the future: SEA and the policy process. *Impact Assessment and Project Appraisal* 15 (3): 219–334.

Barrow, C.J. 1997. *Environmental and social impact assessment: An introduction*. London, UK: Arnold.

Bartlett, S. 2001. Systems thinking for learning-oriented SEA. Paper presented at the International Association for Impact Assessment Annual Conference, Cartagena, Colombia.

Bina, O. 2007. A critical review of the dominant lines of argumentation on the need for strategic environmental assessment. *Environmental Impact Assessment Review* 27:585–606.

Bond, A.J., and D.J. Brooks. 1997. A strategic framework to determine the best practicable environmental option (BPEO) for proposed transport schemes. *Journal of Environmental Management* 51 (3): 305–21.

Boothroyd, P. 1995. Policy assessment. In F. Vanclay and D.A. Bronstein (Eds), *Environmental and social impact assessment*, 83–128. Chichester: Wiley.

Brown, A.L., and R. Therivel. 2000. Principles to guide the development of strategic environmental assessment methodology. *Impact Assessment and Project Appraisal* 18 (3): 183–9.

Buckley, R.C. 2000. Strategic environmental assessment of policies and plans: Legislation and implementation. *Impact Assessment and Project Appraisal* 19 (3): 209–15.

CEA Agency (Canadian Environmental Assessment Agency). 1999. *Strategic environmental assessment: The 1999 Cabinet Directive on the environmental assessment of policy, plan and program proposals: Guidelines for implementing the Cabinet Directive*. Hull, QC: Her Majesty the Queen in Right of Canada.

CEARC (Canadian Environmental Assessment and Research Council). 1990. *The environmental assessment process for policy, plan, and program proposals*. Ottawa: Minister of Supply and Services Canada.

CEC (Commission of European Communities). 1993. *The European high speed train network*. Environmental Impact Assessment Executive Summary. Brussels: Research and Consulting Mens en Ruimte.

Clark, R. 2000. Making EIA count in decision-making. In M.R. Partidário and R. Clark (Eds), *Perspectives on strategic environmental assessment*, 15–27. New York: Lewis Publishers.

CNSOPB (Canada–Nova Scotia Offshore Petroleum Board). 2005. Strategic environmental assessment of the Misaine Bank area. Report to the CNSOPB prepared by CEF Consultants. Halifax: CNSOPB.

Connelly, B. 2000. Notes from a speech to the Ontario Association for Impact Assessment, 1 June, Ottawa.

CSTA (Council of Science and Technology Advisors). 1999. *Science advice for government effectiveness*. Ottawa: SAGE.

Dalal-Clayton, B., and B. Sadler. 2005. *Strategic environmental assessment: A sourcebook and reference guide to international experience*. London, UK: Earthscan.

Davey, L. 1999. Adopting a SEA framework for the development of legislation in Nova Scotia, Canada: Benefits and challenges. Paper presented at the International Association for Impact Assessment Annual Conference, Glasgow, Scotland.

DEAT (Department of Environmental Affairs and Tourism). 2000. *Strategic environmental assessment in South Africa: Guideline document*. Africa: CSIR–DEAT.

FEARO (Federal Environmental Assessment Review Office). 1992. *Environmental assessment in policy and program planning*. Hull, QC: Minister of Supply and Services Canada.

———. 1993. *The environmental assessment process for policy and program proposals*. Hull, QC: Minister of Supply and Services Canada.

Fischer, T.B. 2002. Strategic environmental assessment performance criteria: The same requirement for every assessment? *Journal of Environmental Assessment Policy and Management* 4:83–99.

Gibson, R.B. 2001. Specification of sustainability-based environmental assessment decision criteria and implications for determining 'significance' in environmental assessment. Report prepared under a contribution agreement with the CEA Agency Research and Development Program. Hull, QC: CEA Agency.

Glasson, J., R. Therivel, and A. Chadwick. 1999. *Introduction to environmental impact assessment: Principles and procedures, process, practice and prospects.* 3rd ed. London, UK: University College London Press.

Harriman, J., and B.F. Noble 2008. Characterizing regional approaches to project and cumulative effects assessment in Canada. *Journal of Environmental Assessment Policy and Management* 10 (1): 25–50.

IAIA (International Association for Impact Assessment). 2002. Strategic environmental assessment performance criteria. IAIA Special Publication Series, no. 1.

Jones, M.L., and L.A. Greig. 1985. Adaptive environmental assessment and management: A new approach to environmental impact assessment. In V.W. Maclaren and J.B. Whitney (Eds), *New Directions in Environmental Impact Assessment in Canada*, 21–42. Toronto: Methuen.

Lee, N., and C. Wood. 1978. EIA: A European perspective. *Built Environment* 4 (2): 101–10.

Marshall, R. 2003. SEA and energy, Part I: The application of statutory and non-statutory SEA in UK electricity transmission and distribution network plans and programmes. Paper presented at the International Association for Impact Assessment meeting, Marrakech, Morocco.

Marshall, R., and T.B. Fischer. 2003. Regional electricity transmission planning and SEA: The case of the electricity company Scottish Power. *Journal of Environmental Assessment Policy and Management* 49:279–99.

NCC (National Capital Commission). 2005. Canada's Capital Core Area Sector Plan. Ottawa: National Capital Commission.

Newfoundland. 1995. *A white paper on proposed reforms to the environmental assessment process.* St John's, NL: Department of the Environment.

Noble, B.F. 2000. Strategic environmental assessment: What is it and what makes it strategic? *Journal of Environmental Assessment Policy and Management* 2 (2): 203–24.

———. 2002a. Strategic environmental assessment of Canadian energy policy. *Impact Assessment and Project Appraisal* 20 (3): 177–88.

———. 2002b. The Canadian experience with SEA and sustainability. *Environmental Impact Assessment Review* 22 (1): 3–16.

———. 2003. Auditing strategic environmental assessment in Canada. *Journal of Environmental Assessment Policy and Management* 5 (4): 127–47.

———. 2004. A state-of-practice survey of policy, plan, and program assessment in Canadian provinces. *Environmental Impact Assessment Review* 24:351–61.

———. 2008. Strategic approaches to regional cumulative effects assessment: A case study of the Great Sand Hills, Canada. *Impact Assessment and Project Appraisal* 26 (2): 79–90.

———. 2009. Promise and dismay: The state of strategic environmental assessment systems and practices in Canada. *Environmental Impact Assessment Review* 29:66–75.

Noble, B.F., and L. Christmas. 2008. Strategic environmental assessment of greenhouse gas mitigation options in the Canadian agricultural sector. *Environmental Management* 41:64–78.

Noble, B.F., and K. Storey. 2001. Towards a structured approach to strategic environmental assessment. *Journal of Environmental Assessment Policy and Management* 3 (4): 483–508.

Partidário, M.R. 1996. Strategic environmental assessment: Key issues from recent practice. *Environmental Impact Assessment Review* 16 (1): 31–56.

———. 2000. Elements of an SEA framework—Improving the added-value of SEA. *Environmental Impact Assessment Review* 20:647–63.

Partidário, M.R., and R. Clark. 2000. *Perspectives on strategic environmental assessment.* New York: Lewis Publishers.

Retief, F. 2007. A performance evaluation of strategic environmental assessment processes within the South African context. *Environmental Impact Assessment Review* 27 (1): 84–100.

Sadler, B. 2005. Canada. In C. Jones, M. Baker, J. Carter, S. Jay, M. Short, and C. Wood (Eds), *Strategic environmental assessment and land use planning: An international evaluation.* London, UK: Earthscan.

Sadler, B., and R. Verheem. 1996. *Strategic environmental assessment: Status, challenges and future directions.* Report 53. Ottawa: CEA Agency.

Sheate, W., J. Richardson, R. Aschemann, J. Palerm, and U. Stehn. 2003. SEA and integration of the environment into strategic decision making. Report to the European Commission, London, no. B4-3040/99/136634/MAR/B4.

SMLP (Saskfor MacMillan Limited Partnership). 1997. Twenty-year forest management plan and environmental impact statement for the Pasquia-Porcupine forest management area. Regina: SMLP.

Therivel, R. 1993. Systems of strategic environmental assessment. *Environmental Impact Assessment Review* 13 (3): 145–68.

———. 2004. *Strategic environmental assessment in action.* London, UK: Earthscan.

Transport Canada. 2001. *Strategic environmental assessment at Transport Canada: Policy statement.* Ottawa: Transport Canada.

UNECE (United Nations Economic Commission for Europe). 1992. *Application of environmental impact assessment principles to policies, plans, and programs.* New York: United Nations.

Verheem, R., and J. Tonk. 2000. Strategic environmental assessment: One concept, multiple forms. *Impact Assessment and Project Appraisal* 18 (3): 3–23.

Vicente, G., and M.R. Partidário. 2006. SEA—enhancing communication for better environmental decisions. *Environmental Impact Assessment Review* 26:696–706.

Chapter 7

Social Impact Assessment and High-Level Radioactive Waste Disposal: The Canadian Concept and Aboriginal Responses

Ron Pushchak and Ann Marie Farrugia-Uhalde

Introduction

All changes to the natural environment are inherently social as well as ecological. Not only does each new project have an impact on its physical and biological surroundings, it also affects the social and economic relations among people and the social and cultural values held by communities. Communities and social groups can be directly affected by changes in social resources that limit or expand their opportunities for social activities. They can also be affected indirectly by changes in the resources they depend on; for example, a change in available food or economic opportunity can significantly affect a community's feelings of cohesion and alter its social exchanges.

The process to determine the acceptability of new impacts on the environment, the environmental assessment (EA) process, is fundamentally a social as well as an environmental exercise. EA is a social process in that it engages the public and its institutions in making a decisions about projects and distributes benefits and costs to stakeholders, social groups, and communities (Francis and Jacobs 1999). The consequences of an environmental assessment approval are widely felt, and often an EA is the only opportunity stakeholders have to consider the value of large social and cultural changes before they happen. In terms of guiding major social change, the EA process has emerged as one of the most important instruments in Canadian public policy to protect human environments.

In few instances have the social impacts of a proposed development been more significant than in the attempt by the Canadian government to develop an approved and accepted high-level radioactive waste disposal facility. A concept was proposed that would dispose of radioactive wastes from nuclear reactors by having them permanently buried deep in the igneous rock of the Canadian Shield. This idea was subjected to an environmental assessment that highlighted the many social concerns that Canadians have about the disposal of nuclear wastes. That EA looked into the value Canadians put on nuclear power and the potential social as well as physical impacts of the waste disposal facility.[1]

The approval of a radioactive waste facility also raised the issue of the potentially significant cultural impacts on First Nations peoples. Since the facility was likely to be sited in the Precambrian Shield region, it was expected to have its most significant social and cultural effects on one or more of the many Aboriginal communities in the North. A northern environmental assessment noted the potential for intercultural impacts and the inequitable distribution of benefits and risks when southern economic interests came into conflict with northern social and cultural values. Resolving the social and cultural conflicts posed a great challenge for the EA process (Mulvihill and Baker 2001).

This chapter outlines the current process of assessing the social impacts of large public projects in Canadian federal environmental impact assessments (EIAs); in particular, it looks at the specific role that social impacts may play in the Canadian nuclear waste disposal EIA. The basic concepts of social impact assessment (SIA) are discussed, as well as the means used to estimate the social effects of large physical changes in northern environments. The chapter focuses on the nuclear waste disposal concept and the impacts of such a facility on Aboriginal communities and cultural groups; it includes the responses of Aboriginal people to the social impacts as recorded in the testimony and submissions to the EIA public hearings.

Assessing Social Impacts

Social impacts include a broad range of outcomes that follow a project; they include 'all social and cultural consequences to human populations of any public or private actions that alter the ways in which people live, work, play, relate to one another, organize to meet their needs and generally cope as members of a society' (Burdge and Vanclay 1995, 32). In the Canadian North, the cultural impact of a radioactive waste disposal facility on Aboriginal communities is particularly significant. Generally, cultural impacts include 'changes to the norms, values and beliefs of individuals that guide and rationalise their cognition of themselves and their society' (Burdge and Vanclay 1995, 32).

While social groups, community interactions, and cultural practices are highly valued parts of the environment, it is nonetheless difficult to accurately predict social impacts in a context of continuing social change; this is a major weakness in social impact assessment. The social reality that we inhabit is a complex and multifactored environment, with a plurality of actors, social groups, and communities, each with a variety of interests—economic, political, cultural, and moral. Not only is the social environment complex, it is also constantly experiencing change in response to a number of internal and external stimuli. Social systems are richly dynamic, continually adapting to a variety of influences, from large-scale market fluctuations to changing methods of communication to the adoption of new technologies.

Vanclay (2002) argues that it is important to distinguish social changes and social impacts. He defines a *social change* as an ongoing change in social conditions that tends to happen in all communities, while a *social impact* is 'an actual experience of an individual or community' in response to a project—it is a change that must be experienced or felt by the people affected rather than a normal change in social conditions that occurs as a community grows (201). For example, a natural increase in the working population of a community is a social change, whereas crowding resulting from a lack of adequate housing because of a sudden population increase when a project's workforce arrives is a social impact. Social impacts can include a shortage of social services, fluctuating property values, or a loss in community cohesion. In each case, social impacts are the experiences of change imposed by the project, and they can be positive as well as negative. Positive changes include improved sanitation, increased educational opportunity, higher community employment, and increased opportunities for social interaction, although the practice in SIA has largely been to predict negative social impacts.

Beyond the task of separating social impacts from ongoing changes in social conditions, the purpose of SIA is to predict impacts beforehand. An environmental assessment, as well as the SIA associated with it, is an anticipatory form of decision making (Interorganizational Committee 1995, 17). An EA forces proponents to predict impacts before damage is done to natural or social environments, at a stage where significant environmental damage can be avoided or minimized. Consequently, predicting impacts is a basic part of SIA (Barrow 2000, 2). As noted above, however, it is difficult to make reasonably accurate predictions of social changes and impacts. While SIA requires that the effects of a development be predicted before a project is approved and irreplaceable resources committed, the means of predicting social impacts are lacking: Predicting what will happen in any social context cannot be done without models of cause and effect, and at present, few predictive models are available. Lawrence (1997) suggests that the difficulty in predicting social impacts has its roots in the rapid development of environmental assessment—that is, EA evolved before a conceptual basis for predicting social impacts was established. He suggests that because theories about how the social environment behaves are fragmented, with different theories being used to explain different social phenomena, it is hard to estimate impacts on the social environment in advance in most cases. Predicting social impacts is also difficult because people can change their behaviour and adapt to new social conditions on the basis of their knowledge of the project and its likely impacts. In this light, Lawrence (1997, 88) suggests that the social environment is 'trans-scientific'; in other words, it is an environment of great uncertainty where impacts cannot be determined through scientific analysis. In many cases, predicting impacts in the social environment may be impossible or impractical, particularly for projects that are infrequent, complex, or involve new technologies.

Because of the uncertain nature of the social environment, SIA involves four challenges:

- To assess non-quantifiable as well as quantifiable social components
- To conduct a comprehensive social analysis
- To predict social change
- To evaluate social impacts

A community or social group is more than the sum of its population and physical resources. The social environment is made up of two kinds of elements, those that social scientists can measure and quantify (the size and composition of the population, its employment, community resources, and economic activities) and those intangible elements that elude measurement (the cohesion of a community, its perceived character, the satisfaction people have with their surroundings) but are nonetheless important components of the social context. It is a paradox in SIA that the things most important to our social well-being are often the most difficult to define and measure.

Assessing impacts in any social environment requires a comprehensive knowledge of all the components of the particular environment and their interactions, a level of understanding that is beyond the ability of social impact assessors to achieve. Thus, in such cases, SIA relies heavily on a consultation process so that it can be decided beforehand what social issues must be addressed and the degree to which they should be investigated.

Predicting the impacts that may be caused by a proposed development in the social environment poses an even greater dilemma for social impact assessors. It seems to require theories for the ways individuals, social groups, and communities behave when confronted by a development. However, social science has so far been unable to provide a reasonably reliable set of theories that can be used to predict social impacts (Lawrence 1997). In the absence of such theories, an SIA process is used to assess social impacts. A process approach to social impact assessment is inevitable for two reasons: first, there is no alternative, and second, failing to assess impacts is intuitively unacceptable, since our experiences with large-scale projects have taught us that social changes do happen, they are significant, and they lead to real social losses and gains.

Adding to the dilemma in assessing social impacts is the fact that there is limited theoretical guidance as to which aspects of our natural, economic, social, and cultural environments we value most. In order to judge whether impacts are acceptable, we need to apply a value to each one. Multi-criteria analysis methods have been developed that assign values to social and cultural resources through the use of surveys and participatory techniques (Massam 1993; Malczewski 1998). However, there are limitations in our ability to determine the nature of social values across a community, including those values involved in the selection of survey samples and groups that purport to represent the values of a community. Moreover, community values are not absolute or unchanging; they vary in each community and social group over time. This makes the assigning of value to social changes uncertain and open to bias. Nonetheless, after social impacts have been predicted, they must be assigned a value and evaluated together with all other impacts to determine on balance whether the environmental impacts are acceptable given the social costs.

Assessing Social Impacts in Environmental Impact Assessment (EIA)

The need to assess the social impacts of a project has been a part of the environmental assessment process since it began more than 30 years ago in the United States. The National Environmental Policy Act (NEPA 1969) defined environment impacts broadly to include both physical changes and major actions that significantly affect the quality of the human environment. The environmental impact statement under the Act called for the use of the social sciences in assessing the impacts of projects on the human environment, with the human environment to be 'interpreted comprehensively' to include 'the natural and physical environment and the relationship of people with that environment' (CEQ 1986, 1508.14). Although social impacts were not the first concern of the EIA under NEPA, given that economic or social effects were 'not intended by themselves to require the preparation of an environmental impact statement,' it was clear that where the economic and social effects of a project are interrelated with natural or physical effects, 'the environmental impact statement will discuss all of these effects on the human environment' (CEQ 1986, 1508.14). These effects included not only direct impacts caused by the project but also the 'aesthetic, historic, cultural, economic, social, or health effects, whether direct, indirect, or cumulative' (CEQ 1986, 1508.8). In the US process, the whole social environment was to be assessed and impacts were to include social effects that were indirectly caused by the project and could accumulate over time. This provided a broad, comprehensive scope for examining social impacts in the United States.

In Canada, the Federal Environmental Assessment and Review Process (FEARP) began in 1973 as a directive issued by the federal Cabinet and not through a law passed by Parliament. It operated as a federal policy procedure, and the government believed its EA policy was a discretionary one that it could choose to apply, if necessary. Federal departments were directed to consider the social as well as the environmental impacts of their projects, but the biophysical environment was the primary concern and social impacts included only those social effects directly related to the project. Federal panels, appointed to issue guidelines for environmental assessments, were the main sources of social impact investigation in the early years of the FEARP. Panels typically directed federal departments to consider the social impacts of projects in their EIAs.

In fact, in two federal EIA cases, the *Rafferty-Alameda* and *Oldman River Dam* cases, the social impacts were so significant that the federal courts decided that the non-enforceable guideline was, in effect, a legally enforceable law. In the *Oldman River Dam* decision, for example, the EA panel found that the major impacts reported in the EIA were effects on fisheries, wildlife, and riparian forest ecosystems and those (social) changes affecting the Peigan Indian Band (FEARO 1992, i). The panel noted that the social effects on the Peigan band included impacts on the band's food supply and potential mercury contamination that could affect their health, could mean losses of cultural resources, and constituted a threat to their claim of Aboriginal rights to the river valley. It determined that the project had important adverse consequences for the social and cultural economy of the Peigan and that great weight should be placed on these consequences for a fair and equitable decision to be reached (FEARO 1992, 26). The panel observed that

> the consequences of the project in these four areas impose substantial environmental, so-
> cial and economic costs . . . and when considered in the context of the small economic
> benefits to be derived from increased irrigation agriculture and other uses of the dam,
> they lead the panel to its preferred, but not unanimous, decision that the dam should be
> decommissioned by opening the low level diversion tunnels and permitting unimpeded
> flow of the river. (FEARO 1992, i)

Clearly, the social impacts on the Aboriginal community were so significant that the panel recommended reversing the effects of a recently completed dam. This decision imposed an EA duty on all federal agencies and ultimately led to a change in the federal process from policy to law in the Canadian Environmental Assessment Act (CEAA 1992).

Although the federal EA process still has the natural environment as its primary focus, given the definition of environment in the Act, proponents are directed to consider social impacts. The Act describes the *environment* as being 'the components of the Earth' and as including '(a) land, water and air, including all layers of the atmosphere, (b) all organic and inorganic matter and living organisms, and (c) the interacting natural systems that include components referred to in (a) and (b) while environmental effects are to include "any effect of any such change on health and socio-economic conditions"' (CEAA 1992, 2(1)).

In practice, the environmental effects of a federal project should include changes in the biophysical environment caused by the project as well as

> certain effects that flow directly from those changes, including effects on
> • human health;
> • socio-economic conditions;
> • physical and cultural heritage, including effects on things of archaeological, paleonto-
>   logical, or architectural significance; and
> • the current use of lands and resources for traditional purposes by aboriginal persons.
> (CEA Agency 1994, 28)

A federal EIA, then, does not consider the direct social impacts of a project, rather only those social impacts associated with biophysical effects, including impacts on the human environment and, indirectly, the social effects on Aboriginal peoples.

At the provincial level, by the late 1970s most governments had adopted some form of EA. Ontario was the first to pass a separate Environmental Assessment Act

(1975) that required EAs for all public sector projects. The Ontario Act specifically defined the environment to include not only the biophysical but the social, economic, and cultural conditions that might be affected by a project (Environmental Assessment Act, R.S.O., 1996, 1(c)).[2] Ontario established not only a broad and inclusive definition of the environment that specifically noted social environments but also an EA process that required that human environments be described, that social impacts be predicted, and that they be evaluated together in the EA.

As a result, it is Canadian policy at both levels of government to conduct an EA that considers social impacts. An EA demands that the effects of a project on the social well-being of affected peoples and communities be estimated and taken into account in the decision of whether to approve a development.

The SIA Process

Given the lack of theory guiding environmental assessments and the limitations in predicting impacts and establishing their value, a broadly accepted process for assessing social impacts is not available. In practice, two major approaches have been used to estimate the social conditions in a community and predict impacts— a technical approach and a more participatory process.

The technical approach, which was the predominant approach early in SIA, is based on a scientific model of social environments. It assumes that assessors can gain an understanding of the valued components and dynamics of a community quantitatively by obtaining data on the community's social attributes—its population size, demographic composition, income, education, occupations and employment, and other elements. Under this approach, it is also assumed that assessors can survey a community in order to determine its preferences for types of impacts. By analysing these data, assessors can draw conclusions about the effects a community may experience when a project is implemented. However, as Barrow (2000, 3) suggests, many social impacts are not open to scientific assessment, and the detached observation of a community alone is not enough if one is seeking to understand its social context.

The technical approach to SIA has been largely unsatisfactory because it has failed to anticipate changes in the unquantifiable but significant aspects of the social environment, to reflect community values, or to establish the significance of cultural resources or the degree of cohesion and vitality that exists in a community through the interpretation of statistical information and community surveys. The typical outcome of the technical approach where communities were poorly understood has been vigorous opposition to and protracted conflicts about proposed projects. The social assessment literature is filled with examples of failure to estimate social impacts accurately when the technical approach was used (Lawrence 1993; Barrow 2000; Vanclay 2002).

Recent SIAs have generally moved away from attempting to predict social impacts mechanically and quantitatively, and have increasingly adopted a model based on public participation to estimate social effects, one that shares more with anthropology than with other social sciences. The participation process takes as its starting point the premise that social reality is complex and hard to assess scientifically and that impacts cannot be predicted with much certainty. The participatory approach calls for assessors to become familiar with the affected community by observing its social and cultural practices and to learn about the significance of human and cultural resources by engaging in dialogue with community members. They make predictions about social impacts after they have asked those most likely to experience social changes to estimate what the impacts are likely to be and to explain their significance. By the same token, this process involves social learning on the part of the affected community, since the sharing of knowledge about related projects and their impacts helps the community learn about the nature of the project and its likely effects (Webler et al. 1995; Saarikoski 2000; Becker et al. 2003). This approach is also reflected in the *action research tradition,* where the active collaboration and mutual learning of the social researcher and the affected community produce relevant information that leads to social action (Franklin 1994; Winter 1996; Huizer 1997).

A participatory model for social impact assessment assumes that having stakeholders take part in the assessment process can resolve some of the dilemmas in SIA (Sinclair and Diduck 2001). Participants can set the scope of the investigation by identifying the appropriate set of impacts and concerns, by collectively projecting the likely impacts, and by assigning values to social and cultural components for evaluation. In embracing the value of participants' knowledge and their qualitative judgments, SIA becomes as much an art as it is a science, and 'it relies a great deal on the professional judgement of researchers, so that qualitative measurements are useful' (Barrow 2000, 5).

Concepts in SIA

Because there is no universally accepted process for impact assessment, the practice of estimating social effects has varied, with proponents selecting methods that best fit their assessment situation and needs. In order to develop a more consistent understanding of SIA practice, a group of leading scholars and practitioners known as the Interorganizational Committee on Guidelines and Principles for Social Impact Assessment outlined a set of guidelines and principles for assessing social impacts. The guidelines have been adopted by the US government to assist project planners and decision makers in meeting the requirements of the National Environmental Policy Act for social impacts (Interorganizational Committee 1995). The six elements outlined below are reasonably consistent with the best accepted practice in assessing social impacts:

1. The scope of a social analysis
2. The profile or baseline conditions of the social environment
3. The projection of social impacts
4. The evaluation of social changes
5. Mitigating measures
6. A monitoring program

The Scope of a Social Analysis

The social environment of any community is a complex network of social groups, economic relations, and cultural activities. It is impossible for any social impact assessment to investigate comprehensively all of the social components in an environment and identify all of the social and cultural activities that might be a part of it. In addition, it is difficult to know the concerns that an affected community might have about a project. If a social assessment cannot be comprehensive, it has to be focused on a set of likely social impacts and concerns. In order to identify the relevant set of probable social impacts, the assessor has to 'scope' the study to include those social and cultural effects likely to be caused by the project, and establish the depth of study for each element in the SIA. Scoping, according to Kennedy and Ross, is 'an EIA activity in which a process is followed to identify the attributes of the environment for which there is concern (public and scientific) and a plan is provided that enables the EIA to be focused on these attributes' (1992, 476).

The scoping step usually takes place after an understanding of the proposed project has been gained and before the start of the EA investigation. Even though it occurs early in the EA process, the scoping stage is critical because the social variables selected for study and the impacts associated with them are essentially those factors that constitute the basis on which the final decision is made. Impacts that are omitted at the outset do not influence the outcome.

A participatory scoping process in SIA typically uses open scoping hearings to give all the participants a chance to raise the social issues that concern them. In public scoping discussions, the study boundaries are decided, social issues are raised, and the likely impacts are identified by the participants. The public scoping process tends to expand the range of social issues and concerns to be addressed. In the end, the objective is to reach a consensus on the scope of the study and to have it reflected in the EIA guidelines at the federal level.[3] Mulvihill and Baker (2001) suggest that the current practice of public scoping hearings, particularly for northern projects where intercultural conflicts frequently occur, adds to the overall success of the social impact assessment by bringing forward social and cultural issues, including the ecosystem and cultural features in a region, local knowledge, cultural symbols, and values that might not have been considered under a more restricted set of concerns.

The Profile or Baseline Conditions of the Social Environment

Social impacts are deviations from existing social conditions that are caused by the project and that would not have occurred as a result of ongoing social changes in the community itself. Thus, in order for assessors to estimate the degree of change from current social and cultural norms, they must take the very important step of developing a profile of the community and establishing its baseline conditions, without the project. Baseline conditions are the existing conditions and past trends found in the human environment in which the proposed activity is to take place (Interorganizational Committee 1995, 26).

The physical impacts of a project such as a nuclear waste facility are likely to oc-cur in a limited geographical area, but the baseline social conditions that will be af-fected by a new disposal technology are harder to define or establish (Murphy and Kuhn 2001). The relevant social or cultural environment is likely to be a widely dispersed set of interested publics, cultural groups, institutions, and organizations (Interorganizational Committee 1995). This was certainly true for the social con-ditions identified in the SIA for the Canadian nuclear waste disposal concept; in-terested Aboriginal groups, organizations, and institutions were found across the country, and more immediately affected groups came forward in four provinces.

Baseline social conditions are most effectively identified through consultations with affected communities, and often focus on

- the population and demographic characteristics of the community;
- relationships that communities have with their physical environment;
- past patterns of social development that indicate the current nature of so-cial change;
- social, economic, and cultural resources that exist in the community;
- social activities that are ongoing, such as kinship, friendship, and social ex-changes; and
- cultural groupings, behaviours, and values that are associated with each culture.

Baseline social conditions are meant to express the resources and activity pat-terns of an affected community, but they also should capture the attitudes people have towards the project and the cultural values affected by it. Often, commu-nity attitudes stemming from existing social conditions shape the community's responses to the actual facility. People's perceptions of the risks a project poses to their health and the environment are also part of the baseline social conditions. Their trust in the proponent's willingness to eliminate or manage those risks will to a large degree affect their responses to the proposal.

Determining the baseline social conditions requires substantial information gathering by the proponent. Ideally, this is an anthropological exercise where the assessors spend enough time in the affected communities gathering published in-formation, conducting community surveys and field studies, and talking to and observing those who will be affected to identify with confidence both the existing conditions and the nature of ongoing social change. In many cases, however, the speed of development limits the time that can be spent on baseline studies.

The Projection of Social Impacts

After the baseline conditions are described and the physical requirements of the project are understood, the next step is to predict the likely social and cultural impacts of the project. However, because most social environments are extremely complex, with a rich web of interactions and a large capacity to adapt to physical and social changes, the social environment is to a large degree *trans-scientific—*

that is, it is not subject to scientific analysis and it has a high level of uncertainty attached to its responses to any project (Berkes 1989). The impacts on the social perceptions, behaviours, and values of affected communities are to some extent unknowable; they can only be estimated, not accurately predicted.

In an SIA, researchers estimate social impacts by comparing the expected future state if the project were to go ahead with the state that would exist without the project and only as a result of social change. The difference between the two states is considered to represent the *actual social impact*. Making projections of future changes in the social environment lies at the heart of social impact assessment (Interorganizational Committee 1995, 28). To the extent that the methods used to do this are borrowed from the physical, economic, or social sciences, some quantitative predictions of population change, employment growth, housing demand, social services needs, and other physical parameters can be made.

Social impact projection methods include comparative analysis (estimating the impacts of a new project based on the outcomes of previous similar projects), extrapolations of existing social trends, and the use of expert social opinion, among others (Interorganizational Committee 1995, 28). However, for the most part, the non-quantifiable social changes that are vitally important to communities are not open to scientific analysis. Often, the most significant outcomes people experience are not determined quantitatively; such outcomes have to do with the impacts of risk, perceptions of unfairness in the distribution of risks and benefits, people's satisfaction with place, the degree of social cohesion they enjoy, and the effects on their cultural practices and values. In the absence of a real ability to predict effects, the practice of SIA has tended to favour public participation methods for estimating impacts, employing consultations with stakeholders to discover their expectations about likely impacts in their communities. Asking those who will be affected to estimate the social changes they expect to happen, given the project and its physical changes, is an increasingly common means of projecting social impacts. This method has employed a number of techniques, such as surveys and interviews, citizen advisory groups, scenario development, workshops, and focus group meetings of community members (Barrow 2000, 89). Using consultation methods to estimate impacts can produce subjective and potentially biased pictures of social change, particularly if communities perceive high levels of risk. In such cases, the estimates of effects may be projections of the fears the community has about the project. On the other hand, members of affected communities, particularly Aboriginal populations, can apply their *traditional environmental knowledge* (TEK), which is based on generations of observation of natural and physical phenomena, as well as their intimate knowledge of their own community's behaviour, to project the likely social impacts of a project (Paci et al. 2002).

The Evaluation of Social Changes

The evaluation step in an EIA is the point at which social impacts are integrated with the physical environment and economic impacts to offer an overall judgment of a project's acceptability. In evaluating whether the expected social changes are acceptable, one must understand the significance of each impact to the community and the interested parties, as well as their willingness to accept them. In many SIAs, assessors have used several quantitative evaluation methods (cost-benefit analysis, cost-effectiveness analysis, simple additive weighting methods, and the like) to judge the acceptability of a project, but with generally unsatisfactory results (Lawrence 1993).

SIA studies have, in recent years, tended to favour broader participatory strategies in determining the acceptability of project impacts. These strategies have been referred to as *social learning models.* Social learning approaches engage all of the stakeholders in learning about each participant's needs and interests (Webler et al. 1995; Sinclair and Diduck 1995; Saarikoski 2000). Communities and groups with different interests are helped to gain this knowledge through presentations, workshops, and investigations. Participants gain an understanding of the significance of social impacts to other groups and communities through exchanges of information and dialogue. They also discover that their own needs can be understood by fellow participants and that it is possible for different stakeholders to reach an area of agreement, where needs can be mutually satisfied and an acceptable action taken to resolve, say, a siting dilemma or a social impact problem. In both the federal and Ontario EA processes, opportunities for negotiation and mediation in impact assessment have increased in recent years, helping groups resolve disputes and come to agreement on the acceptance of social impacts.

Mitigating Measures

In each SIA, not only are impacts projected, but measures are identified to prevent or reduce adverse effects. Mitigating measures are identified steps that reduce or prevent the impacts that are expected to occur as a result of the project. Some social impacts can be reduced or eliminated if the design of the facility or its operation is altered or if additional steps are taken to compensate those affected to offset the social or economic effects they will have to bear. In some cases, the risk and health-related impacts of a project can be reduced if the control of the facility is shared with the community or if there are procedures in place whereby community members are notified in an emergency and the health and social consequences of a sudden event can be adequately managed (Castle 1993). In current SIAs, mitigating measures play a significant role in the process by altering the social impacts of the project on the affected community.

In a participatory approach to SIA, the mitigating measures are developed through citizen advisory groups, workshops, and public meetings that examine community preferences for different measures and determine the balance to be struck between design or operational changes and compensation, management, or control-sharing measures. Through consultation with affected communities, proponents are likely to discover the most appropriate set of measures—one that achieves the highest level of acceptance.

A Monitoring Program

Given the uncertainty that exists in the social environment, it is likely that many of the impact projections in an SIA will not be accurate. In some cases, deviations from the expected outcome may be sufficiently great to call for corrective action. Monitoring provides for the repeated and ongoing measurement of social conditions after the project is in operation. In recent SIAs, a monitoring program has become an essential step in determining whether a proponent complies with the mitigation measures required as part of the project's approval and, more broadly, whether social conditions have been affected in ways that were projected to happen.

A monitoring program confirms predicted outcomes. In such cases, it serves as a check on mitigating measures to make sure they perform as planned. The program can also detect unanticipated social impacts that might call for corrective actions. Further, as Armour suggests, 'community resistance to facility siting proposals regarded as threatening in some way is often rooted in the concern that the really serious problems are the ones that creep up gradually—such as contamination of groundwater or loss of way of life—and that these may go undetected until it's too late' (1988, 249). In many cases, individuals and communities affected by the project have called for a monitoring program as part of their acceptance of the project, so that they might maintain some control over the project's effects after it has been approved.

A monitoring program is an important means of verifying that social impact predictions are accurate and of addressing expected social effects. Monitoring is very useful when the project takes place in an environment where little detailed information is available or where high levels of variability or uncertainty exist (Interorganizational Committee 1995, 32). In the Canadian EIA process, a monitoring or follow-up program must be a component of a comprehensive environmental assessment study, and in cases where the project is actualized (after a comprehensive study, mediation, or a panel review), the proponent must ensure that a follow-up program is designed and implemented. In Ontario, there is no specific provision for monitoring the impacts of a project, although there is an opportunity for the minister of the environment to impose monitoring as a condition for approving the project. The extent of the monitoring effort is, however, most often not in the control of the affected community.

Steps in the SIA Process

The SIA process includes each of the key components listed above and follows a sequence of steps that is typical in environmental assessment practice (Table 7.1). While these steps are common in many SIAs, a more participatory approach has meant that most steps in the process are carried out through consultations with people in affected communities. Greater community participation in SIA has tended to follow a set of participatory principles. The Interorganizational Committee (1995, 35) has identified a number of principles to achieve high levels of participation:

Table 7.1   The social impact assessment process

SCOPING	Potentially impacted individuals, groups, and communities and their concerns are identified. The type, scale, and focus of the assessment is decided. The boundaries and limits of the assessment, and the social indicators to be used, are decided. Assessment methods and the sources of data are also decided.
ALTERNATIVES	Reasonable alternatives to the proposal are developed based on the needs of the project and the social needs of the community.
PROFILING	Existing social conditions of the affected communities are established. A profile is built of the characteristics and trends in the community before the start of the project.
PROJECTION	Estimates are made of what is expected to happen both without the project and with the project once it is built. Those who are likely to be affected by the expected changes are identified, including those who benefit and those who do not.
ASSESSMENT	Magnitudes of expected impacts are estimated for both the project and its alternatives. The social impacts that are likely to be most significant are identified. Also, the impacts that can be avoided or mitigated are noted.
EVALUATION	Trade-offs between social impacts and benefits are noted. Those who benefit and those who lose are identified. The overall social impacts of the project and its alternatives are assessed to determine if they are acceptable.
MITIGATION	Measures available to reduce or offset the expected impacts are identified.
MONITORING	A plan for ongoing monitoring is established to indicate if corrective action is needed. The actual impacts are measured and compared with the projected impacts.

Source: Adapted from Barrow 2000, 38.

Principles for Participation in Assessing Social Impacts
- Involve the diverse public (identify and involve all potentially affected groups and individuals).
- Analyse impact equity (clearly identify who will win and who will lose, and emphasize the vulnerability of under-represented groups).
- Focus the assessment (deal with the issues and public concerns that 'really count', not those that are 'easy to count').
- Identify methods and assumptions, and define significance in advance (define how the SIA was conducted, what assumptions were used, and how significance levels were selected).
- Provide feedback on social impacts to project planners (identify problems that could be solved with changes to the proposed action or alternatives).

- Use SIA practitioners (trained social scientists employing social science methods will provide the best results).
- Establish monitoring and mitigation programs (manage uncertainty by monitoring and mitigating adverse impacts).
- Identify data sources (use published scientific literature, secondary data, and primary data from the affected area).
- Plan for gaps in data.

These principles show that the ideal SIA is strongly based on public consultation in many of its steps and that an SIA is expected to seek out and include throughout the process a broad spectrum of groups and individuals from the community. In addition, an SIA should attempt to deal with the fairness of the project's outcomes, indicating what the distribution of benefits and costs will be to members of the community and whether an under-represented group without much political or economic power is vulnerable to the project's impacts. Further, it is clear that, in its scoping step, the ideal SIA is meant to take into account the unquantifiable social impacts that are difficult to count but that 'really count' in the community.

The Canadian Radioactive Waste Disposal Concept: EIA and Aboriginal Social Impacts

SIA emerged as a central component in the assessment of environmental impacts at about the same time EIA began in Canada. In the years following the start of the EIA policy, several major northern developments with significant impacts on the social and cultural lives of Aboriginal peoples demonstrated the importance of SIA. Fortunately, a model process for consultation, scoping, and examining the social impacts of large-scale northern development had been established by Justice Berger's inquiry into the effects of the Mackenzie Valley Pipeline, proposed in the early 1970s. That process set a benchmark for assessing social effects on Aboriginal communities and economies that is reflected in current SIA practice. The social and economic impacts of the pipeline on northern communities were estimated to be so significant that the pipeline's approval was delayed for 10 years (Berger 1977; Gamble 1978). Other projects that illustrated the significance of social impacts on Aboriginal communities include the James Bay Great Whale project, proposed in 1990 (Berkes 1981; Niezen 1993, 1998), and the proposal in the mid-1980s to expand low-level military flight training over Labrador and eastern Quebec, the traditional Aboriginal homeland of the Innu (Pushchak 2002).

However, the environmental assessment for the Canadian high-level radioactive waste disposal concept, proposed in 1989, was an exceptional event in social impact practice. This EIA examined the impacts of a concept rather than the impacts of a project tied to a site, and it was not linked to any single Aboriginal community. Since the concept did not specifically affect one group, it potentially affected all groups and promised broad social effects across the entire Aboriginal culture. The EIA exposed the conflicting interests of two cultures, the southern non-Native culture that wanted to dispose of the radioactive wastes in a northern remote site and the Native culture that feared that the site would be located in one of its traditional regions. Beyond this, the EIA dealt more with intangible and unquantifiable social impacts, including questions about the fairness of the project to Native people, the effects of the concept on their cultural values, the impacts on future generations, and the effects on Aboriginal and treaty rights.

The project began 35 years ago, when the Canadian government and Atomic Energy of Canada Limited (AECL) undertook the task of finding the most appropriate method for the disposing of nuclear fuel waste. Since the beginning of the nuclear age, several solutions had been suggested, ranging from burying the waste under Antarctic ice, to injecting it into the seabed, to launching it into space. None were practical in the Canadian context. The method deemed the most suitable for Canada was first suggested in 1972 by a committee that included AECL, Ontario Hydro, and Hydro-Québec. The committee developed the geological disposal concept that called for burying the waste deep within the earth's crust.[4] In 1974 the government decided, after consultations between the Department of Energy, Mines and Resources (now Natural Resources Canada) and AECL, that further research on nuclear fuel waste would focus on disposal deep in the plutonic rock of the Canadian Shield and that other alternatives would not be considered.[5] For technical reasons, plutonic rock was preferred as the disposal medium, and the shield's vast geographic area offered several opportunities for a site because of the many plutonic formations within it (Dormuth 1996).

In 1978 AECL was given the responsibility of developing the geological disposal concept, and in 1981 the federal and provincial governments decided that, rather than selecting a disposal site, they would first seek the public's acceptance of the disposal concept. Given the broad public opposition to nuclear power plants and waste disposal facilities at the time, gaining public acceptance was believed to be critical to the success of any nuclear waste management strategy (Slovic 1987; Slovic et al. 1991; Shrader-Frechette 1993). Therefore, the search for a site would not be attempted until the disposal concept was found to be environmentally and socially acceptable to Canadians.

The disposal concept had to undergo an environmental impact assessment and public review. In 1989 the minister of the environment appointed a federal Environmental Assessment Review Panel (EARP), also known as the 'Seaborn Panel' after its chairman, Blair Seaborn, to provide guidelines for the EIA and to review the environmental impact statement (EIS) after it was completed. The panel conducted two sets of nationwide public hearings with broad public and Aboriginal participation to gain as much input as possible from the affected Canadian public. The first hearing entailed a series of scoping meetings and open houses where social issues and concerns were raised. Much of the scoping testimony was incorporated into the final set of EIA guidelines, issued to AECL in 1992, that set the scope of the EIA (EARP 1991).

Scoping the Concept SIA

Aboriginal people are expected to be the group most affected by the geological disposal of nuclear fuel waste because the potential sites in plutonic rock are found in Native lands on the Canadian Shield. Although many people would assume that the Canadian Shield is largely uninhabited, it does include a significant portion of tribal lands and its environment is an important part of their livelihood. The Native participants raised concerns about effects on their tribal lands and social practices at the scoping hearings, which examined a number of social issues:

- The potential impacts on future generations
- The general social impacts on Native economies and ways of life
- The failure to consider alternatives
- The Aboriginal participants' lack of trust in the proponent and the EIA process

Aboriginal participants were also concerned about the narrowness of the panel's mandate, which did not allow a broader discussion about the social acceptability of Canada's nuclear power and energy policy, and they raised such social concerns frequently. Throughout the scoping hearings, Native participants expressed strong opposition to the concept and said they would continue to oppose siting it in their communities (Pushchak and Heisey 1992).

Following the scoping hearings, the panel made sure that the social concerns raised by the Native participants were included in the EIA. The panel required in its guidelines that AECL consider the social and cultural impacts of the concept. To 'assure safe and acceptable management of nuclear fuel waste,' the panel's scoping guidelines directed AECL to consider 'ethical and moral perspectives, along with various social issues [to be] . . . *as important as* scientific, technical and economic considerations' (FEAP 1992, 1). In addition, the panel told AECL that in its examination of social impacts it must pay specific attention to Aboriginal communities (FEAP 1992, 2). Despite these demands, AECL's EIA produced a very technical environmental impact statement, with much of its content focused on the design of the facility and the waste containers rather than on the social effects of the project.

The EIA of the Deep Disposal Concept and SIA

In 1994, within two years of receiving the guidelines, AECL produced its environmental impact statement on the geological disposal concept. Its proposal involved placing nuclear fuel waste in corrosion-resistant metal containers, designed to last at least 500 years, depending on the material used, and burying them in vaults excavated deep in a plutonic rock formation 500–1000 metres below the earth's surface. The containers were to be placed in boreholes drilled in the floors of rooms that extended from a main disposal vault. A buffer of sand and clay would surround each container of waste, and once they were all filled, the rooms and the entire vault would be backfilled with clay and granite (Figure 7.1).[6]

The EIA Review

The second set of national hearings took place after the EIA was completed in 1994, and for almost one year, the panel conducted public hearings on the concept. In

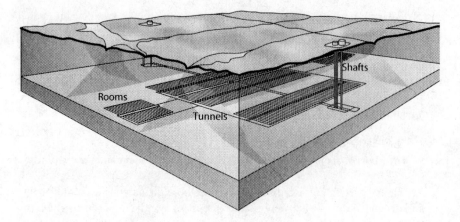

Figure 7.1   Artist's Rendering of a Disposal Vault

Source: AECL 1994.

1996 and 1997, the panel visited 19 different cities and communities, travelled to the provinces of Ontario, Quebec, New Brunswick, Saskatchewan, Manitoba, and Alberta, and conducted 54 days of hearings. The hearings were divided into three phases. Phase I focused on the broad societal issues related to long-term management of nuclear fuel waste. Phase II focused on the safety of the AECL geological disposal concept from a scientific and engineering point of view. The final phase involved community hearings, which gave the public a chance to voice its concerns about the safety and acceptability of the proposed concept (FEAP 1998). The panel also accepted written submissions on the concept.

The Seaborn Panel made a substantial effort to ensure that Aboriginal people had an opportunity at the hearings to express their concerns about the social and cultural impacts predicted in the EIA. The panel invited tribal chiefs to speak at the hearings and organized Native community visits to hear submissions from individual community members. Three of the 16 communities visited were Aboriginal (Sagkeeng First Nation in Manitoba, Ginoogaming First Nation in Ontario, and Serpent River First Nation, also in Ontario), and 58 elders, Native individuals, and groups from these three communities participated. In addition, 36 Native peoples travelled to other hearing locations to participate. The testimony at the public review hearings provides the single most extensive record of Aboriginal social and cultural views on nuclear fuel waste impacts in Canada (Farrugia-Uhalde 2003). The hearing record reveals that the concerns of the First Nations people were the same ones they had raised in the scoping hearings eight years earlier and were consistent with those of other Native peoples faced with similar nuclear siting projects.

The EIA included an outline of the steps AECL promised to take during the site selection process, including the most significant step—to search for a community willing to accept the facility voluntarily. AECL indicated that only communities willing to accept the facility freely would be considered as potential host sites for the waste. It was hoped that Aboriginal communities could be found that believed the social impacts of the project were acceptable and would thus accept the project willingly. However, if a community willing to accept the nuclear waste could not be found, the search for a willing host site would fail.

AECL's prediction of social impacts in its EIA was specifically criticized for failing to consider the ways Aboriginal people would be affected by the disposal concept. This was the view of the Seaborn Panel and many hearing participants, both Native and non-Native (FEAP 1998).[7] In its final report, the Seaborn Panel concluded that 'the AECL concept did not take place within the context of a comprehensive social and ethical framework' and that 'the EIA gave little indication that AECL had attempted to . . . pay special attention to the viewpoints of Aboriginal people . . . or [to] how the traditional knowledge and experience of Aboriginal people might be incorporated into any analysis of the effects of the facility' (FEAP 1998, 60). The panel's finding was reflected in the opinions of Native participants, who noted that the amount of energy and time spent considering Aboriginal concerns and the ways the concept would affect their lives was completely inadequate, considering that Native peoples would be the ones most significantly affected by the concept that AECL had proposed. Chief Richard Kahgee explained that the review included little information about the potential impacts of the proposal on Aboriginal people. This was seen as unacceptable considering that First Nations were the ones to be most directly and dramatically affected by the proposal and that proper, respectful, and culturally and politically appropriate consultation with First Nations people by First Nations people should have been undertaken (Kahgee 1997).

The Aboriginal Social Impacts

The testimony in the EIA review hearings provided an understanding of Aboriginal views and concerns about the impacts of the disposal concept (Farrugia-Uhalde 2003). Judging by the number of times issues were raised by Aboriginal participants at the hearings and how long they were discussed, there were three social effects that Aboriginal people were most concerned about:

- Effects on Aboriginal spiritual, cultural, and social values
- Failure to respect treaty and Aboriginal rights
- The limited Aboriginal role in planning and decision making

In addition to these three, a number of other social issues were discussed by many Native participants at considerable length:

- Aboriginal involvement and the poor communication of information (this discussion focused on the barriers to communication)
- The protection of future Native generations and the threat to their cultural well-being and preservation
- Impacts on the health of the environment and indirectly on the health and safety of Aboriginal communities

- Criticisms about the EIA's deficiencies
- Past and present hardships, including the social impacts of large min-ing projects, logging, and hydro development, among other projects (this discussion usually involved issues about trust in government and large corporations)
- Lack of consideration given to opposition to a waste disposal or storage system and to the disposal options preferred by Aboriginal peoples
- The inequity of the outcome: the geological disposal concept unfairly plac-es a nuclear fuel waste repository a great distance from the producers of the waste and close to communities that have not used nuclear energy (the dis-cussion also involved intergenerational inequity—i.e., present generations benefit from the energy, while future generations will have to deal with the consequences of its waste).
- The use of traditional environmental knowledge, the importance of its preservation, and its significance in finding the most appropriate way to manage the nuclear waste
- Concerns about the ethics of voluntary siting and compensation (Aborigi-nal peoples did not support this site selection method)

Several of the identified impacts revealed the complexity of the social environ-ment and the challenges it poses to SIA. Many of the Aboriginal issues could not be assessed quantitatively, and this meant that the evaluation of social changes relied heavily on the stories First Nations people told about their activities and ways of life. For example, the extent of the impact on traditional Native ways of life was hard to estimate because of the subjective value of the Native economy and lack of knowledge about how much First Nations people depended on it. Impacts such as these also demonstrated the challenges involved in predicting social change; it was very difficult to forecast how a large development project would influence the stability of a culture whose dynamics were not clearly understood by assessment authorities.

Despite the limitations in the SIA, the hearings demonstrated that Aboriginal concerns about social change had remained consistent over time. The same con-cerns raised in the scoping hearings were repeated in the hearing process. Four issues stand out:

- Effects on First Nations culture
- Effects on future generations
- Fairness
- Voluntary site

Effects on First Nations Culture

Protection of cultural well-being was an issue that received considerable atten-tion from First Nations participants during the public hearings on the concept of geological disposal. Despite the difficulty in quantifying cultural practices and beliefs, the Aboriginal participants made it clear that their cultural resources and practices were the aspects of First Nations society that 'really counted' in their way of life. Cultural impacts have continued to be central concerns for Native peoples in Canada, the United States, and elsewhere (Hanson 1995).

Effects on Future Generations

Aboriginals suggested that it was unfair that future generations would be compelled to participate in managing a nuclear waste facility they had not created. These participants expressed their moral obligation to prevent harm to the generations that would follow. The same was found to be the case in the United States, where Native individuals expressed their concern for future generations and their worry about the significant risks to their lives posed by a development such as this (Gowda Rajeev and Easterling 2000).

Fairness

During the hearings on geological waste disposal in Canada, First Nations people continually voiced concerns about the unfair distribution of risks and benefits imposed on them by the project. They suggested that the waste facility would create risks in Native communities, while the benefits would be enjoyed by others. They suggested that this was unfair because Aboriginal peoples had not played a part in making the decision to generate nuclear wastes and they had not benefited from a power source whose wastes would be stored in their community. The Aboriginal participants submitted that it would be fairer if the responsibility to dispose of the wastes was borne by the communities that derived the benefits.

Voluntary Siting and Compensation

Native Canadians expressed the fear that divisions among Aboriginal communities would result if a voluntary siting method included offers of compensation to encourage communities to volunteer. They feared that the communities that received compensation would be resented by those who did not. Such divisions were observed between traditionalists and business-oriented tribal officials in a US nuclear waste siting strategy. For this reason, many Native leaders have strongly opposed voluntary sites and compensation offers by the nuclear industry.[8]

The EIA Decision: Rejection of the Disposal Concept

In the hearings, the First Nations participants consistently expressed their rejection of the concept of nuclear fuel waste disposal on tribal lands. Given the significant social impacts that were expected and the strong opposition to the concept, it appeared to the panel that the possibility of finding a site for the facility would be low, even if a search were undertaken for a Native community willing to host it. Collectively, the First Nations communities responses indicated that the concept was not socially acceptable, and this led the panel, in its final report, to recommend that the concept not move to the siting stage: '[T]he AECL concept for deep geological disposal has not been demonstrated to have broad public support. The concept in its current form does not have the required level of acceptability to be adopted as Canada's approach for managing nuclear fuel waste' (FEAP 1998). The panel recommended, among other suggestions, that a participation process be initiated to involve Aboriginal people, since the likely site would be in a location that would affect them as a group specifically.

In December 1998 the federal government issued a response that addressed several issues in the panel's report (Government of Canada 1998). With respect to Aboriginal communities, the government made a commitment to begin a dialogue with First Nations peoples to better assess the social impacts of a nuclear waste facility on their communities. There then followed something of a standstill in the process. This was, however, a familiar turn of events in the realm of nuclear facility siting as it has been over the last few decades, with many projects stalled because of public opposition and uncertainty. Four years later, the government introduced the Nuclear Fuel Waste Act (2002) to guide the long-term management of nuclear fuel waste.[9] However, unlike the old concept proposal, this Act will consider not only geological disposal, but also above- and below-ground centralized storage alternatives and above-ground storage options at reactor sites. Clearly, new disposal alternatives have been added, and many of them may not affect Native lands, particularly those at existing reactor sites. In such cases, nuclear wastes will remain in southern regions and social impacts in northern communities will not be an issue.

The Concept EIA Outcome

In the end, expectations about the social impacts of the deep disposal concept, together with the input of Native participants in the SIA scoping and hearing processes, changed the course of nuclear fuel waste management in Canada. Aboriginal people established a position opposing the social effects of the disposal concept at the outset, a position that was shared by First Nations groups across the country and that continued unchanged throughout the EIA process. Despite the concept's endorsement by recognized scientific and public bodies, Aboriginal representatives suggested that alternatives that might not impose social and cultural impacts on them were not being considered. They successfully presented the case that their social and cultural values were important in the EIA process and should be taken into account in the decision. They also argued that the EIA had failed to address a great deal of social unfairness in the proposal, both to existing First Nations communities and to future generations.

The social impacts of the disposal concept were the determining factors in the reversal of more than 25 years of nuclear waste disposal policy, despite the limited abilities in SIA to predict the social impacts of change expected. A participatory SIA process ultimately resulted in a decision that was more democratic than a non-participatory process because the affected populations took part in making the decision. Moreover, the outcome dealt more directly with the conflict between southern and northern First Nations cultures by allowing the consideration of more equitable alternatives for nuclear waste disposal.

Conclusion

Assessing social impacts is now a central part of the Canadian environmental assessment process, and a participatory approach that involves socially and culturally affected populations has emerged as the paradigm for assessing the social impacts of major developments. An SIA process that can take into account the cultural and social differences between northern Aboriginal and southern cultures has been essential in assessing the effects of large development projects on Aboriginal communities, and it is evident that social impacts have played an important role in determining the outcomes of these projects. This was particularly true for the proposed concept for the disposal of high-level radioactive wastes, where the decision was largely a result of the anticipated social and cultural impacts of the concept. However, the assessment process has been carried out without clear and accepted theoretical guidelines for predicting or assessing impacts; many social impacts were identified through participatory means rather than through science-based predictions. The principles of public involvement have been important in shaping an SIA process that can take into account the unquantifiable social impacts on cultures— those that are difficult to count but that count in the social and cultural lives of Aboriginal people.

Notes

1. Atomic Energy of Canada Limited and Ontario Hydro, the proponents of the disposal concept, did not initially identify a site or outline a search process to find a location for the nuclear waste facility; rather, the disposal concept was proposed with the understanding that a site would be found after the technical concept was approved through the EA process.

2. *Environment* in the Ontario Act means (i) air, land, or water, (ii) plant and animal life, including human beings, (iii) the social, economic, and cultural conditions that influence the life of human beings or a community, (iv) any building, structure, machine, or other device or thing made by human beings, (v) any solid, liquid, gas, odour, heat, sound, vibration, or radiation resulting directly or indirectly from the activities of human beings, or (vi) any part or combination of the foregoing and the interrelationships between any two or more of them.

3. In the Ontario process, recent changes to the provincial EA Act require that the proponent prepare terms of reference (ToR) to set the study scope. Terms of reference are intended to be drafted in consultation with affected stakeholders, but in practice, the opportunities for input by stakeholders are much narrower. The terms of reference are written by the project developer and tend to be more restrictive than in the federal scoping process.

4. The deep disposal concept was supported by a number of Canadian authorities: Ontario's Royal Commission on Electric Power Planning (1978), the House of Commons Standing Committee on Energy, Mines and Resources (1988), and a 1977 study group chaired by F.K. Hare (Aikin et al. 1977).

5. Plutonic rock is formed deep in the earth by the crystallization of magma and/or by chemical alteration and is often referred to as intrusive igneous rock. It has many characteristics that are favourable to waste disposal; it is found in many locations across the Canadian Shield in large, consistent rock formations that can withstand geological stresses, allow heat to dissipate, minimize access to ground water, and retard or prevent the movement of contaminants (AECL 1994, 97–103).

6. In the AECL concept, several man-made and natural barriers were included to provide long-term protection for humans and the environment (EARP 1994). It was assumed that once the vault was sealed, it would be perpetually safe, even if people were not able to monitor the site for the thousands of years the waste remained toxic. A key objective was that the disposal site should be able to exist safely without constant supervision once it was closed so that future generations would not have to be burdened with looking after waste that was not their own.

7. The critics included the Joint First Nations' Submission to the Federal Environmental Review, the Federation of Saskatchewan Indian Nations, the Assembly of Manitoba Chiefs, the Assembly of First Nations of Quebec and Labrador, and the Grand Council of the Crees of Quebec (Orkin 1995; FEAP 1998).

8. Views on voluntary siting and compensation were expressed by Wendell Chino, an adamant promoter of development on Native lands, and Grace Thorpe and Rufina Laws, both well-known Native activists who clearly opposed the promises of compensation promoted by the nuclear industry (NECONA 1993; Hanson 1995; Thorpe 1995, 1996, 1997; Chino 1996; Gray-Kanatiyosh 1997).

9. The Act also called for the creation of a Nuclear Waste Management Organization (NWMO), which was established at the end of 2002. The NWMO, formed and funded by the nuclear energy industries (Ontario Power Generation, New Brunswick Power, Atomic Energy of Canada Ltd, and Hydro-Québec), will 'propose to the government of Canada approaches for the management of nuclear fuel waste and implement the approach that is selected' (Nuclear Fuel Waste Act 2002). The government also established the Nuclear Fuel Waste Bureau (NFWB) in the Department of Natural Resources to oversee the NWMO, the Government of Canada, and the minister of natural resources under the 2002 Act.

References

AECL (Atomic Energy of Canada Ltd). 1994. *Environmental impact statement on the concept for disposal of Canada's nuclear fuel waste.* AECL–10711, COG–93–1. Ottawa: Scientific Document Distribution Office.

Aikin, A.M., J.M. Harrison, and F.K. Hare. 1977. *The management of Canada's nuclear wastes.* Report EP 77–6. Ottawa: Minister of Energy, Mines and Resources, Canada.

Armour, A. 1988. Methodological problems in social impact monitoring. *Environmental Impact Assessment Review* 8:249–65.

Barrow, C. 2000. *Social impact assessment: An introduction.* London, UK: Arnold.

Becker, D., et al. 2003. A participatory approach to social impact assessment: The interactive community forum. *Environmental Impact Assessment Review* 23:367–82.

Berger, T. 1977. *Northern frontier, northern homeland: The report of the Mackenzie Valley Pipeline Inquiry.* Ottawa: Government Supply and Services.

Berkes, F. 1981. Some environmental and social impacts of the James Bay Hydroelectric Project, Canada. *Journal of Environmental Management* 12:157–72.

———. 1989. The intrinsic difficulty of predicting impacts: Lessons from the James Bay hydro project. *Environmental Impact Assessment Review* 8:201–20.

Burdge, R., and F. Vanclay. 1995. Social impact assessment. In F. Vanclay and D. Bronstein (Eds), *Environment and Social Impact,* 31–65. Chichester: John Wiley and Sons.

Castle, G. 1993. Hazardous waste facility siting in Manitoba: A case study of a success. *Journal of Air and Waste management* 43:963–9.

CEAA. 1992. *Canadian Environmental Assessment Act.* Bill C–13. Ottawa: Government of Canada.

CEA Agency (Canadian Environmental Assessment Agency). 1994. *The responsible authority's guide.* Ottawa: Minister of Supply and Services Canada.

CEQ (Council on Environmental Quality). 1986. *National Environmental Policy Act Regulations.* Section 40, CFR 1508. Washington, DC: CEQ.

Chino, W. 1996. The Mescalero Apache Indians monitored retrievable storage of spent nuclear fuel: A study in environmental ethics. *Natural Resources Journal* 36 (4): 673.

Dormuth , K. 1996. *Nuclear fuel waste management and disposal concept public hearings,* AECL. General Session/Public Hearing, Toronto, 11 March. Ottawa: CEA Agency http://www.ceaa-acee.gc.ca/0009/0001/0001/0012/0002/transcripts_e.htm.

EARP (Environmental Assessment Review Panel). 1991. News and views from the Chairman: An interview with Blair Seaborn. *Dialogue* 3 (Spring): 1.

———. 1994. Recap of the concept. *Dialogue* 6 (Spring): 1.

Farrugia-Uhalde, A. 2003. Nuclear fuel waste and aboriginal concerns. Canada's nuclear fuel waste management concept public hearings: A content analysis. Master's of science thesis, Ryerson University, Toronto.

FEAP (Federal Environmental Assessment Panel). 1992. *Final guidelines for the preparation of an environmental impact statement on the nuclear fuel waste management and disposal concept.* Ottawa: FEAP.

———. 1998. *Report of the Nuclear Fuel Waste Management and Disposal Concept Environmental Assessment Panel.* Ottawa: FEAP.

FEARO (Federal Environmental Assessment Review Office). 1992. *Old Man River Dam: Report of the Environmental Assessment Panel.* Ottawa: FEARO.

Francis, P., and S. Jacobs. 1999. Institutionalizing social analysis at the World Bank. *Environmental Impact Assessment Review* 19:341–57.

Franklin, B. 1994. Grassroots initiatives on sustainability: A Caribbean example. In D. Bell and R. Keil (Eds), *Human society and the natural world: Perspectives on sustainable futures,* 1–10. Toronto: York University.

Gamble, D. 1978. The Berger Inquiry: An impact assessment process. *Science* 199 (3): 946–52.

Government of Canada. 1998. *Government of Canada: Response to the Nuclear Fuel Waste Management and Disposal Concept Environmental Assessment Panel.* Ottawa: Natural Resources Canada.

Gowda Rajeev, M., and W. Easterling. 1998. Nuclear waste and native America: The MRS siting exercise. *Risk-Health Safety and Environment* 9:229–58.

———. 2000. Voluntary siting and equity: The MRS facility experience in native America. *Risk Analysis* 20 (6): 917–29.

Gray-Kanatiyosh, B. 1997. Online book review of J. Weaver (Ed.), *Defending Mother Earth: Native American perspectives on environmental justice.* Maryknoll, NY: Orbis Books.

Hanson, R. 1995. Indian burial grounds for nuclear waste. *Multinational Monitor* 16 (9): 21–6.

Huizer, G. 1997. Participatory action research as a methodology of rural development. Food and Agriculture Organization, *SDdimensions*, posted May 1997, http://www.fao.org/ WAICENT/ FAOINFO/SUSTDEV/PPdirect/PPre0021.htm. .

Interorganizational Committee on Guidelines and Principles for Social Impact Assessment. 1995. Guidelines and principles for social impact assessment. *Environmental Impact Assessment Review* 15:11–43.

Kahgee, R. 1997. *Federal environmental assessment review of the environmental impact statement on the concept for disposal of Canada's nuclear fuel waste*. Saugeen First Nation, 24 February, vol. 44. Ottawa: CEA Agency.

Kennedy, A., and W. Ross. 1992. An approach to integrate scoping with environmental impact assessment. *Environmental Management* 16:475–84.

Lawrence, D. 1993. Quantitative versus qualitative evaluation: A false dichotomy. *Environmental Impact Assessment Review* 13:3–11.

———. 1997. The need for theory building. *Environmental Impact Assessment Review* 17:79–107.

Malczewski, J. 1998. *GIS and multicriteria decision analysis*. New York: John Wiley and Sons.

Massam, B. 1993. *The right place: Shared responsibility and the location of public facilities*. New York: John Wiley and Sons.

Mulvihill, P., and D. Baker. 2001. Ambitious and restrictive scoping: Case studies from northern Canada. *Environmental Impact Assessment Review* 21:363–84.

Murphy, B., and R. Kuhn. 2001. Setting the terms of reference in environmental assessments: Canadian nuclear fuel waste management. *Canadian Public Policy* 27 (3): 249–66.

NECONA.1993. *National Environmental Coalition of Native Americans*. http://www. alphacdc. com/necona/necona.html.

NEPA. 1969. *The National Environmental Policy Act*. Washington, DC.

Niezen, R. 1993. Power and dignity: The social consequences of hydro-electric development for the James Bay Cree. *Canadian Review of Sociology and Anthropology* 30 (4): 510–29.

———. 1998. *Defending the land: Sovereignty and forest life in James Bay Cree society*. Boston: Allyn and Bacon.

Nuclear Fuel Waste Act. 2002. *An Act respecting the long-term management of nuclear fuel waste*. Bill C-27. http://www.parl.gc.ca/37/1/parlbus/chambus/house/bills/government/C-27/C-27_3/90140bE.html.

Orkin, A. 1995. *Joint First Nations' submission to the Federal Environmental Review Panel of the Nuclear Fuel Waste Management and Disposal Concept*. Federation of Saskatchewan Indian Nations, Assembly of Manitoba Chiefs, Assembly of First Nations of Quebec and Labrador, Hamilton, Grand Council of the Crees (of Quebec). http://www.ccnr.org/AECL_Andy.html.

Paci, C., A. Tobin, and P. Robb. 2002. Reconsidering the Canadian Environmental Impact Assessment Act: A place for traditional environmental knowledge. *Environmental Impact Assessment Review* 22:111–27.

Pushchak, R. 2002. Environmental justice and the EIA: Low-level military flights in Canada. *International Journal of Public Administration* 25 (2/3): 169–91.

Pushchak, R., and A. Heisey. 1992. Canada's novel approach to siting a high level radioactive waste facility: A greater chance of success? *Journal of the International Association of Impact Assessment* (Washington, DC), August.

Royal Commission on Electric Power Planning. 1978. *A race against time: Interim report on nuclear power in Ontario*. Toronto: Queen's Printer.

Saarikoski, H. 2000. Environmental impact assessment (EIA) as a collaborative learning process. *Environmental Impact Assessment Review* 20:681–700.

Shrader-Frechette, K. 1993. *Burying uncertainty: Risk and the case against geological disposal of nuclear waste*. Berkeley: University of California Press.

Sinclair, J., and A. Diduck. 1995. Public education: An undervalued component of the environmental assessment public involvement process. *Environmental Impact Assessment Review* 15:219–40.

―――. 2001. Public involvement in EA in Canada: A transformative learning perspective. *Environmental Impact Assessment Review* 21:113–36.

Slovic, P. 1987. Perception of risk. *Science* 236:280–5.

Slovic, P., M. Layman, and P. Flynn. 1991. Risk perception, trust, and nuclear waste: Lessons from Yucca Mountain. *Environment* 33 (3): 6–11, 28–36.

Standing Committee on Energy, Mines and Resources. 1988. *Nuclear energy: Unmasking the mystery.* Tenth Report of the Standing Committee on Energy, Mines and Resources; B. Sparrow, Chairman, Ottawa, Second Session of the Thirty-third Parliament, 1986–1988.

Thorpe, G. 1995. Radioactive racism? Native Americans and the nuclear waste legacy. *Indian Country Today* 14:A5.

―――. 1996. Our homes are not dumps: Creating nuclear free zones. *Natural Resources Journal* 36 (4): 715.

―――. 1997. Our homes are not dumps: Defending Mother Earth. In J. Weaver (Ed.), *Defending Mother Earth: Native American perspectives on environmental justice.* Maryknoll, NY: Orbis Books.

Vanclay, F. 2002. Conceptualizing social impacts. *Environmental Impact Assessment Review* 22:183–211.

Webler, T., H. Kastenholz, and O. Renn. 1995. Public participation in impact assessment: A social learning perspective. *Environmental Impact Assessment Review* 15:443–63.

Winter, R. 1996. Some principles and procedures for the conduct of action research. In O. Zuber-Skerritt (Ed.), *New directions in action research,* 13–27. London, UK: Falmer Press.

Chapter 8

The Cheviot Mine Project: Cumulative Effects Assessment Lessons for Professional Practice

Roger Creasey and William A. Ross

Introduction

The authors of this chapter were both involved, in different capacities, with the approval process for the Cheviot coal mine. One of us (R.C.) was, at the time, a staff member with the Alberta Energy and Utilities Board and directly involved in both of the hearings on the project (1996 and 2000). The other (W.R.) was a cumulative effects assessment specialist on the panel for the second set of hearings in 2000. Throughout this document, we draw on our experiences and observations at the hearings, but we also rely on the two panel reports (Bietz et al. 1997, 2000) and on the environmental impact assessments (EIAs) (CRC 1999a, 1999b).

The Cheviot mine project was proposed for the Rocky Mountains in west-central Alberta. The proposed open-pit (surface) coal mine permit area is approximately 23 km long and 3.5 km wide. Overall, the proposal includes the construction, operation, and decommissioning of a coal processing plant and the development, operation, and reclamation of a large open-pit coal mine.

In Alberta, the development of a coal mine is based on a two-stage regulatory approval process (a brief description of the Alberta EIA process is provided in Alberta Environment 2003a, as well as in chapter 16 of this book). The initial approval stage deals primarily with the conceptual plans for the mine project as a whole. The second stage allows for site-specific changes to the conceptual plans approved during the initial stage of review. The two-stage approval process for coal mine projects is designed to look at the full range of likely environmental and technical issues associated with a project on a broad-scale basis. In the case of the Cheviot mine project, a federal authorization from the Department of Fisheries and Oceans was also required.

From the aspect of environmental impact assessment, the Cheviot mine project involved both provincial and federal requirements. The EIA for the project had to meet the requirements of both the Alberta Environmental Protection and Enhancement Act (provincial EIA legislation), described in chapter 16 of this book, and the Canadian Environmental Assessment Act (federal EIA legislation), described in chapter 14. In addition, specific requirements for cumulative effects assessment (CEA) came to be problematic for the project. The basic principles of CEA are succinctly described in the next section of this chapter. The public hearing process and court challenges initiated by environmental groups opposing the project complicated the eventual regulatory approvals. The major lessons learned from this case study have significant consequences for other projects undergoing assessment, especially in terms of the requirements for CEA. These lessons deal with such interesting issues as the treatment of future human activities in CEA, availability of information related to other industrial activities, and the management of cumulative effects.

This chapter will document principles of cumulative effects assessment, the review process for the Cheviot Mine, and the compilation of the environmental impact and CEAs, and will describe how legal and regulatory requirements affected the practice of impact assessment in this case. Emphasis will be on the lessons learned from what we believe is the most important CEA case study in Canada.

Cumulative Effects Assessment

A detailed description of cumulative effects assessment is provided in Ross 1998, especially as it pertains to environmental impact assessments for projects, as was the case for the Cheviot review. Ross (1998) also developed requirements for project CEAs, and these were the criteria he used to evaluate the Cheviot proponent's CEA submitted for the 2000 hearings. A summary of this material follows.

Cumulative effects assessments may be classified as regional cumulative effects studies and project cumulative effects assessments. The former do not focus on any one project, but instead examine how the various human activities in a region combine to cause cumulative effects. The latter focus on a single (proposed) project and are assessments of the impacts of that project in combination with those of other (past, present, and future) human activities. As noted in chapters 14 and 16, project CEAs are required under both Canadian and Alberta EIA processes.

It is worth noting that cumulative effects are the effects that people notice, and as such, they are the effects of concern to people. People care about the quality of water and air, not about whether one source or another is responsible for contaminating water or air. That is why it is essential for project decision makers to pay attention to cumulative effects rather than to single project effects. In turn, this is why good EIA legislation requires that cumulative effects be assessed before project decisions are made.

Ross (1998) developed four requirements for doing project cumulative effects assessments and used them to evaluate the Cheviot mine CEA in 2000:

1. Identify important impacts (scoping).
2. Identify other human activities that contribute to the same impacts.
3. Predict cumulative effects (and determine their significance).
4. Suggest appropriate means of managing the cumulative impacts.

Other material dealing with cumulative effects assessment in Canada is found in Hegmann et al. 1999, Spaling et al. 2000, and Baxter et al. 2001.

Project Description

The Cheviot coal mine is a proposal of Cardinal River Coals (CRC) and entails the construction, operation, and eventual reclamation of an open-pit coal mine, the restoration of the Mountain Park subdivision rail line, the upgrading of the existing access road into the Cheviot mine area, and the installation of a new transmission line and substation to supply electrical power to the Cheviot mine.

The project is located in the Rocky Mountains of west-central Alberta 320 km west of the city of Edmonton (Figure 8.1). The western edge of the mine project lies 2.8 km east of the boundary of Jasper National Park. The area of the proposed mine is characteristic of the foothills region of Alberta, which includes rolling hills and mountainous terrain with subalpine and alpine vegetation. The mine area itself is 23 km long and 3.5 km wide and is located along an east-west trending valley.

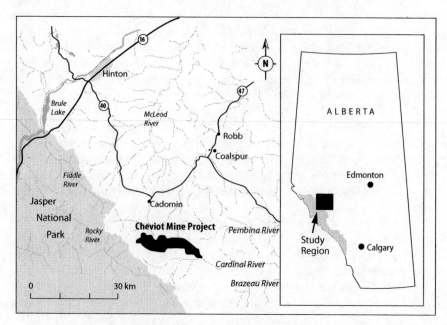

Figure 8.1   Location plan of the Cheviot mine project

Source: Alberta Energy and Utilities Board.

Mining activity was carried out within the proposed mine permit boundary from the early 1900s until the 1950s and was centred on the former mining town of Mountain Park. Mining during this period was primarily underground, although some minor surface mining activity was also conducted. The town of Mountain Park was abandoned with the introduction of diesel-powered train engines, and no residents or operating facilities exist there now. The coal product is a high-quality ore intended for metallurgical applications in offshore markets.

Environmental Impact Assessment Process

The Cheviot mine project presents some unique problems in the area of the environmental impact assessment process followed by CRC. The project includes both a surface mine producing a projected 3.2 million tonnes of coal per year and a coal processing plant. As a result, it is a project requiring a mandatory EIA as described in the Environmental Assessment Regulations of the Alberta Environmental Protection and Enhancement Act (AEPEA) (Alberta Environment 2003c). This Act and its associated regulations govern and describe the components in the EIA process in the province of Alberta. As part of the EIA process, the government works with the proponent to prepare draft terms of reference (TOR) describing the issues and topics that should be addressed within the EIA analysis. Once compiled, the draft TOR is open for public review and comment, during which time the general public, environmental groups, and other interested parties have the opportunity to identify other aspects of the environment that they believe should be considered in the impact assessment process, or otherwise to propose changes to the TOR. After considering those public comments, the government issues a final TOR for the EIA, which in the Cheviot case was done on 23 January 1995 (Bietz et al. 1997). The EIA for the Cheviot mine was submitted by the applicant in March 1996. Following review of the document in relation to the TOR, the director of assessment advised the Energy and Utilities Board (EUB), the regulator for energy projects in Alberta, that the EIA was complete and met the requirements of the environmental assessment provisions of AEPEA.

Once deemed complete under AEPEA, the EIA becomes part of the overall application to the Energy and Utilities Board. This agency is responsible for conducting a review of energy development projects, including coal mines, and renders approvals based on the need of the project and on whether it is 'in the public interest'. Depending on the nature of any objections made by interested parties, the EUB may conduct public hearings to review evidence related to the application, and it ultimately renders a decision on the acceptability of the proposal. It did so for the Cheviot mine project, a reflection of the considerable interest in the proposal. For example, the proponent held over 160 meetings with various public participants in the process and with regulatory officials. The original hearing was held in Hinton, Alberta, from 13 January to 20 February 1997, plus one day in April. During the 1997 hearings, 40 persons participated on behalf of CRC and 46 on behalf of governments and regulators; 67 others were active participants. The hearings in 2000 commenced on 1 March, continued to 10 March, and, after an adjournment, were completed on 25 to 27 April. At the 2000 hearings, 19 persons participated on behalf of CRC, 43 represented governments and regulators, and 50 others were active participants. The participants represented Aboriginal groups, local communities, environmental groups, mine workers, and other interested industries and individuals. (Information concerning the hearings was extracted from Bietz et al. 1997, 2000.)

An interesting complication arose in that another regulatory environmental impact assessment, one by the Government of Canada (CEA Agency 2003), was also required in the overall approval process for the Cheviot proposal. This was triggered by the need for an authorization under the Fisheries Act, a federal statute. When a federal approval is related to a project, the Canadian Environmental Assessment Agency (CEA Agency) coordinates a referral process within federal agencies and departments. In the Cheviot project, the Department of Fisheries and Oceans asked that the application be referred for panel review. Since both the province and federal EA processes recommended a panel review, a joint panel was struck to deal efficiently with the project approval.[1]

For a project as large and extensive as the Cheviot coal mine, it is not surprising that the company originally identified 99 valued ecosystem components (VECs); that is, 99 items were originally identified during the issues scoping process (in 1995). CRC defined VECs as being 'those environmental attributes associated with the proposed project development, which have been identified to be of concern by either the public, government, or the professional community' (CRC 1996). However, through a staged process of consolidation, that number was, after consideration of regional or cumulative effects, reduced to 28 VECs and then, after those with a moderate or high significance were considered, further reduced to 11 key VECs for consideration for cumulative effects assessment (CRC 1999a). These are identified in Table 8.1. This final number included the issue of surface water quality, but at the time of the assessment analysis, one specific element, selenium, was not considered to be important. As it turned out, that element would quite suddenly come to be considered important to the regional water quality, yet at the time of the scoping process, the proponent could not have anticipated how significant it would be.

Cumulative Effects Assessment within the EIA Process

An interesting aspect of the Cheviot Project involved the manner in which cumulative environmental effects were, and were not, described and addressed. Under both the provincial and federal environmental assessment processes described above, an analysis of the cumulative effects caused by the proponent's project and other human activities is required.

Of particular interest was the analysis and conclusions reached during the review of the project's cumulative effects on local carnivore species. The EIA identified significant adverse cumulative effects (as identified by CRC's expert consul-

Table 8.1   CRC summary of key regional valued ecosystem components

Surface water flow	Flow alteration, groundwater contribution, effects on fish and fish habitat
Water quality (3)	Alteration of water quality due to (1) nutrient enrichment, (2) sedimentation, and (3) effect on fish and fish habitat
Significant plant communities	Loss of significant communities or species
Elk (ungulates)	Habitat alteration and loss; habitat effectiveness
Grizzly bear (carnivores)	Habitat alteration and loss; habitat effectiveness
Harlequin duck	Alteration and loss of habitat
Neotropical birds	Alteration and loss of habitat
Traditional land access (First Nations)	Alterations to access for traditional uses
Public access (recreation)	Alterations to access for recreation, recreation sites, and features

Source: CRC 1999a.

tant) on carnivores, specifically the grizzly bear. The interesting feature is that the impact was clearly cumulative, caused by the coal mines, the oil and gas activity, the forest harvesting, and the recreational activities in the region. Moreover, the model that showed that cumulative effects would be significant with the Cheviot mine also showed that the cumulative impact of existing human activity, without the Cheviot mine, was significant. In effect, the CEA of the proposed mine had identified an existing problem, one that the proponent, CRC, would need to deal with in order to obtain approval for the mine.

Because the impacts were identified as significant, there was concern that the project might not receive approval. Accordingly, CRC proposed a 'carnivore compensation program' to manage these effects, making them acceptable (Herrero and Jacob 1996). The carnivore compensation program involved several government agencies (both federal and provincial), several industries active in the area, and recreational users of the area. The principle here is that cumulative effects require cumulative solutions. The suggestion was that the existing significant adverse impacts would be reduced through the carnivore compensation program, even with the new project added. The review panel had to determine what was in the public interest given these circumstances.

In its report, the EUB–CEA Agency Joint Review Panel commented on the difficulties faced by a proponent that attempts to address regional cumulative effects that are to a large degree created by activities outside the boundaries of its project. In this regard, the panel noted that

the ultimate success of [the carnivore compensation program] will depend on active participation of a range of parties. The Panel notes from the evidence provided at the hearing that
the level of proactive participation by companies in such processes tends to be directly tied
to the degree that a program may affect either their present operations or future approvals.
Government, on the other hand, while wishing very much to participate in a comprehensive manner, often has difficulty in identifying adequate resources. (Bietz et al. 1997, 90)

The panel went on to suggest that for regional environmental management to
occur in a manner that would address the cumulative effects, the government regulators (EUB and Alberta Environment particularly) may have to reconsider how
they approve activities in the region. This is so even though those activities (other
coal mines, oil and gas exploration and development, forestry, and recreation—
see Figures 8.2, 8.3, 8.4, and 8.5) are not part of the Cheviot mine project. The
panel recognized that cumulative effects require integrated action by several government regulatory bodies and that action should be directed towards looking at
approvals on a regional basis. Project-by-project assessments can, unfortunately,
tend to ignore regional (comprehensive and holistic) approaches to environmental management. For this reason, the panel also noted the need for all jurisdictions
to share the management of the land base and development approvals for which
they are responsible.

This aspect was reiterated by the federal government in its response to the panel's 2000 report. In relation to the need for regional grizzly bear management, it
stated: 'Given the scale of current and projected future land use activities identified by the joint panel, Canada believes that timely implementation of scientifically based landscape condition targets is required with or without a Cheviot Coal
project' (Canada 2001).

Figure 8.2   Luscar coal mine just north of proposed Cheviot mine

Source: Alberta Energy and Utilities Board.

Figure 8.3   Luscar coal mine just north of proposed Cheviot mine

It is worth noting the condition 'with or without a Cheviot Coal project'. The point is that the cumulative effect on carnivores (as indicated by the effects on grizzly bears) was cumulatively significant even without the Cheviot project. The proposed project would make an existing significant adverse effect worse (or better if the cumulative effects management program, the carnivore compensation program, worked).

Subsequent to the release of the first panel report in June 1997 (Bietz et al. 1997), the decision to approve the mine proposal was challenged in federal court by a coalition of environmental groups. The most important claim, from our point of view, was that the review did not deal, as required by the Canadian Environmental Assessment Act, with future human activities, those that would act cumulatively with the Cheviot mine in the future. That challenge was successful in that the court noted that there were four deficiencies within the review of the project that needed to be considered before the proponent could be considered to be in compliance with the requirements of the Canadian Environmental Assessment Act and the panel's terms of reference (Campbell 1999). Specifically, the four points noted to be in non-compliance with federal environmental assessment law were the need to

- obtain all available information about likely forestry in the vicinity of the project, consider this information with respect to cumulative environmental effects, and reach conclusions and make recommendations about this factor;

- obtain all available information about likely mining in the vicinity of the project, consider this information with respect to cumulative environmental effects, and reach conclusions and make recommendations about this factor;
- do a comparative analysis between open-pit mining and underground mining at the project site to determine the comparative technical and economic feasibility and comparative environmental effects of each; and
- consider documents submitted by one specific intervenor to the project.

As a result of the successful legal challenge, the approvals rendered by the panel were effectively quashed and the panel was reconvened to hear evidence related to the four aspects found lacking by the courts. To assist the panel, Cardinal River Coals conducted a more comprehensive cumulative effects assessment (CRC 1999a, 1999c) and also compiled additional information on the project's effects, specifically the future cumulative effects (CRC 1999b).

During the second set of hearings, in the year 2000, the CRC submission (CRC 1999c) was evaluated, in terms of its treatment of cumulative effects, against the following requirements: specifically that a CEA should (Ross 1998)

- identify valued ecosystem components (VECs) affected by the proposed project (scoping);
- determine what other past, present, and future human activities have affected or will affect these VECs;
- predict the impacts on the VECs of the project in combination with the other human activities, and determine the significance of the impacts; and
- suggest how to manage the cumulative impacts.

Figure 8.4   Forestry activity near site of proposed Cheviot mine

Figure 8.5   Petroleum activity near site of proposed Cheviot mine. Well pad is connected with a road access and seismic line.

It was determined that the first two requirements (scoping and identification of other human activities) had been well done but that there were problems with the prediction of cumulative effects and there was very little to guide the panel on how to manage the cumulative impacts (Ross 2000). The selection of future human activities was based on the proponent's knowledge of the coal mining industry and on good consultation with other industries in the area. The problems of impact prediction were mainly driven by inconsistencies in the future activities assumed by the several sub-consultants (different futures were assumed for different VECs). This was cleared up during the hearings, and in the end, a satisfactory set of predictions for the various VECs was obtained.

The final requirement, management of cumulative effects, proved to be the most difficult. To determine whether a project is in the public interest, a panel needs to know whether the cumulative impacts of the project and other activities can be managed. Pure self-interest tends to drive a proponent to suggest that such impacts can be managed, but this may require the proponent to go far beyond its own authority and suggest how governments can regulate others. Certainly, Cardinal River Coals was reluctant to engage in such suggestions. Again, in the end, enough suggestions were offered at the hearings to satisfy the panel. The most important of these was the carnivore compensation program, needed to overcome the potential for significant adverse effects on grizzly bears. We believe that the management of cumulative effects is the most challenging aspect of cumulative effects assessment.

Lessons Learned

The review of the Cheviot mine project and the environmental assessment that was conducted for the project prompted several observations on the state of the EIA process for a large project conducted under the laws of the province of Alberta, those of Canada, or those of both jurisdictions. These observations are noted and expanded upon here.

1. *Cumulative effects require cumulative solutions.* Project impact assessment, as illustrated in this example, has moved from a relatively narrow examination of the project site to a broader analysis of how activity on that site can affect the environment some distance away or can have an impact on environmental components that move through or from the project site, such as wide-ranging wildlife species. This means that effects beyond the lands controlled, leased, or owned by the proponent have to be considered in the environmental management plan. It also means that the effects of other human activities (past, present, and future) must be considered in assessing (cumulative) impacts. While complicating the management options, this approach has been successful in provoking a cooperative multi-sectoral approach to regional environmental management.

The cumulative effects in the Cheviot case were truly cumulative in that they were caused by a variety of human activities: coal mines (existing and planned), forest harvesting, recreational use of the area, and oil and gas activity. Effective management of the cumulative effects not only required modifying the Cheviot mine, but also modifying the full variety of human activities contributing to the cumulative impact. In other words, cumulative effects require cumulative solutions. Implicit in the effective management of cumulative effects are regional co-operation and collaboration.

Project-based cumulative effects assessments identify impacts that need management; the consequence is cumulative effects management (usually regional), which, in turn, leads to better information for future project decisions. In the case of this particular mine proposal, one carnivore species became important to the assessment and to the broader regional management of the environmental effects—the grizzly bear. This animal is widely recognized as an indicator of ecosystem vitality in this part of the Rocky Mountains. The grizzly ranges widely throughout the area of Alberta's eastern slopes and is also found in the adjacent Jasper National Park just west of the proposed mining area. Expert testimony presented to the review panel (both in 1997 and in 2000) acknowledged the sensitive nature of the grizzly bear and the ecosystem it inhabits:

> CRC (the proponent) noted that carnivores, especially the large bodied ones, can be considered both as indicator and as umbrella species for impact assessment purposes. An indicator species in this context is a species that is particularly sensitive to the effects of development and human activities. Measurements of the effects of development on such species provide a measure of the success of impact mitigation programs. For umbrella species, the presence of declines in populations and habitat for such species are taken to

indicate not only stresses on the species itself, but also on other species and on the eco-system to which they belong. Protection of the umbrella species, on the other hand, will generally result in the preservation of adequate ecological conditions for other species. (Bietz et al. 1997, 75–6)

In order to address this species, the proponent was instrumental in arranging for a grizzly bear management framework involving several jurisdictions and interested parties. In addition, from the time of the first panel review in 1997 to the second in 2000, the Alberta government implemented a regional, sustainable, integrated resource management process. These processes, and others operational in the region, help to address cumulative effects through adaptive planning and resource management. They serve to provide the context within which individual development proposals can be considered by approval agencies. We believe that such measures must be a significant part of cumulative effects management.

2. *Don't be too conservative in seeking and incorporating information on other projects that may aggravate your cumulative effects assessment.* In the case of Cheviot, the successful argument used in the courts was that the applicant did not document how the effects of other activities in the area acted cumulatively with the direct effects of the mining project, especially for future projects. Whether by oversight, difficulty in accessing the information, or expediency, the applicant failed to take a broad approach to identifying those other future activities. That failure resulted in the protracted legal and regulatory reviews that spanned about five years from the beginning of the impact assessment process. Ultimately, the project was approved again in 2000, but by then the proponent had lost the business opportunities it had arranged. The project was indefinitely suspended in the fall of 2000, several years after the original EA documentation was compiled and more than three years after the first comprehensive panel review. To this date (early 2009), the project has not been started.

Herein lies one aspect of the Cheviot EA review that became a priority issue. Specifically, the difficulty in addressing the information requirements of an appropriate cumulative effects assessment led to the federal court challenge noted earlier. That successful appeal to the courts caused a three-year delay in the project and also necessitated the reconsideration of certain parts of the original EIA review before the same review panel. In fact, the delays and resulting uncertainty associated with the project's approval is cited by CRC as the main reason that the current status of the project is described as being 'suspended'.

3. *Remain flexible to accommodate new or changing aspects of the impact prediction.* Despite a comprehensive scoping process that led to the identification of just under 100 topics as important to those canvassed, an indication that selenium was important to the consideration of regional water quality was not raised until well into the review process. This forced the proponent, as well as everyone involved in the review, to readjust to a new issue. Evidence that the element selenium may be occurring at levels above that considered harmless to the environment was submitted to the second version of the panel review in the spring of 2000. This concern was based on preliminary chemical analysis of fish samples from water bodies in the vicinity of the existing mining operations. CRC provided a technical specialist at the hearings to address this new issue. It also stated that these indications were indeed serious and warranted investigation. Responding to the issue, CRC helped to form a Selenium Working Group, whose aim was to develop an understanding of the sources of the selenium, what guidelines would be appropriate if action was warranted, and what impact selenium would have on fish and other aquatic species. It was predicted that this working group could take up to 18 months to complete its studies.

On this specific issue the panel concluded that '[s]elenium levels in the aquatic environment, while warranting ongoing monitoring and research, do not current-ly represent a significant risk of adverse effects on regional water quality' (Bietz et al. 2000, 68). It is important to note that by properly addressing the selenium issue when it first appeared, CRC was able to handle the issue to the satisfaction of the re-view panel. In this way, a commitment to adaptive management (i.e., to addressing issues as information becomes available) was the appropriate course of action.

4. *Don't underestimate the legal and procedural sophistication of those who oppose the project.* The Cheviot project was opposed by a number of parties, but most aggressively by a coalition of environmental groups. These groups objected pri-marily to the proposal of a new strip mine adjacent to a national park with World Heritage Site designation. They participated in the lengthy public panel review of the proposal and sponsored the successful federal court challenge to the decisions and recommendations reached by the panel. One can argue that the opportunity for judicial review was presented in the proponent's evidence and EIA documents and that a proper impact statement would be robust enough to sustain a legal challenge, no matter what legislative framework was used.

5. *Management of cumulative impacts is both difficult and essential.* For this proj-ect, it was challenging to get the proponent to identify means of managing cumu-lative impacts. But, at the same time, the introduction of the carnivore compensa-tion program was a great benefit to the project proposal. For the one potentially significant adverse effect, one that was explicitly cumulative, this program offered the panel a demonstration that cumulative impacts can be managed and perhaps, with mitigation measures in place, made even less significant with the project than without the project.

Conclusion

The Cheviot environmental assessment process was found to be fraught with problems and procedural difficulties, many of which were exacerbated by the complications brought on by cumulative effects assessment and the nuances of the federal and provincial legislation governing the process. Lessons can be learned from the experience nonetheless.

- Do not be too conservative in seeking and incorporating information that may aggravate the cumulative effects assessment of the project.
- To manage cumulative effects, use cumulative solutions.
- Remain flexible to accommodate new or changing impact predictions.
- Do not underestimate the legal and procedural sophistication of those opposing the project.

Environmental impact assessment in both Canada and the province of Alberta are mature processes that have been shown to be excellent tools for project planning and adaptive management. Nonetheless, the experience with complex development proposals such as Cheviot has indicated where an awareness of certain aspects of EA can improve professional practice.

Note

1. Joint panels of this nature are conducted under the Harmonization Agreement between Canada and Alberta. This agreement is intended to meet the needs of each jurisdiction and yet allow for efficient processing of the proposal and the participation of the public and other interested parties (Alberta Environment 2003b).

References

Alberta Environment. 2003a. Legislation website http://www.qp.gov.ab.ca/documents/acts/E12. cfm, visited 2003.

———. 2003b. http://www3.gov.ab.ca/env/protenf/ccme/, visited 2003.

———. 2003c. http://www3.gov.ab.ca/env/protenf/EPEA/eiaproce.html, visited 2003.

Baxter, W., W. Ross, and H. Spaling. 2001. Improving the practice of cumulative effects assessments in Canada. *Journal of Impact Assessment and Project Appraisal* 19 (4) (December): 253–62.

Bietz, B.F., G.J. Miller, and T. Beck. 1997. *Report of the EUB-CEA Agency Joint Review Panel*. Calgary: Alberta Energy and Utilities Board and the CEA Agency, June.

———. 2000. *Report of the EUB–CEA Agency Joint Review Panel, Cheviot Coal Project, Mountain Park Area, Alberta, Cardinal River Coals Ltd.* Calgary: Alberta Energy and Utilities Board and the CEA Agency.

Campbell, D.R. 1999. *Decision of Court in Appeal of Cheviot Decision.* Ottawa, April.

CEA Agency (Canadian Environmental Assessment Agency). 2003. Legislation website, http://www.ceaa-acee.gc.ca/0011/act_e.htm, visited 2003.

CRC (Cardinal River Coals). 1996. *Cheviot mine project application.* Hinton, AB: CRC, May.

———. 1999a. Letter to panel regarding Cumulative Effects Assessment. Hinton, AB: CRC, September.

———. 1999b. *Cheviot mine supplemental information.* 3 vols. Hinton, AB: CRC, October.

———. 1999c. *Cheviot Mine Cumulative Effects Assessment.* Hinton, AB: CRC, November.

Canada, Government of. 2001. Federal government response to the September 12, 2000 Environmental Assessment Report of the EUB–CEA Agency Joint Review Panel on the Cheviot Coal Project. Available at http://www.dfo-mpo.gc.ca/media/backgrou/2001/ hq-ac29_e.htm.

Hegmann, G., C. Cocklin, R. Creasey, S. Dupuis, A. Kennedy, W.A. Ross, H. Spaling, and D. Stalker. 1999. *Cumulative effects assessment: A practitioners' guide.* Prepared by Axys Environmental Consulting Ltd and the CEA Working Group for the Canadian Environmental Assessment Agency. Hull, QC.

Herrero, S., and H. Jacob. 1996. Cheviot mine project: A proposed carnivore compensation program. Submitted on behalf of Cardinal River Coals to the panel, December.

Ross, W.A. 1998. Cumulative effects assessment: Learning from Canadian case studies. *Journal of Impact Assessment and Project Appraisal* 16 (4).

———. 2000. Cumulative effects assessment: An evaluation: The Cheviot Coal Mine Project Joint Review. Prepared for the Cheviot Coal Mine Project Review Panel.

Spaling, H., J. Zwier, W. Ross, and R. Creasey. 2000. Managing regional cumulative effects of oil sands development in Alberta, Canada. *Journal of Environmental Assessment, Policy and Management* 2 (4): 501–28.

Chapter 9

Multi-jurisdictional Environmental Assessment

Patricia Fitzpatrick and A. John Sinclair

Introduction

The management of the environment in Canada is subject to complexity stemming, in part, from the absence of clear reference to it in the Canadian Constitution.[1] The division of powers between the provinces and the federal government in the Constitution does not assign absolute control over environmental issues to either level of government (Harrison 1996). Consequently, the 'constitutional authority over environmental protection is shared by the federal and provincial governments' (McKenzie 2002, 106). This jurisdictional situation has ultimately led to overlapping legislative responsibilities (see Fafard 2000; Hessing et al. 2005; Valiante 2002).

Environmental assessment (EA) provides a good example of a process for environmental decision making that has multi-jurisdictional implications. In Canada, the federal government and all of the provincial and territorial governments have their own legislated EA processes. Some recent land claims agreements may also include provisions for EA. In practice, this means that a project could be subject to at least two EA processes—one administered by the province (or territory) and one administered by the federal government. If more than two jurisdictions are affected by a project and/or another regulatory process applies, it is possible that multiple assessment processes apply to one project. This is viewed as problematic because multiple assessments can cause duplication of effort, lead to process inefficiencies, and reduce the effectiveness of the assessment exercise.

The purpose of this chapter is to discuss how government in Canada is currently addressing issues associated with multi-jurisdictional EA and to illustrate the complexities of such approaches through a case study of the Wuskwatim generating station and transmission lines projects in Manitoba (Wuskwatim projects). The first section reviews the issues associated with assessing projects under different EA processes and the different efforts employed to reduce duplication. The second section provides an examination of the harmonization process for the Wuskwatim projects.

The final section reviews key issues associated with harmonization and reiterates why this is an area of particular importance to the system of environmental management in Canada and abroad.

We used a number of methods in gathering data. First, we completed a review of pertinent EA legislation, agreements, policies, and literature. Second, we conducted semi-structured interviews with individuals involved in multi-jurisdictional EA, representing government, industry, and the public. Finally, we collected and analysed data surrounding our case study, including the literature on the public registry and interviews with assessment participants.

Harmonization and EA

There are four standard approaches to addressing overlap in EA employed in Canada: standardization, substitution, harmonization, and exemption (Fitzpatrick and Sinclair, 2009). *Standardization* involves the adoption of the same EA process in multiple jurisdictions. Thus, regardless of how many assessments are triggered, the EA process remains the same. Early efforts to standardize EA processes in Canada were quickly abandoned owing to the complexity of power sharing (Kennett 2000).

Substitution occurs when any other federal process is used in place of the federal EA. This approach was recently tested when the government substituted the National Energy Board review process for the federal EA process in the Emera Brunswick Pipeline case. While the outcomes of this case are still being reviewed, a report by Schneider et al. (2007) indicates that a key concern of the public was the differing mandates of the two agencies. The Canadian Environmental Assessment Agency (CEA Agency) is charged with ensuring that EAs are conducted in such a way as to ensure informed decision making for sustainable development; the National Energy Board, however, is mandated 'to promote safety, security, environmental protection and economic efficiency' with respect to energy developments. Moreover, participants expressed concern about the formal hearing processes used for the review. The reviewers and case participants were left with significant concerns about EA substitution.

Harmonization, the third approach, involves the rationalization of EA processes (Valiante 2002), accomplished through the coordination of the legislative frameworks prescribed by different jurisdictions, so that a project undergoes a single review. A project is therefore subject to one assessment that meets the needs of perhaps many assessment legislations. This process is also referred to as the 'one-window approach' to EA. By fostering a one-window approach, harmonized EAs are thought to minimize duplication, reduce process complexity, and strengthen the use of resources with respect to efficiency and effectiveness (Fitzpatrick and Sinclair 2005; Kennett 1993, 1997; Lawrence 1999).

The system of environmental management that is used across Canada also gives rise to a newer, fourth approach to overlapping EA requirements: exemption. An example of *exemption* would be when, in the presence of a newer claims-based EA process, the federal government excludes the application of the Canadian Environmental Assessment Act 1992, c. 37 (CEAA) within a specific region. In essence, the claims-based EA replaces the CEAA.[2]

Harmonization is overwhelmingly the most common approach to minimizing process overlap in Canada, and thus it is the focus of this chapter. Historically, harmonization has taken place on a project-by-project basis. In this way, when a proponent's project triggers the EA process of more than one jurisdiction, the representatives of different government agencies meet together to negotiate how the assessment can be completed effectively and efficiently with minimal duplication and overlap. Relevant ministers then sign a memorandum of understanding that outlines the assessment process to be followed. More recently, however, some jurisdictions have formalized the harmonization process through bilateral agreements. The Canadian Council of Ministers of the Environment (CCME) signed the Canada-Wide Accord on Environmental Harmonization in 1998.[3] The purpose of this agreement is to improve environmental protection through the promotion of sustainable development, foster inter-jurisdictional cooperation, and streamline environmental management. The accord includes a Sub-agreement on Environmental Assessment that outlines how the signatory jurisdictions 'are seeking to provide the public, proponents, and governments with greater consistency, predictability and timely and efficient use of resources where two or more Parties are required by law to assess the same proposed project.' This sub-agreement establishes, among other things,

- the required components of a harmonized environmental assessment (sec. 4.1);
- the necessary stages of the assessment (sec. 4.2);
- the minimum components of the public participation program (sec. 4.2);
- the lead party for facilitating an environmental assessment (sec. 5.6); and
- the need to develop specific bilateral agreements to implement the sub-agreement (sec. 5.9).

Eight bilateral agreements between the Government of Canada and the provincial/territorial governments are in place today (see Figure 9.1): Yukon Territory (2003), British Columbia (2005), Alberta (2005), Saskatchewan (2005), Manitoba (2007), Ontario (2004), Quebec (2004), and Newfoundland and Labrador (draft, 2004). These cooperative agreements transfer the principles of the 1998 accord into specific operating plans for each jurisdiction.

In the first edition of this book, we indicated that there was significant variation among these agreements (Fitzpatrick and Sinclair 2005). In other words, no two agreements covered the same range of issues. However, since that time, three agreements have been negotiated and three more renewed. This new set of bilateral agreements is more consistent. Most of the agreements cover

- administrative aspects of coordinated EA (e.g., scope of the agreement, objectives, designated offices, lead-party identification, agreement duration, and renewal process);

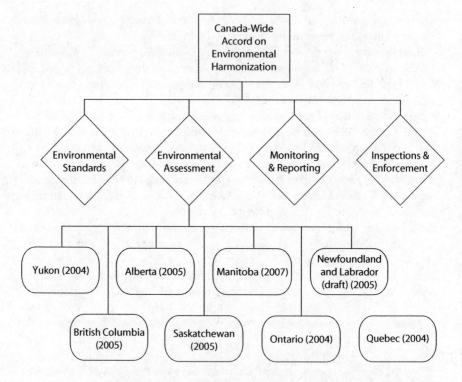

Figure 9.1   The system of EA coordination negotiated between the federal and provincial governments. As noted in the text, Quebec is not a signatory to the Canada-Wide Accord on Environmental Harmonization, thus the bilateral agreement negotiated with the federal government.

Source: Fitzpatrick and Sinclair, 2009

- preliminary process (e.g., consultation between parties, notification, determination of interest);
- cooperative assessment process (e.g., administering team, development of the environmental impact statement [EIS] guidelines, EA determinations, follow-up);
- reference to public participation;
- provisions for a joint panel;
- Aboriginal considerations;
- accommodating interests; and
- transboundary provisions.

Furthermore, six agreements now include a process for dispute resolution between parties, and five include reference to coordinating decisions. Thus, the current collection of bilateral agreements is apparently more uniform among jurisdictions.

There remains, however, significant variability in how each agreement addresses aspects of public participation in the EA process (e.g. when the public should get involved, how the public should get involved; see chapter 4). Table 9.1 identifies the requirements for public participation in EA provided for in each agreement. A quick survey of the table reveals the degree to which these bilateral agreements vary in terms of these common requirements. Although the agreements are re-markably more consistent than they were in the past, this table illustrates that differences remains. For example, only six of the eight agreements include specific reference to access to information. Furthermore, each of the public participation activities is addressed in a different manner in each agreement.

Table 9.2 provides an example of these differences by comparing the clauses on participation found in the Yukon and Alberta agreements. While it is reasonable to expect that the agreements would differ, just as provincial legislation differs, these differences add another layer of variability and bring into question how clear the process is to the public, to the proponents, and to the regulators.

A second important exception to improved consistency involves the EA legisla-tion itself. Although the bilateral agreements are more consistent, the triggers for an assessment to start and the EA process and decision points vary significantly. In other words, because the assessment processes vary among jurisdictions, so too

Table 9.1   Opportunities for public participation identified in existing harmonization agreements

	Yukon Territory	British Columbia	Alberta	Saskatchewan	Manitoba	Ontario	Quebec	Newfoundland
General reference	38					13(2), 13(3)	11(c)	
Notice		14(1)(c)	10.2(b)	23(2)(b)	38(b)		11(b)	
Access to information			9.0	23(2)(a)	38(a)	13(2)(a)	11(a)	17(a)
Participant assistance	53	15(3)(d)	10(3)	29(3)(d)	52	18(4)	Annex B	22(3)
Public comment		14(1)(c)	10(2)(c)	23(2)(e)	38(c)			17(b), 17(c), 17(d)
Public hearings	39	15–19	10(2)(l) Section 7	29–33	46–52	18	14	22–26

Source: Fitzpatrick and Sinclair, 2009.

Table 9.2　Provisions for public participation in two bilateral agreements

	Public Participation
Yukon Territory	38. The Parties will coordinate the timing of their determination for the need for public participation. Each Party will determine its need for public participation in as timely a way as possible for the cooperative environmental assessment and will communicate this need to the other Party. If both Parties determine that public participation is appropriate, they will consult on the establishment of a joint process.
Alberta	**9.0 PUBLIC REGISTRY** 9.1 For projects subject to a cooperative environmental assessment, the Parties agree to cooperate in meeting their respective public registry requirements. Both Parties will maintain public registries in accordance with the requirements of their respective legislation. 9.2 The public, proponent and the other Party will be provided access to the registries maintained by the Parties in accordance with their legislated requirements. 9.3 The Parties will continue to evaluate opportunities to provide the public with more convenient access to information about cooperative environmental assessments, including linking the websites of the Parties. **10.0 PUBLIC INVOLVEMENT** 10.1 The Parties involved in a cooperative environmental assessment will facilitate public participation, where consistent with their policies and legislation, which may include providing access to information, technical expertise, and participation at public meetings. 10.2 Provision for public participation in cooperative environmental assessments shall include where applicable, but not be limited to: a. public notice that a cooperative environmental assessment will be conducted; b. public disclosure and an opportunity to comment on a proposed terms of reference for the environmental assessment report; c. an opportunity to participate in public consultations required by the terms of reference, as part of the preparation of an environmental assessment report; d. public notification that the final terms of reference for the environmental assessment report has been issued; e. an opportunity to be consulted in the comprehensive study process in accordance with section 21 of the Canadian Environmental Assessment Act; f. an opportunity to participate in the comprehensive study in accordance with section 21.2 of the Canadian Environmental Assessment Act; g. an opportunity to comment on the comprehensive study report in accordance with section 22 of the Canadian Environmental Assessment Act;

Continued

Table 9.2   Continued

h.   public notification of the availability of the environmental assessment report;

i.   an opportunity for members of the public to comment on the environmental assessment report provided by the proponent;

j.   public notification of the opportunity to comment on the need for a public hearing;

k.   public notification of the Parties' intention to appoint a joint panel in accordance with section 58(3) of the Canadian Environmental Assessment Act; and

l.   if a public hearing is held, an opportunity for members of the public to participate in the hearing.

10.3   Consistent with the Sub-agreement, each Party will provide participant funding based on its law or policy.

10.4   Notices of filing of environmental assessment documents, placed by the proponent or Lead Party in the cooperative environmental assessment process, will indicate the involvement of both Parties. If the other Party later withdraws from the cooperative environmental assessment because it has determined it has no environmental assessment responsibility, that Party shall notify the proponent and the public of the change using equivalent notification procedures. The Parties agree that when a joint panel review is undertaken, public notices placed by either Party will identify the involvement of both Parties in the hearing.

---

does the implementation. For example, significance is an important concept in the EA process (see chapter 3), as it establishes the magnitude of the projected impacts of a project or activity. Under the CEAA, it is the proponent who establishes significance after having considered relevant mitigation measures. A project may not proceed if it is determined to have significant adverse impacts that cannot be justified, and thus very few impacts are ever deemed to be significant under this assessment process. However, under the Mackenzie Valley Resource Management Act 1998, c. 25, significance is determined by the arm's-length, government-appointed board charged with reviewing the developer's assessment report (see chapter 11). Under this legislation, determining whether the impact is significant is necessary so that mitigation measures can be applied. Thus, most impacts are deemed significant in the Mackenzie Valley. This example serves as one illustration of the differences between EA processes. Thus, regardless of the comment element (the federal EA process), harmonized EAs vary across jurisdictions.

Notwithstanding the questions and issues surrounding harmonization, there is a need for jurisdictions to cooperate when more than one assessment process is triggered. To deepen our understanding of the issues surrounding multi-jurisdictional assessments, it is instructive to consider a case study. In the first edition of this book, we reviewed the assessment of the Sable Gas project. This project was a joint venture among several energy companies to extract, process, and transport natural gas from offshore Nova Scotia, through New Brunswick, to markets in the United States (Fitzpatrick and Sinclair 2005). The proposal was subject to numerous regulatory regimes and triggered reviews under five different regimes: the National Energy Board (NEB), the Canada–Nova Scotia Offshore Petroleum Board, the Province of Nova Scotia, the Province of New Brunswick, and the CEA Agency.

The environmental assessment for the Sable Gas project illustrated the trade-offs associated with harmonizing disparate processes (Fitzpatrick and Sinclair 2005). The strengths of the harmonized assessment included a stronger public registry for participants, more financial resources for public participants, and a more extensive scope of review than under any one of the individual EA processes. Data suggested that the harmonized review increased access to assessment information through a single, harmonized registry. EA participants, who constitute a key component of the federal EA but are absent from the provincial counterparts in this region, received participant funding. Furthermore, the scope of the EA review went beyond the EA requirements of any of the five individual review processes triggered by the process. The key weakness of the process was the quasi-judicial hearing format. In this case, the informal, non-adversarial public hearing venue utilized under the federal EA process was replaced by the quasi-judicial, adversarial approach used by the National Energy Board. This has become a commonplace trade-off when harmonizing the federal process with other EA frameworks; consequently, the spirit of informal hearings becomes lost in a sea of lawyers. Finally, as the review becomes more legalistic, participants are compelled to engage legal representation, which entails a significant financial commitment that is not provided for under current participant funding guidelines in EA legislation. We concluded, however, that the Sable Gas case study did show that a one-window or cooperative approach to EA can be achieved and that this may not be a bad thing for certain provisions, such as public involvement.

The Wuskwatim Generating Station and Transmission Lines Projects

In this edition, we offer a more recent case study—that of the Wuskwatim generating station and transmission lines projects (Wuskwatim projects); this case allows us to consider the progress in EA harmonization that has occurred since the Sable case. The Wuskwatim projects involved the construction of a low head, modified run of the river dam, to produce 200 megawatts of electricity, and three 230 kV transmission line segments, totalling 247 km (CEC 2004a). The projects triggered reviews by three separate bodies. The Department of Fisheries and Oceans and Transport Canada conducted a comprehensive study of the generating station under the terms of the Canadian Environmental Assessment Act (see chapter 2). This assessment was founded on the projects' need for Fisheries authorization (this involves permission to modify fish habitat) and on their location within a navigable water. The provincial EA was undertaken under the terms of the Environment Act SM 1987–88, c. 26 of Manitoba (see chapter 18). The generating station triggered a class three assessment, and the transmission lines triggered a class two assessment. As part of this review process, the Clean Environment Commission (CEC) was directed to gather public comment on the assessment guidelines and to later hold public hearings about the impact statement. Finally, the provincial Public Utilities Board was charged with reviewing the justification of, need for, and alternatives to the projects as stipulated in the Public Utilities Board Act, C.C.S.M. P280.

Harmonization was pursued through a two-tiered approach. First, the federal and provincial EA processes were harmonized under the terms of the 2000 Canada-Manitoba Agreement on Environmental Assessment Cooperation. As per this agreement, the two jurisdictions coordinated the common steps in their respective EA processes, such as the development of EIS guidelines and a technical review of documentation. For example, since an EIS was required under both assessment processes, the federal and provincial governments both contributed to the development of the impact statement guidelines, which indicate what the proponent had to consider during the EA process. Both levels of government also contributed to a technical review of the project. However, because this assessment was not deemed a joint panel review (see chapter 2), there was no project-specific agreement outlining the roles and responsibilities of each party, nor was there an obligation for the federal or provincial governments to participate in EA activities required by the counterpart's legislation.

To complicate matters further, two provincial public decision processes also had to be satisfied—the EA process and the Public Utilities Board process.[4] Coordination of these two processes (which both include the opportunity for hearings) was less systematic than the coordination of the federal and provincial EA processes, and occurred much later in the review. At the pre-hearing conference, the chair of the CEC described the decision to combine the two provincial reviews as follows:

> [T]he Minister also asked the Commission to review the justification, need for and alternatives for the proposal that is traditionally carried out by the Public Utilities Board. The Commission, I think, was very frank in saying we didn't feel we had the depth of experience, either in our Commission or in the technical support that is usually provided. . . . I went to see the Chair of the Public Utilities Board, and that is where the idea was proposed that . . . we seek to use at least two of their Commissioners. (CEC 2003, 6)

Indeed, it was a full month after the final EIS guidelines were issued by the federal and provincial governments (9 April 2003) and fifteen days after the EIS was submitted (30 April 2003) that the province announced, on 15 May 2003, that the two provincial public hearings would be held together. As noted by three participants, merging these two provincial processes following the finalization of the assessment guidelines resulted in unclear wording relating to the 'justification for the project' and the 'need for the project' components of the assessment (Interviews 21, 23, and 25). Participants questioned whether these aspects of the CEC hearing met the requirements of the Public Utilities Board legislation. One commented: 'Frankly, I think that [the merging of the processes] complicated things in my view. Certainly it provided for a lot more information than would normally be dealt with. And whether or not the CEC, even with the addition of the Public Utilities Board members to the Commission, were the appropriate body to talk about the economics of it, I don't know' (Interview 22). In sum, the provincial harmonization of the EA and Public Utilities processes in one hearing contributed to process uncertainty and general dissatisfaction with the review.

As noted earlier, our specific discussion of the strengths and weaknesses in the harmonization of the three decision processes focuses on public participation. The harmonized process for the Wuskwatim projects included the following opportunities for public participation (Fitzpatrick 2006):

- Written and verbal submissions regarding the scope of the assessment
- Written comments related to the conformity of the impact statement to the guidelines
- Written interrogatory (IR) process (a multiple series of written questions and answers to supplement the impact statement)
- Verbal presentations, through 32 days of public hearings, that were supported by written material
- Written comments about the draft comprehensive study report

In addition, participants in the provincial process were eligible for intervenor funding, as per section 13.2 of the Environment Act.

The opportunities for public participation described above reflect aspects of both the provincial and the federal legislation. For example, provincial processes allocated participant funding, recognizing it as an important aspect of the project review in that it allowed participants to prepare for the hearings. Unfortunately, money was made available for only the provincial hearings process, not the federal comprehensive study process.[5] However, only the federal comprehensive study process allowed for public comment on the draft report prior to the decision by the government minister.

The combining of the activities for public participation in the harmonized EA went beyond what was required under any of the three processes involved. In other words, the harmonized approach might offer the public more opportunity to be involved in the decision process. Nevertheless, the harmonized EA of the Wuskwatim projects was not without problems. Five key issues emerged concerning the strengths and weaknesses of the EA process in this case: government involvement, scope, process, access to information, and timing of the decision.

Government Involvement

The role of government in the hearings process was one of the most significant concerns raised by participants in the harmonized EA of the Wuskwatim projects. Except in the case of a joint review panel, a harmonized assessment under the terms of a bilateral agreement requires only that the two levels of government coordinate the steps in the assessment process, wherever possible. Thus, while there was collaboration during the scoping and technical review stages of the Wusk-watim EA, when it came time for the hearings, both levels of government took a hands-off approach.

The provincial government viewed the hearings as an opportunity for the public to consider and debate the projects. Representatives of the Government of Manitoba attended all hearings, and gave testimony on two occasions, but did not engage in cross-examination during this process. Correspondence provided to the Clean Environment Commission prior to the hearings explained this deliberate strategy: 'As you are aware, in order to protect the validity of the CEC independent review, provincial decision-makers cannot take an active role in the hearing process or comment in any substantive way on issues that will ultimately be the subject of their licensing decisions' (Strachan 2004). The provincial government pointed out that it had completed its analysis prior to the hearings, as part of its technical review of the EIS, and therefore had decided to be primarily an observer at the public hearings.

Since the hearings were part of the provincial EA, the federal government had an even smaller presence at the hearings. While the federal government decided to consider the findings of the public review in preparing its comprehensive study, it chose not to participate in or attend the hearings. This role was particularly problematic for hearing participants: 'There is a clear absence of a federal presence in this room and throughout this hearing process and clearly there are federal jurisdictional issues here that are not being addressed' (CEC 2004b, 7554); and 'Why has the CEC not asked for Federal Government experts?' (CEC 2004b, 7237).

The lack of active government involvement in the public review was problematic for a number of reasons (Fitzpatrick 2006). First, active government involvement in public hearings increases the range of issues that are considered publically. Other participants, including non-governmental organizations and the general public, while well funded for this review (see below), lack the human and financial capacity to canvass all issues. Thus, while an inactive government may preserve the independence of the CEC, it also detracts from efforts to ensure that alternative perspectives are explored during the hearings. Second, although the government completed its review of the assessment prior to the public hearings, its lack of active involvement gave the impression that it was forgoing its duty. The technical review was not open to the public, and only summary notes were posted on the public registry. Since the technical review occurred before the hearings, it was unclear how issues arising from the public hearings would be addressed by government. Finally, an inactive government presence during the hearings created process uncertainty. There was some belief that the federal government would, at a minimum, testify at the hearings in front of the CEC. When this did not occur, participants became confused and made statements like those noted above.

In the end, the absence of the federal government was noted by the provincial CEC in its final report: 'The Commission agrees that the cooperative assessment process in Manitoba is not easily understood and found little evidence of its practical application during the review of the Wuskwatim Projects. The Commission realized little benefit from the cooperative approach that was apparently undertaken in connection with this review' (CEC 2004a, 7–8).

Scope

Participants also expressed concerns associated with the scope of the review. As discussed in chapter 3, the scope of the assessment lays out which aspects of the project must be considered in an EA. That is to say, the scope identifies the activities to be assessed (e.g., the dam, the transmission lines, etc.) and the components to be considered (e.g., the purpose of, need for, and alternatives to the project, the environmental impacts, the social impacts, etc.).

That this assessment is referred to as the 'Wuskwatim projects [plural] ' EA is deliberate, as the project was split into two components: the generating station and the transmission lines. The federal responsible authorities exacerbated this split when, in scoping, they made the determination that the federal review would only focus on the generating station aspect of the project. The transmission lines component of the projects was considered only inasmuch as the federal government assessed cumulative effects (Canada 2005). The rationale for this decision was that the Fisheries authorization under section 35(2) of the Fisheries Act and authorization under section 5(1) of the Navigable Waters Protection Act (which had triggered the review) applied to the generating station. The responsible authorities determined that they had no regulatory triggers pursuant to the transmission lines aspect of the projects.

Fortunately, in this case, the provincial EA process included both project components. However, the different scopes contributed to the belief, by some, that this assessment was not, in fact, harmonized (Fitzpatrick 2006). For the public, having a harmonized process that only considered one aspect of the project was very confusing—after all, how can you have a power generating station without hydro lines to move the power?

Process

The hearings format was a third area of concern. Although every assessment process to which a given project is subject can involve hearings, the characteristics of the hearings process required by each framework are varied. Hearings can be held in public or in camera; they may follow a legalistic (quasi-judicial) process or be informal in tenor (similar to a question-and-answer activity) (see chapter 4).

Hearings under the Environment Act in Manitoba are generally characterized as informal opportunities for the public to present information about a project. The commission is required to make information available to the public and to hold hearings in a manner that offers the public an opportunity to participate, which generally involves a relaxed, non-adversarial environment. The hearings of the Public Utilities Board, however, are more formal in nature. This formality is expressed through very specific additional procedures in the hearings process, including interrogatories (IRs) and quasi-judicial hearings.

The Public Utilities Board utilized a more formal process for the Wuskwatim project hearings:

> It became a quasi-judicial process, at least it looked like it did. It wasn't supposed to; it was supposed to be about the public participating, but it seemed to become a really legal thing, and so if you are a participant in that kind of thing, it became very clear to me that it wasn't about presenting in a way the participant felt comfortable with, or just get their voices out, or to bring communities' information in . . . it was about fitting within a particular set of rules and fitting within the scope. . . . And I thought, within a court of law this makes perfect sense, but this isn't [a court of law]. (Interview 30)

It is becoming more common in harmonized EAs to replace an informal public hearings environment with a legalistic, formal process (e.g., the Sable Gas and Mackenzie Gas projects). Federally, hearings under the Canadian Environmental Assessment Act are informal. However, when this process is harmonized with a more legalistic review, like the National Energy Board Act, R.S. 1985 CN-7 or the Environmental Protection and Enhancement Act, E12 of Alberta, informal hearings are sacrificed as one trade-off.

A quasi-judicial review has some strengths (Fitzpatrick 2006; Fitzpatrick and Sinclair 2003). Some participants, particularly those represented by counsel, have suggested that evidence is subject to more scrutiny in a quasi-judicial process. The IR process is seen to make the impact statement a more iterative document, reflecting changes in the project design, in the predicted impacts, and in prescribed mitigation as the project is reviewed. IRs are also seen as a good tool for evaluating evidence prior to the public hearings. One participant in the hearings for the Wuskwatim projects suggested the 'the Public Utilities Board process is better at having a rigorous examination for the issues . . . and I think there is a better role for public participation in the Public Utilities Board' (Interview 23).

These strengths, however, can be considered weaknesses in terms of engaging the public. A legalistic process can alienate those who are not represented by counsel and who are less knowledgeable about how to behave in a court of law. This format can increase the costs associated with participation because legal representation can be expensive. A legalistic process may also alienate the general public (see Fitzpatrick and Sinclair 2003). Finally, while the IR process may serve to improve the information presented to the proponent, it nevertheless creates another layer of information with which the public must become familiar.

In the review process for the Wuskwatim projects, the IRs generated over 500 questions, which led six participants to question how valuable the exercise was to the process, particularly when balancing input versus return (Interviews 14, 15, 19, 22, 28, and 30).

> We were quite concerned with the Commission engaging us in the IR process, that we saw pretty much as a duplication to what we had done through our review process by making documents available, answering questions, addressing questions to the extent that we thought they needed to be addressed, and then the Commission starting it all over again and trying to reengage us in the process. (Interview 22)

Four people, however, felt that IRs serve an important function in the review process. For example, one commented: 'Well the IRs are useful in the sense that you can get as much information as possible ahead of time, and they are cheaper. That is just the basic efficiency' (Interview 16).

The relative value of IRs is illustrative of the larger debate surrounding the value of informal versus quasi-judicial processes. It is unlikely, however, that an informal hearings process can satisfy EA legislation based on a quasi-judicial hearings format. Thus, this trade-off will continue in the future.

Access to Information

Numerous authors, including Sinclair and Diduck (see chapter 4) and Sinclair and Doelle (2003), have examined the importance of a public registry system for the collecting and maintaining of information about projects undergoing assessment. A public registry allows participants to be confident that they have access to the information they need to review the project documentation. Following the completion of the EA, the public registry serves as a permanent record of the events that transpired throughout the review.

The review process for the Wuskwatim projects included an extensive set of documents that were intended to be available to the public, but, unfortunately, accessing these records was difficult. In the case of a federal-provincial harmonized EA, each government manages its own public registry, in accordance with its own legislation. The Canadian Environmental Assessment Act and the Environment Act of Manitoba, however, have different definitions of material that must be included in such a registry (see Table 9.3). Of particular importance is how legislation treats correspondence. The CEAA (subject to areas of third-party information described in s. 55 (4)) includes a record of public comments related to the EA; as such, this public registry includes the significant body of correspondence that takes place throughout an EA. Provincially, however, this correspondence is placed on the public record only at the discretion of the director of environment approvals.

Furthermore, all material submitted to the CEC may be put on its own record of process. However, this record is separate from the provincial public registry, since evidence submitted to the CEC is only placed on the provincial public registry at the discretion of the director of environmental approvals. Little evidence presented at the hearings was posted on the provincial public registry (Fitzpatrick 2006), and it was therefore important to consult the CEC office, to get a comprehensive understanding of the hearings, and the federal and provincial registries, to see other case documentation.

In essence, three registries had to be consulted to obtain a comprehensive picture of the EA. Each site had a unique set of documentation related to the EA, none of which was complete. Consequently, although participants had access to EA material, this material was not readily available. Furthermore, all central repositories were housed in Winnipeg. Residents of northern Manitoba, where the projects are to be situated, needed to travel south to access each complete official record.

Table 9.3   Description of material available on the public registries

The Canadian Environmental Assessment Act (1992)	The Environment Act of Manitoba (1987)
55(3) Subject to subsection (4), a public registry shall contain all records produced, collected, or submitted with respect to the environmental assessment of the project, including (a) any report relating to the assessment; (b) any comments filed by the public in relation to the assessment; (c) any records prepared by the responsible authority for the purposes of section 38; (d) any records produced as the result of the implementation of any follow-up program; (e) any terms of reference for a mediation or a panel review; and (f) any documents requiring mitigation measures to be implemented.	17 Subject to section 47, the director shall maintain or cause to be maintained a public registry, containing for each proposal received (a) a summary, prepared by the proponent in form and detail approved by the department; (b) the disposition and status of each proposal; (c) a copy of the environmental licence, where applicable; (d) a copy of the assessment report; (e) justification for not accepting the advice and recommendations of the commission, where applicable; and (f) justification for refusing to issue an environmental licence, where applicable; and (g) such other information as the minister or director may from time to time direct.

Timing of the Decision

Perhaps the greatest area of concern surrounding the harmonized EA process involved the timing of the EA decision(s). The CEC released its report in September 2004, three months after the hearings had ended. The CEC 'determine[d] that, if the appropriate mitigation and monitoring regime is put in place and the Projects are constructed and operated as proposed, the adverse effects on the biophysical, socioeconomic and cultural environment will not be significant'(CEC 2004, 1). Normally, the province issues a licence to proceed with a project shortly after the hearings conclude and the CEC has reported. The federal government, however, did not release its draft comprehensive study report until November 2005, 14 months after the province appeared poised to make its decision. The draft report summarized in two tables the concerns raised by the public during the provincial hearings (Canada 2005). After collecting comments on the report, the federal government made its final decision on 21 June 2006. Like the CEC, the federal government also concluded that the project was not likely to cause significant impacts. At this time, the provincial government issued licences under the Environment Act of Manitoba. There were clearly two assessment processes in the mind of the public.

In accordance with the federal process in effect at the time, the federal government could have, in fact, bumped its EA up to a panel review at any time prior to the decision. Were this to have occurred, a (second) independent federal panel would have been struck. This aspect of the EA caused significant confusion among the public participants and reinforced ongoing concerns about the lack of harmonization in the case.

Discussion and Conclusion

The growing number of empirical studies related to harmonized assessments illustrate mixed results. As noted above, the EA of the Wuskwatim projects offered increased opportunities for public participation. Thus, strictly from the perspective of the number of participants, harmonization can be seen to strengthen the EA. Nonetheless, the Wuskwatim EA and other cases show that the complexities associated with harmonization have an effect on public participation (Fitzpatrick and Sinclair 2005, 2009; Kwasniak 2008). However, with the lack of involvement of all government parties in all aspects of the assessment (e.g., the Public Utilities Board in scoping, the federal government in the provincial hearings), one could argue that the assessment, while meeting the terms of the bilateral agreement, was not harmonized. In fact, except in cases of joint review panels, all bilateral agreements negotiated under the auspices of the Canada-Wide Accord on Environmental Harmonization provide for coordinated steps in the EA, rather than a one-window hybrid assessment approach. Thus, the entire process might better be termed 'cooperative' (as the bilateral agreements are titled) rather than 'harmonized' (as the sub-accord is titled). Participants in EA for the Wuskwatim projects, in fact, expected a one-window approach. They expected not only that the federal and provincial EAs would be coordinated, but also that parties would be active in each stage and would come to common, timely decision points. However, while increased federal-provincial harmonization was sought for and attained to some extent, the same cannot be said for the two provincial reviews. Harmonizing the provincial EA and the review by the Public Utilities Board was significantly less successful, particularly since harmonization was initiated so far along in the review process and the 'new' process included quasi-judicial hearings.

Support mechanisms can be put into place to mitigate many of the negative consequences for public participation. For example, participant funding could be expanded to include all aspects of the EA, although this would require federal involvement (in terms of both funding and administering) in the program. Second, a workshop on how to be an intervener should be offered by the government team involved in facilitating the EA (see, for example, the program presented for the Sable Gas project (Fitzpatrick and Sinclair 2005)). Third, an improved and shared (across jurisdictions) issues-tracking system should be put in place so that the public could more easily access all the information around an issue and its final status. Finally, and perhaps most importantly, decisions about how to streamline the EA should involve public participants and result in significantly more education of the public following each of the decision points.

Multi-jurisdictional EAs are becoming increasingly important across Canada. In our chapter in the first edition of this book, we noted that that the majority of EAs are not subject to harmonization (Fitzpatrick and Sinclair 2005; Lawrence 1999). While this is true, it is because the majority of EAs at the federal level involve screenings for small localized projects. It is the large projects, which include some screenings, comprehensive studies, and panel reviews, that are increasingly subject to some form of harmonization. Between October 2003 and May 2008, for example, 13 of the 17 panel reviews, and 19 of the 24 comprehensive studies involved multi-jurisdictional EAs. This is a significant investment of both human and financial resources.

The growing importance of multi-jurisdictional coordination is evident through two recent initiatives pursued by the federal and provincial governments. First, in April 2007, the federal government issued a Cabinet Directive on Streamlining Regulation (Canada 2007). The purpose of this directive is to increase the efficiency and effectiveness of regulation, minimize duplication, and increase cooperation and coordination among federal departments and between federal and provincial/ territorial governments. These objectives, of course, reflect those associated with EA coordination.

The second initiative, spearheaded by the Canadian Council of Ministers of the Environment, focuses on federal-provincial/territorial EA coordination. This project is designed to strengthen efforts to streamline EA processes. In the program proposal, it is recognized that 'there are process inefficiencies at every level—provincially, territorially and within federal departments—and these are exacerbated when more than one process applies to a project' (CCME 2008). This being so, the CCME has undertaken to evaluate the current system that is used to coordinate EAs in Canada.

It is clear that in Canada we still have to face many challenges before we will reach the stated goal of EA harmonization, as established by the CCME. At present, redress is most actively pursued through the negotiation of bilateral agreements between jurisdictions; however, four such agreements remain outstanding (those with Prince Edward Island, New Brunswick, Nova Scotia, and the Inuvialuit), and one remains in draft form (with Newfoundland). There is no doubt that drafting bilateral agreements presents difficult legal challenges. Provincial and territorial jurisdictions are reticent about creating what amounts to a new EA law through such agreements, and all jurisdictions want to maintain their current decision-making authority. Governments do not give up the power to make decisions easily, and in many EA situations the public does not want to see decision-making powers eroded, particularly those of the federal government. Thus, the extent to which EA laws have become harmonized through these agreements is questionable; rather bilateral agreements tend to be cooperative plans that are focused on how to proceed. However, that there are now eight bilateral agreements in place is an improvement from where things were at in 2003.

With increasing application of bilateral agreements, the difficulties of this approach are becoming clearer. First, although the agreements are more consistent, significant variability remains surrounding public participation. Second, the EA processes that are harmonized with the federal Canadian Environmental Assessment Act continue to be markedly different. Furthermore, the differences between what the law states and what the agreements provide for make it difficult for proponents and the public to know what to expect. In practice, the federal and provincial governments share information on EA cases under consideration through their respective legal regimes all the time. Problems only arise when a decision may be made under two or more EA laws. Ultimately, experience (by any or all of the participants) plays a significant role in creating process certainty, and the experience of working with a harmonized review varies by jurisdiction. Harmonization of the EA process thus continues to add further layers of complexity to the review of primarily large-scale projects.

While the Wuskwatim case highlights some of the issues surrounding harmonization in Canada, it is worth noting that many other jurisdictions are grappling with similar issues. Internationally, the issue of harmonization is gaining considerable interest and attention (Bjørnøy 2006; Connelly 2006). Three reasons underpin this interest. First, with the almost global adoption of EA as a planning tool, internationally funded projects may trigger multiple processes. In fact, the experience of overlap and duplication in Canada is becoming more common at the international level. Second, there is increasing recognition that large-scale projects may lead to transboundary concerns. The Convention on Environmental Impact Assessment in a Transboundary Context (commonly called the Espoo Convention) establishes the responsibility of nations to assess the international impacts of projects. This convention outlines the obligations of the 'host' country to notify the impacted neighbouring nation of projects and to include its participation in EAs (Connelly 1999). Third, political bodies, such as the European Union, are making efforts to standardize policies across member states. The Aarhus Convention, ratified primarily by countries in Europe, creates a standard for signatory countries about access to information, public participation in decision making, and public recourse (Hartley and Wood 2005).

As we have illustrated, Canada's efforts to avoid duplication and overlap in EA processes through harmonization have had mixed results, particularly with respect to public participation (Fitzpatrick and Sinclair 2005, 2009; Kwasniak 2008; Schneider et al. 2007). What is becoming clear is that the most successful aspects of any effort centre on meaningfully engaging the public in process decisions early on and ensuring that the roles and responsibilities of government authorities are clearly communicated through the review. As demonstrated by our analysis of the EA of the Wuskwatim projects, a bilateral agreement is not enough to achieve these aspects. The assessment might have been more successful had a project-specific agreement been negotiated as well, not only among the federal and provincial regulators, but also with the participation of the public at large. This would ensure better outcomes, for it seems there is always the need for negotiation, even if there is a federal-provincial agreement in place.

Acknowledgements

The authors gratefully acknowledge the participants of the EA of the Wuskwatim projects for sharing their insights and experience. We also thank the three anonymous reviewers, whose comments strengthened the presentation and analysis. Funding for this research was provided by the Social Sciences and Humanities Research Council of Canada and the University of Winnipeg.

Notes

1. *Constitution Act, 1867*, UK, 30 and 31 Victoria, c. 3
2. Exemption has been implemented in two areas: the Mackenzie Valley of the Northwest Territories (whereby the EA process outlined through the Mackenzie Valley Resource Management Act reigns supreme); and Nunavut (whereby the EA process outlined in parts 5 and 6 of the Nunavut Final Agreement reigns supreme). As this is a more recent and more geographically (and politically) focused approach to overlapping EA, with potentially significant implications, it will be discussed in a future publication.
3. The CCME is comprised of the 14 environment ministers, from the federal, provincial, and territorial governments. All members, with the exception of the member from Quebec, approved the accord (Valiante 2002).
4. The Public Utilities Board is involved in regulating aspects of public utilities in Manitoba in the interest of preserving the public good. With respect to Manitoba Hydro, the Public Utilities Board is charged with such responsibilities as setting rates, regulating energy export, etc. As noted earlier, in terms of the Wuskwatim projects, the Public Utilities Board was charged with considering the need for and alternatives to the project—that is, with examining the economic case for development.
5. The EA of the Wuskwatim projects was triggered prior to the 2003 revisions to the Act. Thus, while the revised federal process allows for participant funding for comprehensive studies, this was not part of the federal process under which the Wuskatim projects were assessed. Consequently, no participant funding was provided for this comprehensive study.

References

Bjørnøy, H. 2006. *Impact assessment: A tool for sustainable development*. Norway: Stavanger.

Canada, Department of Fisheries and Oceans, and Transport Canada. 2005. *Wuskwatim Project Comprehensive Study*. Winnipeg: Department of Fisheries and Oceans Canada.

Canada, Government of. 2007. Cabinet directive on streamlining regulation. Ottawa, ON.

CCME (Canadian Council of Ministers of the Environment). 2008. *CCME action on environmental assessment*. Winnipeg, MB: Report to the Regulatory Advisory Council.

CEC (Clean Environment Commission). 2003. Wuskwatim generation and transmission project pre-hearing conference transcript. Winnipeg, MB: CEC.

———. 2004a. *Report on public hearings: Wuskwatim generation and transmission line projects*. Winnipeg, MB: CEC.

———. 2004b. Wuskwatim generation and transmission project hearing transcript. Winnipeg, MB: CEC.

Connelly, R.G. 1999. The UN convention on EIA in a transboundary context: A historical perspective. *Environmental Impact Assessment Review* 19:37–46.

————. 2006. Acceptance speech for the Rose-Hulman Award. Norway: Stavanger.

Fafard, P. 2000. Groups, governments and the environment: Some evidence from the Harmonization Initiative. In P.C. Fafard and K. Harrison (Eds), *Managing the environmental union: Intergovernmental relations and environmental policy in Canada*, 81–101. Kingston, ON: School of Policy Studies, Queen's University.

Fitzpatrick, P. 2006. The environmental assessment process-learning nexus: A Manitoba case study. *Prairie Perspectives* 9 (1): 1–30.

Fitzpatrick, P., and Sinclair, A.J. 2009. Multi-jurisdictional environmental impact assessment: Canadian experiences. *EIA Review*. doi:10.1016/j.eiar.2009.01.004b.

————. 2005. Multi-jurisdictional environmental assessment. In K. Hanna (Ed.), *Environmental impact assessment: Process and practice*, 160–84. Toronto, ON: Oxford University Press.

————. Forthcoming. Multi-jurisdictional environmental impact assessment: Canadian experiences. *EIA Review*.

Harrison, K. 1996. *Passing the buck: Federalism and Canadian environmental policy*. Vancouver, BC: University of British Columbia Press.

Hartley, N., and C. Wood. 2005. Public participation in environmental impact assessment: Implementing the Aarhus Convention. *Environmental Impact Assessment Review* 25 (4): 319–40.

Hessing, M., M. Howlett, and T. Summerville. 2005. *Canadian natural resource and environmental policy: Political economy and public policy*. 2nd ed. Vancouver, BC: University of British Columbia Press.

Kennett, S. 1993. Interjurisdictional harmonization of environmental assessment in Canada. In S. Kennett (Ed.), *Law and process in environmental management*, 297–318. Calgary, AB: Canadian Institute for Resources Law.

————. 1997. Boundary issues and Canadian environmental legislation. In L.K. Caldwell and R.V. Bartlett (Eds), *Environmental policy: Transnational issues and national trends*, 131–55. Westport, CT: Quorum Books.

————. 2000. Meeting the intergovernmental challenge of environmental assessment. In P.C. Fafard and K. Harrison (Eds), *Managing the environmental union: Intergovernmental relations and environmental policy in Canada*, 107–31. Kingston, ON: School of Policy Studies, Queen's University.

Kwasniak, A. 2008. Harmonization in environmental assessment in Canada: The good, the bad and the ugly. Unpublished manuscript, Ottawa, ON.

Lawrence, D. 1999. *Multi-jurisdictional environmental assessments*. Ottawa: Canadian Environmental Assessment Agency.

McKenzie, J.I. 2002. *Environmental politics in Canada: Managing the commons into the twenty-first century*. Don Mills, ON: Oxford University Press.

Schneider, G., A.J. Sinclair, and L. Mitchell. 2007. Environmental assessment process substitution: A participant's view. http://www.cen-rce.org/eng/caucuses/assessment/docs/Final%20Substitution%20Paper%20March29.pdf.

Sinclair, A.J., and M. Doelle. 2003. Using law as a tool to ensure meaningful public participation in environmental assessment. *Journal of Environmental Law and Practice* 12 (1): 27–54.

Strachan, L. 2004. Correspondence. In Clean Environment Commission (Ed.). Winnipeg, MB: Public Registry, Province of Manitoba.

Valiante, M. 2002. Legal foundations of Canadian environmental policy: Underlining our values in a shifting landscape. In D. VanNijnatten and R. Boardman (Eds), *Canadian environmental policy: Context and cases*, 3–24. 2nd ed. Don Mills, ON: Oxford University Press.

Chapter 10

Harmonization and Efficacy: What the 2010 Winter Olympics Tell Us about Environmental Impact Assessment in Canada

Dan Kellar and Kevin S. Hanna

The two-week spectacle that is a modern Olympics inevitably has impacts on the economic, social, and environmental systems of the host city. So it might follow that when a nation such as Canada—one with a long history of provincial and federal environmental impact assessment (EIA) application—hosts a winter Olympics, as it will in 2010 at Vancouver and Whistler in British Columbia, the role of EIA would be showcased as a key planning mechanism. After all, EIA helps ensure that the Olympic Games yield the best possible benefits for the host communities.

This chapter briefly outlines the application of EIA to the 2010 Winter Olympic Games. The games will require the development or redevelopment of 18 specific sites and two large-scale linear facilities. Here we use the 2010 context to reflect on challenges in Canadian EIA. While we draw from a recent study of EIA and the 2010 Winter Olympics, our aim is to provide a critical reflection on the efficacy of an EIA grounded in a present-day large-scale undertaking, one with unique time and political pressures.

Ideally, impact assessment could help make the lasting legacy of the Olympics one of environmental responsibility, positive social impacts, sustainable community benefits, and sustainable design. Indeed, EIA could provide the venue through which planning and the environment intersect and the opportunities to improve Olympic projects are realized. But, if anything, the 2010 Olympic experience illustrates, several pervasive challenges in EIA—the tendency to see it as perfunctory—a sort of green stamp of approval; or on the other hand as a process that stands in the way of development—something to be avoided if at all possible. Many of the issues evident in the 2010 example are hardly new. Others have noted that differences in federal and provincial regulations for environmental assessment contribute to bureaucratic gridlocks, projects being questionably approved or delayed, and confusion within agencies, and among industrial stakeholders, and ordinary citizens (e.g., Marsden 1998; Diduck and Sinclair 2002; Gibson and Hanna 2005).

The Olympics bring several interesting and unique variables into a development process—a non-negotiable timeline, global attention, and national pride. The International Olympic Committee (IOC) has articulated an environmental vision (the 'third pillar') and expects it to be reflected in event venues. But after the games are awarded and planning and development begin, the IOC has no real power to enforce its environmental vision. The IOC statement of environmental quality objectives is in part a reflection of larger political and social pressures. The Vancouver Olympic Organizing Committee (VANOC) has sought to present itself as seeking sustainability in its development decisions, an interesting objective for a temporary event and a momentary entity.

Hosting the Winter Olympics was weakly approved by Vancouver voters, and the planning process has seen some environmental controversy. But in many respects environmental issues have not been the primary focus of the criticism of the games. Instead opposition has focused on the social impacts of hosting the event. Other issues have also emerged, such as increasing overall costs, the impacts on government services as British Columbia attempts to pay for the games, municipal loans to pay for Olympics-related developments, and now the high prices and the controversial allocation of tickets for the events. As for EIA, its role has in many respects been part of a process of building public good will. Even though an EIA was not legally required for all 2010 projects, Olympic organizers made the commitment to use EIA to assess all of the projects. The task of EIA in the preparation for the 2010 Winter Games has been to provide the environmental review function, and this is an important for the games' public image, since the Vancouver organizers have pointed out that 2010 will be a 'green games'.

High achievement in sport, culture, and environmental sustainability are objectives of the Olympic movement. This came about gradually, perhaps in no small part due to the controversy over the 1992 Albertville Winter Games in France. There, some venues were contentious because of the destruction of alpine habitats. Certainly the IOC has also become cognizant of the growing global interest in environmental issues and the prominence of sustainability as a policy concept. The adoption of the IOC's third pillar—the environment—is a response to the greater global social concern for the state of the environment worldwide. The Lillehammer Winter Games, in Norway, gave the environment a prominent place in planning largely because of the importance residents placed on environmental quality (Myrholt 1996). In some respects, Lillehammer set new expectations for venue planning. But it was Sydney, in 2000, that hosted the first declared 'green games', and subsequent games have worked to promote this image. In 2008, reports from the Summer Games in Beijing often made reference to that city's legendary poor air quality. While China had promised to 'green' Beijing in time for the Olympics, many might say its efforts fell short, and this, along with political issues, will be part of the Beijing Olympic memory.

Our Approach

In 2007, Kellar interviewed 14 stakeholders involved in the 2010 EIA process to help in the general understanding of the role of EIA in the 2010 Winter Games. These people represented federal and provincial government agencies, VANOC, First Nations interests, local governments, private sector consultants, and environment and social equity organizations. This was a small elite group, intimately involved in the Olympic process, who worked as advocates, project managers, and EIA administrators, or served as critics and observers. Each individual was well connected to the EIA or the impact contexts, and all were involved in, or keenly aware of, the Olympic planning process. Since our respondents provided candid observations and information, we do not identify their comments by employment category.

We also examined specific EIA reports and documents, as well as site-planning documents. We considered factors such as the role of harmonization and interjurisdictional collaboration, cumulative effects, the place of EIA in assessing broadscale events, and the efficacy of EIA in planning for a time-sensitive project or for a collection of projects. Each issue relates to the broader practice of EIA across Canadian jurisdictions. The role of EIA described here provides a snapshot of practice in Canada. In this book, however, chapter 14 explains the federal system and chapter 15 describes the BC process in more detail, so we do not dwell on describing those systems.

The Setting

The area between Vancouver and Whistler is called the Sea-to-Sky Corridor, a reference to Vancouver's ocean-side setting and Whistler's mountain location. Most of the Nordic and other snow events will be held at Whistler, while arena-based events will be held in Vancouver. Vancouver is largely a flat area, bounded by the sea and the Fraser River on the west, north, and south, with the Fraser River Valley forming the eastern edge. Whistler is located in a valley at the base of several large mountains about two hours by car north of Vancouver. The rugged landscape accounts for Whistler's tourism-based economy. Indeed, the community has no other economic base; nor does it have a history beyond tourism.

The Vancouver Olympic venues are mainly being constructed at sites that have already been developed, although the athletes' village and the primary media centre require some sea infill to extend their area. One competition venue is located between Vancouver and Whistler—the Cypress Mountain site for snowboard halfpipe and freestyle skiing competitions. The Sea-to-Sky Highway was a relatively narrow road and required expansion in preparation for the Olympics. The light rail system (LRT) is being developed between Vancouver's airport and the city core. These two transportation developments, while not acknowledged as Olympic projects, were certainly hastened, some would say even built, just for the Winter Games.

Highway expansion between Squamish and Whistler has pushed through forests, narrow valleys, and mountain areas. This may be the most significant project, in impact terms, related to the Olympics. The Whistler Nordic Centre, located in the unique Callaghan Valley, is at an elevation that holds deep snow well into the spring months. This valley backs onto the Pemberton Glacier. The other developments around Whistler are at varied elevations and have a range of latent landscape impacts, from minor effects to potentially significant cumulative impacts on wildlife habitat and alpine/subalpine systems. All the developments north of Vancouver involved the removal of some forest, and the Sea-to Sky Highway expansion required blasting works and significant new road infrastructure. The new LRT line is being built through established urban areas and connects to the existing LRT system.

Harmonization

Since both the BC and Canadian governments have a role in the Winter Olympics, both the provincial and federal EIA systems have potential applicability, although, as we will see, one certainly has been applied more than the other. The federal and BC governments have an agreement to harmonize their EIA processes when an undertaking is subject to both the Canadian Environmental Assessment Act (CEAA) and the BC Environmental Assessment Act (EAA). The reasoning behind harmonization is that it saves time, money, and other EIA resources, and that it ultimately results in a more efficient process. A joint panel reviews large applications, and proponents develop a single EIA that adheres to the requirements of each process. The ideal in practice is that the single EIA conforms to both federal and provincial laws while eliminating overlap and potential redundancy. Table 10.1 provides an outline of Olympic venues and the status and jurisdictional application of EIA.

The Canada–British Columbia Agreement for Environmental Assessment Co-operation (1997 and 2004) provides the framework for federal/provincial cooperation in EIA. Five Olympic developments have been assessed under a harmonized process. Four of these related to the EAA and the CEAA, while one was a joint CEAA and BC Parks review, with no formal EAA involvement. Only two of these sites are competition venues—the Whistler Nordic Centre and the Cypress Mountain venue. The Cypress Mountain site was jointly assessed through the CEA Agency and BC Parks. Cypress Mountain is located within a provincial park and hence would have been subject to BC Parks' own EIA process, but since there was a federal 'trigger' (funding), the two levels of government developed a specific and unique harmonization process. Some respondents commented that the BC Parks process has a greater conservation and landscape impact sensitivity than the kind of EIA review process typically done under British Columbia's EAA.

Table 10.1 2010 Winter Olympics Venues and EIA Status

Venue Name and Location	Basic Stats (development type; sport; considered an Olympic development [yes/no])	Impact Issues	EIA Decision; Type; Trigger(s); Requirements; Significance	EIA Process Applied
Richmond Oval (Richmond)	Competition venue; long-track speed skating; yes	Brownfield redevelopment on edge of marine waters	Approved; screening; mitigation funding; effects not likely adversely significant	CEAA
Hillcrest/Nat Bailey Arena (Vancouver)	Competition venue; curling; yes	Reconstruction of existing facilities	Approved; screening; mitigation; funding; effects not likely adversely significant	CEAA
Trout Lake (Vancouver)	Competition venue; hockey; yes	Reconstruction of existing facilities	Approved; screening; mitigation; funding; effects not likely adversely significant	CEAA
Killarny (Vancouver)	Competition venue; hockey; yes	Reconstruction of existing facilities	Approved; screening; mitigation; funding; effects not likely adversely significant	CEAA
Cypress Mountain (West Vancouver)	Competition venue; freestyle skiing, snowboarding; yes	Endangered habitat; provincial park	Approved; screening, harmonized (BC Parks/CEAA); mitigation; funding and BC Parks; effects not likely adversely significant	CEAA and BC Parks
General Motors Place (Vancouver)	Competition venue; hockey; yes	Renovation of existing facility	Not started	Pending

Continued

Table 10.1 Continued

Venue Name and Location	Basic Stats (development type; sport; considered an Olympic development [yes/no])	Impact Issues	EIA Decision; Type; Trigger(s); Requirements; Significance	EIA Process Applied
Whistler Nordic Centre	Competition venue; cross-country skiing, biathlon, and ski jumping (long jump); yes	Endangered species and habitat; cultural space; water quality	Approved; screening, harmonized; mitigation + follow-up; funding, Nav Waters PA, VANOC opt-in to BC EAA; effects not likely adversely significant	Harmonized by opt-in by VANOC
Whistler Sliding Centre	Competition venue; luge, bobsled, skeleton; yes	Water quality; impact on wildlife	Approved; screening; mitigation; funding; effects not likely adversely significant	CEAA
Whistler Creekside	Competition venue; downhill ski racing; yes	Stream issues; impact on wildlife	Approved; screening; mitigation; funding; effects not likely adversely significant	CEAA
UBC Winter Sports Arena (Vancouver)	Competition venue; ice and sledge hockey; yes	Renovation of existing facility	Internal	VANOC self-assessment
Pacific Coliseum (Vancouver)	Competition venue; figure skating and short track; yes	Renovation of existing facility	Not applicable	VANOC self-assessment
BC Place (Vancouver)	Non-competition venue; opening and closing ceremonies; nightly medal presentations; yes	Renovation of existing facility	Not started	Pending
Whistler athletes' village	Non-competition venue; residence for athletes; yes	Water quality; cultural issues; impact on wildlife	Approved; screening; mitigation; funding, Nav Waters PA; effects not likely adversely significant	CEAA

Project	Description	Concerns	Status	Assessment
Whistler ceremony plaza	Non-competition venue; awards and celebration plaza; yes	Impact on wildlife; water quality	Not started; funding	Pending
Whistler media centre	Non-competition venue; media broadcasting; yes	Renovation of existing facility	Not applicable	None needed
Main Media Centre (Vancouver)	Non-competition venue; media broadcasting; venue by VANOC but not development for EIA	Water quality (extends into harbour); impact to water animals and habitat	Approved; screening, harmonized; mitigation; funding, Nav Waters PA, Fisheries Act, Prov Rev Projects Reg; effects not likely adversely significant	Harmonized
Vancouver athletes' village	Non-competition venue; residence for athletes; yes	Water quality (extends into harbour); impact on water animals and habitat; Brownfield redevelopment.	Approved; screening; mitigation (limited); funding, Nav Waters PA; effects not likely adversely significant	CEAA
Whistler Nordic rec trails	For cross-country skiing and hiking purposes; a Legacy Project; EIA was split off from the Nordic Centre	Endangered species and habitat; cultural space; water quality	Approved; screening, harmonized; mitigation; funding, Nav Waters PA; effects not likely adversely significant	Harmonized
Canada Line; Sky Train (Richmond, Vancouver)	Public infrastructure; connects YVR to downtown; not by VANOC	Water quality; impact on wildlife	Approved; screening, harmonized; mitigation; funding, CEPA, Nav Waters PA, Fisheries Act, proponent BC EAA opt-in; effects not likely adversely significant	Harmonized
Sea-to-Sky Highway (North Vancouver to Whistler)	Public infrastructure; main highway route connecting Whistler and Vancouver; not by VANOC	Water quality; impact on wildlife; impact on residents, endangered species, and habitats	Approved; screening, harmonized; mitigation; funding, Fisheries Act; effects not likely adversely significant	Harmonized

The EIA for the Whistler Nordic Centre was not, strictly speaking, subject to the EAA, but it was subject to the CEAA. This was one of the projects that VANOC asked to be subject to both processes. One of the attractions to bringing the BC system into this review was the role of timelines in the EAA. Certainly, VANOC wanted to ensure that the EIA was completed within a set time period, which the federal process could not stipulate, at least formally. However, the Whistler Nordic Centre EIA was eventually split into two parts: (1) the Nordic Centre and access roads and (2) the Legacy Recreational Trails. The rationale for splitting the project was that the recreational trails portion of the project needed more time for consultation during the design and planning stage and more time for information to be collected to ensure an adequate impact assessment, but the federal process had no provision for allowing a staged or postponed assessment when making a judgment about the significant effects of a project (CEA Agency 2005a). Splitting the EIA for the project would allow the Olympic portion (the Nordic Centre) to proceed, so that it could be completed on time, while the recreational trails would be assessed later. Splitting the EIA made sense to some respondents, since the recreational trails would not be used for the Olympics and thus did not have to be completed in time for the games. The trails are considered a 'legacy'. The Nordic Centre is a substantial project with important implications for the Callaghan Valley, the undeveloped area in which it is being built. From a landscape viewpoint, and indeed simply from a best-practice perspective, splitting the project weakens EIA's role in developing the best possible image of collective and cumulative impacts. Beyond that, it illustrates the challenges in harmonization, not uncommonly seen, where processes have different timelines, coverage, and postponement rules.

Whether the recreational trails from the main Nordic Centre will be assessed according to their cumulative effects remains to be seen. Some believe that the EIA for the trails was split from the venue to avoid a 'significant potential adverse environmental impacts' decision, which would have slowed construction. Seen separately, the impact image of the trails appears less substantial, and the separation weakens the potential for developing a cumulative perspective. The Canadian Environmental Assessment Agency (CEA Agency) expressed its concern over this change in a letter to the BC government: 'Changing the scope of a project while under review causes considerable concern for the Agency ... possible environmental effects of the project may be missed if assessment of the LRT [Legacy Recreational Trails] is not included in the CEAA [CEA Agency] screening' (CEA Agency 2005b, 1). In the same letter, the CEA Agency noted that a new EIA for the trails would later be required to 'ensure that environmental impacts including cumulative effects of the LRT in relation to the rest of the project be fully reviewed' (CEA Agency 2005b, 2). The practice challenge here, brought about in part by the nuances of harmonization, is that while we can account for some cumulative impacts after venue construction, the opportunity for a more comprehensive EIA, one that considers the project as whole, is in essence lost.

In our interviews, the opinions expressed about the efficacy and compatibility of the harmonized approach vary. Some respondents believed that the harmonized process was effective in identifying and reducing impacts, since the two systems brought different perspectives and criteria to bear on impact identification. Most held that the two systems have worked well together despite substantial differences in coverage, scoping, and screening approaches. The Cypress Mountain EIA was noted as 'working well across two processes', although this EIA had no stressors. But others did not see harmonization as particularly effective. One respondent suggested that 'if nothing . . . they were clearly two different processes running almost entirely separately with little coordination,' and 'the Nordic Centre stands as an example.'

In Table 10.1, the importance of the CEA Agency is apparent. The CEAA is more likely to be required for 2010 projects. In most instances, federal funding is the primary trigger for the application of the Canadian Environmental Assessment Act. Other federal areas of responsibility include navigable waters, fisheries, and transportation, and these also act to trigger the Act. BC's Environmental Assessment Act, it seems would not have applied to any direct Olympic development. Without VANOC's early commitment to use EIA, it is likely that only through federal participation would there have been any meaningful application of EIA to the 2010 Winter Olympics. Some ancillary projects, such as highway improvements, would likely have been subject to EIA regardless of their Olympic connections, but these are considered by VANOC and the respective governments as not really being Olympic projects.

The complexity in harmonizing EIA is an ongoing challenge for the CEA Agency. Even with a harmonization agreement in place, conflicts emerge, as in the Nordic Centre example. Differences will arise naturally in a nation with 14 EIA systems and 14 divergent points of view on the benefits of harmonization. Approaches to participation, intervenor funding, scoping, coverage, and strategic and cumulative effects assessment differ greatly, and even in harmonized contexts, these can be difficult to merge. The problem is that it becomes easier to meet the lowest common denominator than to strive to practise the best possible approach.

While harmonization was largely seen as positive—indeed, essential given the time imperative in the hosting of the Winter Olympics—there was the view that, at best, harmonization provided a framework for coordinating two systems, that it failed to provide an integrated inter-jurisdictional approach to EIA, and as the Nordic Centre case suggests, even modest stressors make harmonized EIA difficult to maintain.

Effectiveness

Only one of the 13 Olympics-related developments required follow-up, which is better than the national average, and none have gone beyond the screening stage. The screening stage entailed the highest level of assessment ever reached by any Olympic project.

In our conversations, respondents provided a somewhat ambivalent image of EIA efficacy and application. Half of those we spoke with suggested that that EIA was valuable in reducing the environmental impacts of some of the Olympic developments. Agency respondents and Olympic proponents were the most positive about the EIA role—perhaps not surprisingly. One benefit of EIA that was noted was its role in bringing the perspectives of First Nations into the planning process. However, some suggested that even in the absence of EIA this would have happened, given the prominent role of 'a First Nations consultation culture within BC development and planning.' Indeed, some interviewees suggested that the First Nations' greatest influence on Olympics planning remains unseen and is not really at odds with the overall development and growth objectives that accompany the Olympics. It was also noted that this influence was advanced outside the EIA rubric.

Another positive example, at least initially, was the role of EIA in the Whistler Nordic Centre and the related identification of impacts on grizzly bear habitat and subsequent mitigation actions. While some respondents noted that EIA resulted in changes to this site that helped conserve bear habitat, others were a bit more measured in their praise. It was noted that British Columbia's Land Resource Management Plan (LRMP) process had already identified the Callaghan Valley as important to grizzly bears, and that the aim of that earlier process was to see that the area remained relatively untouched. Notwithstanding the evidence of the importance of the area to bears, Olympic development proceeded. Contrary to macro-habitat planning objectives, the Whistler Nordic Centre will bring more people into the area, will result in trail construction, and will see the construction of a permanent facility. The overall development changes are not compatible with long-term bear management objectives and would seem to contravene the objectives of the LRMP, a highly participatory and long-range planning process. As for EIA, some mitigation measures were certainly adopted, and these helped improve the project, but there was 'never any question that the Nordic Centre would be built,' and thus the LRMP objectives have been put aside at this locale. Olympic projects may also displace other established plans, or alter planning objectives already established by the province or by local governments.

The efficacy issue inherent in the Nordic Centre review relates to the relationship between EIA and other planning processes. While the EIA result does not invalidate the LRMP results and recommendations, it might be seen as legitimizing the decision to—in essence—ignore them. On the other hand, the role of EIA can be seen as providing an opportunity to ensure that the conservation objectives of other processes are considered in the planning process, and some would interpret the EIA function in this example as doing just that—essentially identifying the habitat issue (or reinforcing what the LRMP had already noted) and proposing mitigation measures.

Some questioned the practical impacts of EIA in fostering environmental protection, noting that most Olympic developments were small in size, with few if any impacts, and in some instances might actually improve the environment; they could, for example, lead to upgraded structures, replacement of inefficient building systems, energy savings, and reductions in waste outputs. Some respondents suggested that the application of building or development standards such as the Leadership in Energy and Environmental Design (LEED) Green Building Rating System would have improved projects 'perhaps even more so' without the use of

EIA. The efficacy of EIA was questioned in terms of its applicability to the nature and scale of some Olympic projects.

Others characterized efficiency in terms of timelines and the efficient use of EIA resources. About half commented positively on the timelines built into the EAA. This, they suggested, provided predictability, certainly with respect to when an EIA would be completed. If anything, timelines were seen as a particularly a desirable feature of the EAA, and it made harmonization an attractive option for overall EIA application, since the federal process does not have timelines. Time emerged as an important theme for those engaged in planning, and an overall thread in all our conversations was the time imperative of site development and the concern that external processes should not unduly interfere with 'getting things ready on time'. The semantic challenge here is that even though it was VANOC that embraced EIA as the *environmental* planning tool, there was an overall tendency among most respondents to see EIA as an external process imposed on Olympic development, rather than as an *integral* planning tool.

Olympic proponents tended to be the most positive about the overall EIA role. They saw it as providing a useful planning support tool—one that helped ensure that a modicum of environmental consideration would be given to site developments. The Cypress Mountain EIA was noted as providing a good review of impacts and providing mitigation and plan amendments.

Overall, those we spoke to appreciated that EIA provided some certainty with respect to approvals and decisions. In other words, while no one expected EIA to radically change a project—and certainly result in a project not being approved—supporters and detractors alike felt that the 2010 role of EIA was to provide a 'sort of green certification' that would allow work to be completed on time and with minimal interference.

The view of effectiveness was qualified by several respondents because of the 'absence of a cumulative perspective' and 'because a broad approach to EIA application was lacking.' Even proponents commented on this. Thus, the perceived efficacy and envisioned potential for EIA to greatly improve the overall approach to Olympics development was seen by some of our respondents as 'unrealized for [the] 2010 context'. It was suggested instead that the inherent value of EIA was its ability to perform a basic role in issues identification after the sites were chosen and the general framework for development was established. It was then that EIA, in addition to identifying issues, could also guide potential mitigation measures. The role of EIA, then, was seen by some as being 'very much after the fact', reactive to issues rather than anticipatory and certainly 'focused on mitigation'.

Concern was also expressed about 'what comes after Olympics'. Specifically, some mentioned the capacity of 'legacy organizations' to manage follow-up or to monitor the operational needs and impacts of the facilities they will inherit after the games are done. Enforceable follow-up provisions were seen as non-existent, and to be sure, not all venues will require such. But for those facilities with long-term environmental impacts, the need to monitor performance and adjust operations may not be adequately accounted for in the EIA process—but then neither the federal nor the provincial system would have a much better capacity to do so.

In many respects, the idea of efficacy reflects the state of EIA practice where prag-
matism trumps the ideal image of EIA as an anticipatory and comprehensive process. The respondents suggested that EIA in the 2010 context can be only as effective
as the somewhat belated and project-specific application allows it to be. Efficacy
issues emerged across their answers. For example, one respondent commented on
the problem of scale and linked it to the adequacy of governing legislation:

> Scoping an environmental assessment to a minute-area scale, while continuing to treat
> potential impacts in isolation, will ensure that significant adverse environmental effects
> will never be discovered and that best practices and environmental protection will not be
> accomplished. The weaknesses discussed in cumulative effects and strategic assessments
> are two of the leading issues regarding the effectiveness of CEAA and the BC EAA. The CEAA
> and EAA should be rewritten with ideals of sustainability and environmental protection
> engrained in every step of the process. Environmental impact assessment . . . needs to be
> more than a bureaucratic checkmark . . . it needs to be a tool used to protect the environ-
> ment from unsustainable development through an adaptive, collaborative, comprehen-
> sive, credible, and transparent process.

Certainly, pressures exist 'to get the work done' and 'on time', as respondents
emphasized. Olympic projects are time sensitive, and so there is an understandable desire to complete all projects on time to avoid international disgrace. As one
respondent said, 'No one wants an Athens', referring to the unfinished venues that
plagued that city's summer games. This requires a quick turnaround time in the
EIA process. Meeting timelines does not, however, mean that that process and quality of EIA have to be sacrificed. Indeed, 2010 provided stakeholders an opportunity
to develop a state-of-the-art EIA, where efficacy could be defined through practice
as being more than the efficient use of resources and quick turnarounds for reviews and decisions, but also as providing project design and implementation with
the best possible impacts—but this would require a comprehensive, cumulative,
and strategic approach, an opportunity missed here.

Cumulative Impacts and a Strategic Vision

The 2010 Winter Olympics will not take place in a single, confined complex. The
sites are spread over a vast landscape, and events will be held in several communities of varying sizes. Cumulative impacts can be difficult to assess, as they require
a complex understanding of interactions—social, economic, and environmental.
The games' most obvious potential cumulative impacts centre on new or expanded transportation systems (between Vancouver and Whistler and the Vancouver
airport and the city centre); the incremental development stemming from venue
creation, notably in the Whistler region; real-estate prices and the transition of
housing stock spurred by the Olympics; and venue redevelopment, notably in the
Vancouver region.

There would seem to be an obvious role for a strategic assessment and certainly
one for a cumulative impact assessment, but neither is present to a great extent in

the 2010 process. We should note that the absence of such comprehensive perspectives was mentioned by respondents across sectors, but some also commented that neither the BC nor the federal EIA system accounts for these to any great degree. The lack of either a cumulative or a strategic assessment reflects a condition of the respective EIA laws and administrations that transcends the application of the two EIA systems to the Winter Olympics. Rather than reflecting a specific weakness, this condition is indicative of general or systemic challenges.

In general, Canadian EIA does not deal well with cumulative effects. Strategic environmental assessment (SEA) remains largely a concept in search of meaningful implementation. There are exceptions, as Creasey and Ross note in chapter 8. In chapter 16, Creasey and Hanna also note that cumulative assessment in Alberta is one of the stronger aspects of that province's EIA system. Noble and Harriman-Gunn, in chapter 6, describe examples of SEA application in Canada, along with the challenges. In chapter 2, Gibson and Hanna also note that strategic approaches are ultimately needed if we are to effectively integrate sustainability into planning and decision-making processes.

When cumulative effects are considered, it seems that the scope is too narrow—spatially, temporally, and thematically—and application is inconsistent. Strategic assessment fares no better. We tend to deal with it in terms of potentials, 'making it tractable', and 'overcoming implementation challenges'. Both strategic and cumulative assessments will require more resources so that the scope can be broadened to include a larger area over a longer amount of time, and both are linked to the emergent notion of sustainability assessment. However, sustainability is not about looking just a few years into the future or at only site-specific impacts; sustainability includes longer temporal time frames and larger spatial settings. Cumulative effects must be assessed beyond the project site. Temporally, the effects have to be assessed well into the future to ensure that the project will not open the door to unsustainable actions. As some respondents suggested, the attention span of Olympic planners 'ends with the closing ceremonies', while the 'legacy', bad and good, will linger well beyond.

Environmental impact assessment reports for the 2010 Olympic projects include few studies that approach an explicit understanding of the cumulative cultural, social, and economic impacts of the games. In our conversations, the games' impact on housing was mentioned, specifically as a focal point for social effects. In fairness, the Olympics have been but one driver in the Vancouver housing boom—property prices have been rising rapidly in recent decades. Housing access and affordability will be affected by the Olympics, but the impacts will be indirect, perhaps incidental, though certainly felt. And while some interviewees described this as a social impact assessment dynamic, which it is, given the existing inaccessibility of housing and the price climate, the cumulative characteristics are substantial and largely unstudied.

Increased land values have helped fuel a conversion and gentrification boom that has affected some of Vancouver's more depressed areas. In one film documentary (Hamilton and Schmidt 2007), we see developers buying low-cost apartment and hotel complexes; evictions follow, and the buildings are soon replaced by high-cost housing and short-term accommodation. Such actions exacerbate homelessness, and homelessness, far from garnering a strategic social policy response, seems to be treated as a crime, where those without shelter face the threat of arrest and removal, though 'removal to where' remains uncertain (a 'no sit, no lie' law). Arrests, as a response to homelessness, beyond being morally problematic, lead to more tax dollars for enforcement and incarceration (Hamilton and Schmidt 2007). Indeed, there are both social and fiscal costs for such policies. Yet a causal link between the 2010 Winter Olympics and such events is difficult to establish absolutely; the variables are complex and rooted in an ongoing history of redevelopment and gentrification. Vancouver has a notable legacy of social problems that the Olympics simply cannot be held responsible for. Ironically, while a global recession looms and real-estate prices are falling in other Canadian cities, the Winter Olympics may actually help insulate Vancouver from falling property prices, at least in the short term.

The occurrence of cumulative effects over the larger landscape may be delayed until well after the Winter Games are finished. Several respondents described an example of this possibility. The Sea-to-Sky Highway expansion will open up new areas, and with the new roads, development will likely ensue. The extent to which other Olympic venues may spur landscape change, improve human access to wilderness areas, and create unanticipated development opportunities has been largely unaccounted for in the EIA process.

The tendency to treat Olympic venues and the ancillary projects as a series of unrelated and isolated developments instead of as an integrated and interrelated system is in part attributable to existing EIA practices in Canada. The lack of integration is not unique to the Winter Olympics, but the games do provide an example of the ongoing challenges.

EIA Resources and Capacity

The availability of resources increasingly limits the work of federal and provincial EIA agencies. The efficacy of EIA is determined in part by the resources and capacity of EIA agencies. Budget cuts are common, and some agencies have lost significant staff resources in recent years. The looming recession will likely see even more cuts to EIA capacity. There is rarely enough staff to follow mitigation requirements, initiate or manage monitoring programs, or ensure that legislated timelines are met. Advancing best practice in EIA and developing strategic approaches cannot be met unless adequate budgets and sufficient human resources are provided for EIA agencies. Public participation, too, requires adequate support on several levels—supporting legislation, agency personnel who are trained to conduct participation programs, and funding for intervenors and related activities. Without public financing for participation, many groups do not have the ability to adequately contribute to EIA proceedings.

In our conversations, respondents discussed the challenges that non-government and non-proponent interests face with respect to informed participation often dependent on a technical capacity to interpret EIA reports and supporting technical documents, time availability, and the need for expert advice or consultant support in the preparation of responses and the assessment of proponent-generated data. Agency respondents noted that these resource needs are perpetual, and for systems with strict timelines in place and the added pressure to complete work in time for the Olympics, some also felt that oversight or appropriate review was not always, as one put it, as substantive 'as it could be'. Many, however, were quick to note that capacity and resources were pan-Canadian issues, but that in the 2010 Olympics these were exacerbated by unique spatial and temporal urgencies. In this respect, respondents across sectors suggested that 'resourcing' posed particular challenges for monitoring and follow-up, although these issues were also seen as being rooted in the administrative and legal structure of the BC and federal EIA systems. Staff consistency was also noted as a problem. The movement of agency staff meant that good knowledge of places, process, development history, and the content of EIA documents was not consistent. Some suggested that agency staff who were perceived as being too stringent in their reviews were transferred. This, it was noted, slowed the process and limited the efficacy of EIA review.

The issue of efficacy in EIA is ongoing. A challenge emerges because of the different ways that effectiveness is conceptualized. Some respondents measured efficacy in terms of following a process, meeting timelines, and having a project move ahead on time. Others looked more critically at the role of EIA in improving projects and considered whether or not EIA makes a substantial contribution to planning and project implementation, or whether it captures larger complex environmental relationships and provides for a longer-term view of impact management. Most respondents, however, agreed that EIA was beneficial in reducing the potential environmental impacts of Olympic development.

Conclusion

The role of environmental impact assessment in the development of 2010 Winter Olympics venues is in many respects indicative of the challenges and common approaches to practice seen across Canada. Olympic organizers embraced EIA as a tool for helping to integrate environmental considerations into the Olympics' development. This reflects a certain respect that EIA has achieved over decades as an established environmental management tool.

Reflecting the comments of respondents, several issues emerge. Harmonization of the federal and provincial EIA systems works well as long as there are no stressors—certainly no major ones. It can unravel when issues become too problematic for one of the systems to accommodate; both then move back to their individual approaches.

Resource issues, staffing, and budgets were seen as limiting factors in the effective application of EIA, and again this was an issue common across jurisdictions. It was felt that significant strategic and cumulative perspectives were lacking, and there was the sense that a substantial opportunity to demonstrate best EIA practices was lost to the imperatives of time and lack of strategic thought. The EIA approach in the 2010 setting was site specific; it lacked connectivity or an integrated vision across venues. Olympic sites were analysed in isolation, and several large infrastructure developments were not included in the environmental vision for the Winter Games—despite the reality that they were being developed for the games. The perceptions of efficacy were diverse, with about half of those we spoke with seeing EIA as an effective environmental planning tool and the other half not seeing it as particularly so. Those who saw EIA as effective tended to focus on the timeliness of process, while the detractors questioned the effect of EIA on planning and its utility with respect to small-scale venues.

So, what does the role of EIA in the 2010 Winter Olympics tell us about Canadian practice? First, there is the caveat that although the games are a complex project with many venues, they are located in one jurisdiction with a role for just two Canadian EIA systems. Regardless, the 2010 EIA experience is helpful for characterizing efficacy. Overall, though, EIA has not been as effective or influential as it could have been in the planning and development for the 2010 games, largely because it was applied to specific sites and was not used in an integrated or strategic way. In practice, the EIA for the games informed site development but rarely planning. While not all Olympic venues needed full-scale EIA attention, there was a sense that the opportunity for greater consideration of complex environmental impacts was lost to the urgencies of time and national pride. Environmental impact assessment has performed for the 2010 setting as well as the respective EIA laws permit or require. It has done no less. But then it has done no more.

References

CEA Agency (Canadian Environmental Assessment Agency). 2005a. Whistler Nordic Centre and separation of the Legacy Recreational Trails for the purposes of review under the Canadian Environmental Assessment Act. CEA Agency, Vancouver.

———— 2005b. Letter from the CEA Agency to the BC Environmental Assessment Office. Subject: Whistler Nordic Centre and separation of the Legacy Trails for the purposes of review under the Canadian Environmental Assessment Act. CEA Agency, Vancouver.

Diduck, A., and A.J. Sinclair. 2002. Public involvement in environmental assessment: The case of the non-participant. *Environmental Management* 29 (4): 578–88.

Gibson, R., and K.S. Hanna. 2005. Progress and uncertainty: The evolution of federal environmental assessment in Canada. In K.S. Hanna (Ed.), *Environmental impact assessment: Practice and participation.* Don Mills, ON: Oxford University Press.

Hamilton, G., and C. Schmidt. 2007. *Five Ring Circus* (documentary film). Vancouver: Rag Tag Productions.

Marsden, S. 1998. Why is legislative EA in Canada ineffective, and how can it be enhanced? *Environmental Impact Assessment Review* 18:241–65.

Myrholt, O. 1996. Greening the Olympics. *Our Planet* 8 (2). United Nations Environment Program.

PART II

Environmental Impact Assessment in Canadian Jurisdictions

Chapter 11

Environmental Impact Assessment in Canada's Northwest Territories: Integration, Collaboration, and the Mackenzie Valley Resource Management Act

Derek R. Armitage

Introduction

Environmental assessment theory and practice have evolved considerably over the last several decades. No longer focused solely on the biophysical impacts of proposed developments, nor devoid of opportunities for meaningful public participation, jurisdictions in both the North and the South have designed and adopted elaborate environmental assessment systems (World Bank 1991; Wood 1995; Gibson 1993, 2001; Lindsay and Smith 2001). Through such efforts, a common set of requirements, conditions, and principles of effective environmental assessment (EA) have been articulated (see chapter 1). In Canada's Northwest Territories (NWT), however, application of principles and best practices for environmental assessment occurs in a dynamic mix of socio-cultural, political, institutional, and economic change. Intense pressures associated with mineral extraction, oil and natural gas development, infrastructure, and tourism activities are having a profound socio-economic and environmental influence (Wismer 1996; Chase 2000; NRTEE 2001). The political and institutional framework in the NWT has likewise been transformed since the early 1990s as a result of the implementation of comprehensive claims agreements negotiated between Aboriginal groups and the federal government (see Reed 1990; Green and Binder 1995; Peters 1999;

Donihee et al. 2000; White 2000).[1] As a consequence of commitments made in the Gwich'in (1992) and Sahtu Dene and Metis (1993) Comprehensive Land Claims Agreements, a new environmental regime has also been established. Proclamation of the Mackenzie Valley Resource Management Act (MVRMA) in 1998 provides one of the more recent examples of a regulatory and institutional framework for integrated resource management and environmental impact assessment (EIA) in Canada (Figure 11.1) (Canada 1998). As required by the Gwich'in and Sahtu claims agreements, the MVRMA establishes an integrated resource planning, management, and assessment framework for the Mackenzie Valley, and has created

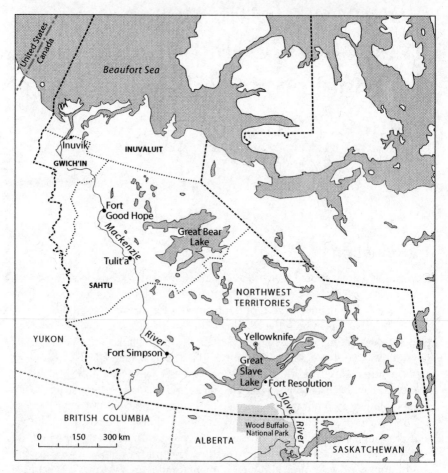

Figure 11.1   Northwest Territories and the MVRMA

In the MVRMA, the Mackenzie Valley is defined as that part of the Northwest Territories bounded on the north by the Inuvialuit Settlement Region (ISR), on the west by Yukon, on the east by Nunavut, and on the south by the 60th parallel, excluding Wood Buffalo National Park. Thus, the geographic extent of the MVRMA includes all of the Northwest Territories, except the ISR and the portion of Wood Buffalo National Park inside the NWT.

new organizational structures and processes to foster greater collaboration among key stakeholder groups in the region.

The purpose of this overview is to highlight key opportunities of and constraints to the implementation of the Mackenzie Valley environmental regime, with a particular focus on the impact assessment process. Although the Mackenzie Valley regime is relatively new, there are two primary reasons why attention to this EIA regime is useful.

First, the analysis illustrates the critical role of environmental impact assessment in shaping the development decision-making process. This is of particular importance in Canada's North because of the intensive resource development pressure and the significant implications for northern communities and ecosystems. The stakes are high indeed (NRTEE 2001). The value of diamond mine production in the Northwest Territories, for example, surpassed $1.7 billion in 2005, nearly three times the value exported in the first year of production in 1999 (Byrd 2006). Diamond production alone is expected to produce billions in taxes and royalties that will accrue to the federal and territorial governments and make significant contributions to growth in GDP per capita in the Northwest Territories. As well, the number of workers employed in the production of diamonds increased by 1,000 in 2003 alone (Byrd 2006). When indirect spinoffs are included in the employment projections, the mining industry is expected to create and maintain approximately 4,000 jobs over the next 25 years (NRTEE 2001). If the proposed Mackenzie Valley pipeline project moves forward, significant economic impacts and expenditures are also expected. Much of this development, however, will take place in sensitive ecosystems and affect the habitat of wildlife (e.g., caribou) and other resources important to First Nations and Inuit.

Second, it is instructive for students, practitioners, and researchers to gain a sense of the opportunities and challenges associated with the development and implementation of an innovative environmental planning and assessment framework. Although the conditions for EA in Canada's North are unique in many respects (see chapter 12), the challenges and opportunities associated with building an EIA process transcend jurisdictional boundaries. This overview of the Mackenzie Valley regime can contribute, therefore, to the development of environmental assessment policy and practice more broadly.

To explore the EA process in the Mackenzie Valley, this chapter is organized into the following sections: (1) an overview of the context within which the MVRMA emerged, with a particular emphasis on the role of comprehensive claims agreements; (2) an outline of the key elements of the MVRMA, highlighting the structures and processes of the EA framework; and (3) an analysis of key opportunities of and constraints to environmental assessment in the Mackenzie Valley and the implications for collaboration and integration. The analysis is based on a review of literature and related documentation pertaining to environmental assessment in the Mackenzie Valley; the results of an EA practitioner workshop held in Yellowknife, NWT, in September 2001 and attended by over 50 individuals (which the author co-facilitated); and eight follow-up interviews conducted with board staff and government officials in 2003.

Comprehensive Claims Agreements and Environmental Assessment in Canada's North

> Having spent the better part of one hundred years pacifying their indigenous peoples and allocating to them what was then perceived as the very worst land . . . the major industrial powers are suddenly finding that these areas . . . cover enormous energy and mineral resources. (Stea and Wisner 1984, 5)

Over the last several decades, governments and industry have increasingly sought to develop resources in areas originally considered unproductive and bereft of significant economic opportunity. Yet, in the 1970s and 1980s, government and industry were not the only groups to recognize the economic potential of northern resources—First Nations communities also recognized the strategic value and importance of natural resources on their lands, and through the difficult and complex establishment of comprehensive claims agreements, they renegotiated the terms under which resource development could take place. For example, the 1975 James Bay and Northern Quebec (JBNQ) Agreement, signed by the Cree and Inuit, established new provisions for environmental assessment and protection. Set in the context of intense hydroelectric power development, the JBNQ environmental regime created mechanisms for greater representation of Cree and Inuit communities in decision making, as well as new opportunities for consultation in the EA process. For the Cree and Inuit, the purpose of these provisions in the JBNQ Agreement was to secure their harvesting rights and protect their way of life (Peters 1999). Then, in 1984, the Western Arctic Land Claim, or Inuvialuit Final Agreement (IFA), led to the establishment of a new EA process, one designed to better represent Inuvialuit concerns and aspirations in the face of oil and natural gas development (see Reed 1990). As articulated in the IFA, a two-stage process for environmental assessment was created, along with the establishment of two 'co-management' bodies to implement the new process—the Environmental Impact Screening Committee (EISC) and the Environmental Impact Review Board (EIRB).[2] This new assessment regime replaced the federal Environmental Assessment Review Process (EARP) and the subsequent Environmental Assessment Review Process Guidelines Order (EARPGO) (see chapter 2) as the primary EIA process in the Inuvialuit settlement area. The environmental assessment process created by the IFA, together with the establishment of co-management boards to implement the process, has provided an innovative model for other EA regimes in Canada's North.

In the Mackenzie Valley, issues and concerns about intensive resource development coalesced around a 1975 royal commission of inquiry on the proposed Mackenzie Valley pipeline, led by British Columbia Supreme Court Justice Thomas Berger (Berger 1977). The proposed pipeline was designed to transport natural gas, and eventually oil, from the Arctic through the Mackenzie Valley, connecting finally with the North American oil and natural gas transportation grid. The scale and scope of the proposed pipeline, however, raised many concerns about the ecological impacts of a major resource development project in areas of exten-sive permafrost and migratory wildlife (i.e., caribou). Aboriginal communities in the Mackenzie Valley also identified many potentially negative impacts of pipeline development on their traditional way of life. A key outcome of the Berger Inquiry was a recommendation not to proceed with pipeline development activities for 10 years, until more comprehensive studies could be undertaken on the potential en-vironmental and socio-economic impacts of the proposed development and, more importantly, until Aboriginal communities in the region had settled outstanding land claims and affirmed their ability to benefit from resource development.

In identifying many issues and incorporating them into the proposed Mackenzie Valley pipeline review process, the Berger Inquiry set high standards for Aborigi-nal participation in resource development decision making and for the integration of economic, socio-cultural, and ecological concerns into environmental assess-ment practice. Yet there remain many unresolved problems associated with the practice of environmental assessment in Canada's North. As Reed (1990) notes, for example, implementation of the IFA environmental review framework was characterized by a number of difficulties, including problems of administrative and jurisdictional overlap, uncertainty about the intent of the screening referral process and of other mechanisms envisioned in the IFA regime, and the limited participation of Inuvialuit communities in the assessment process. Likewise, as Peters (1999) illustrates, experiences with the James Bay and Northern Quebec Agreement environmental regime brought to light a number of difficulties with the establishment of new structures and processes for assessment. For example, the administrative and jurisdictional complexity of the advisory committee struc-tures created to enhance representation and improve participation was not well suited to Inuit or Cree cultural traditions. Peters (1999) also notes that many of the sections of the JBNQ Agreement dealing with environmental assessment were difficult to translate into working principles. Finally, as Wismer (1996) notes with respect to EA experiences in the Mackenzie Valley, the assessment process for the BHP Billiton Ekati Diamond Mine development raised a number of issues, includ-ing the timely provision of documentation (much of it highly technical and in English only) to Aboriginal communities potentially affected by development of the mine; the procedural fairness, rigour, and constructive participation of diverse Aboriginal communities; the degree of emphasis accorded to political, economic, and intergenerational equity issues; the development of subsequent monitoring programs, particularly of social and cultural impacts; and the integration of tradi-tional ecological knowledge in the EA process.

Over the last decade, however, the regulatory, socio-political, and institutional context in which environmental assessment takes place in the Mackenzie Valley has been transformed (Armitage 2005; Fitzpatrick et al. 2008). Comprehensive claims have been settled, while additional 'agreements-in-principle' have been or are being negotiated between the federal government and other First Nations in the Mackenzie Valley. Moreover, two decades of EA experience in the context of increasing pressure for resource development have played an important role in shaping the principles now being applied to assessment in the Mackenzie Valley. To what extent, therefore, does the environmental regime established by the MVRMA facilitate assessment practice? And what are the challenges associated with the implementation of an EA process that seeks to enhance integration and foster collaboration in a complex and dynamic environment?

The Mackenzie Valley Resource Management Act: An Evolving Framework for Integration and Collaboration

The Mackenzie Valley Resource Management Act was proclaimed 22 December 1998 in response to obligations made under the Gwich'in and Sahtu Dene and Metis Comprehensive Land Claims Agreements. Each agreement called for the establishment of new legislation requiring the coordinated regulation of land and water throughout the Mackenzie Valley, both in settlement areas (i.e., those lands covered by the Gwich'in and Sahtu claims) and in adjacent areas in the Mackenzie Valley (i.e., those lands not covered under a comprehensive land claim). Furthermore, clauses in both the Gwich'in and Sahtu agreements called for the establishment of several new decision-making entities or public boards (see Table 11.1) to be responsible for (1) preparing regional land-use plans to guide development decision making; (2) regulating the use of land and water; and (3) carrying out environmental assessments and reviews of proposed developments. Each of these claims-based requirements was incorporated into the design of the MVRMA, creating as a result an integrated and collaborative framework to foster the conservation, development, and utilization of land and water resources; protect the environment from significant adverse impacts; and ensure the social, cultural, and economic well-being of residents and communities in the Mackenzie Valley. The primary components of the MVRMA are summarized below to illustrate how this system is intended to function. More comprehensive reviews of the MVRMA are provided by Donihee et al. (2000) and Donihee (2001) (see also Fitzpatrick et al. 2008). The overview provided here draws upon these reviews but emphasizes only those elements of the Act relevant to this analysis.

MVRMA, Part 1: General Provisions Respecting Boards

Part 1 of the MVRMA outlines the intent, scope, and operational arrangements associated with the various boards responsible for implementing the environmental planning, management, and assessment regime in the Mackenzie Valley (see Table 11.1). Referred to as 'institutions of public government' (IPGs), the boards are independent of the federal government and have many rights, powers, obligations, and privileges of decision making. Moreover, the boards are responsible for hiring their own staff and can make bylaws to govern their internal activities. As stipulated in the land claims process, board members are jointly appointed by the appropriate First Nations and government (federal/territorial) bodies. In total, six boards have been established with the proclamation of the MVRMA, representing both the geo-

Table 11.1   Selected institutions of public government for environmental assessment in the Mackenzie Valley

Board	Primary Mandate	Website
Mackenzie Valley Environmental Impact Review Board (MVEIRB)	Environmental assessment and review throughout the Mackenzie Valley, including the Gwich'in and Sahtu settlement areas	www.mveirb.nt.ca
Mackenzie Valley Land and Water Board (MVLWB)	Issuing land-use permits and water licences for all areas in the Mackenzie Valley outside of the Gwich'in and Sahtu settlement areas; preliminary screening	www.mvlwb.com
Gwich'in Land and Water Board (GLWB)	Issuing land-use permits and water licences within the Gwich'in settlement area; preliminary screening	www.glwb.com
Gwich'in Land Use Planning Board (GLUPB)	Development and implementation of a land-use plan for the Gwich'in settlement area	www.gwichinplanning.nt.ca
Sahtu Land and Water Board (SLWB)	Issuing land-use permits and water licences within the Sahtu settlement area; preliminary screening	www.slwb.com
Sahtu Land Use Planning Board (SLUPB)	Development and implementation of a land-use plan for the Sahtu settlement area	www.sahtulanduseplan.com

graphic scope of the Mackenzie Valley and the First Nations communities in those regions. Because their composition is determined by agreements made within the land claims process, IPGs are often referred to as 'co-management' boards. This implies an arrangement that involves the sharing of power and responsibility among government and First Nation groups and therefore an explicit opportunity for different groups to represent their interests in the decision-making arena. Strictly speaking, however, the boards created by the MVRMA are intended to represent the broader public interest, not the groups that nominate them (e.g., Gwich'in, Sahtu, or government). It is also important to note that the current structure and existence of the boards reflect the outcome of the *settled* comprehensive claims in the Northwest Territories. New IPGs in the North Slave, South Slave, and Deh Cho regions are permitted under the MVRMA and may be established depending on the outcome of ongoing settlement negotiations. The current and future establishment of boards, therefore, is a crucial component of the environmental assessment process in the Mackenzie Valley because of the implications for regional integration and collaborative practice. As Donihee (2001, 5) notes:

The establishment of co-management institutions reflects a deliberate choice made by first nations at land claim negotiation tables. . . . The Gwich'in and Sahtu first nations might have chosen to focus on more complete control over their settlement lands. However, the Mackenzie Valley is also an ecosystem and the first nations in the valley have had long experience with proposals for large scale development which could affect several of the regional settlement areas. The Dene and Metis are also aware that the management of migratory wildlife populations such as barren ground caribou and of water resources, for example, can and will benefit from a broader focus. The MVRMA provides such a focus, in the land and resources context, without threatening first nations governance or the specific management of settled lands.

MVRMA, Part 2: Land-Use Planning

Part 2 provides the framework for the land-use planning process and the establishment of the Gwich'in and the Sahtu land-use planning boards. Both of these boards are responsible for developing a land-use plan in their respective settlement areas and, once these are approved, for ensuring that all proposed developments conform to the plan. Once a land-use plan has been approved by a land-use planning board, it must be submitted for approval to the First Nations authority of the settlement area, as well as to territorial and federal government departments. Part 2 (section 35) of the Act outlines the core principles for land-use planning in both the Gwich'in and Sahtu settlement areas:

(a) the purpose of land use planning is to protect and promote the social, cultural and economic well-being of residents and communities in the settlement area, having regard to the interests of all Canadians;

(b) special attention shall be devoted to the rights of the Gwich'in and Sahtu First Nations under their land claims agreements, to protecting and promoting their social, cultural and economic well-being and to the lands used by them for wildlife harvesting and other resource uses; and

(c) land use planning must involve the participation of the first nation and of residents and communities in the settlement area.

In the Gwich'in settlement area, the land-use planning process evolved from the Mackenzie Delta–Beaufort Sea regional land-use planning initiative. The current Gwich'in Land Use Plan was formally approved in August 2003 (http://www. gwichinplanning.nt.ca/). A draft of the Sahtu Land Use Plan is also available for review (http://www.sahtulanduseplan.com), while the initiation of land-use planning processes in the remainder of the Mackenzie Valley will depend on the outcome of settlement negotiations in the North Slave, South Slave, and Deh Cho regions.

The land-use planning process and the development of a binding land-use plan in each settlement area is a central element of the MVRMA framework and is intended to play an important role in the EIA process. Eventually, all decisions associated with the regulatory and environmental impact assessment process in the

Mackenzie Valley will need to be consistent with an approved land-use plan. As outlined in section 46(1) of the Act,

> [t]he Gwich'in and Sahtu First Nations, departments and agencies of the federal and territorial governments, and any body having any authority under federal or territorial law to issue licenses, permits or other authorizations relating to the use of land or waters or the deposit of waste, shall carry out their powers in accordance with the land use plan applicable in a settlement area.

MVRMA, Parts 3 and 4: Land and Water Regulation

Parts 3 and 4 have established a new land and water management regime in the Mackenzie Valley.[3] Specifically, Part 3 (sections 54 and 56) of the Act established the Gwich'in Land and Water Board (GLWB) and the Sahtu Land and Water Board (SLWB), while Part 4 established the Mackenzie Valley Land and Water Board (MVLWB). The MVLWB has jurisdiction for land and water regulation in those regions of the Mackenzie Valley outside of the Gwich'in and Sahtu settlement areas. Largely because of concerns expressed by First Nations groups in those areas of the Mackenzie Valley with unsettled claims, Part 4 was not called into force until March 2000. Under section 59 of the Act, all land and water boards in the Mackenzie Valley are authorized to issue, renew, amend, suspend, and cancel authorizations for land- and water-use permits in their respective jurisdictions. Land and water boards also have the authority to develop any guidelines and/or policies necessary to support their licensing and permitting activities.

Activities of the land and water boards, however, are not carried out in isolation from the land-use planning or environmental assessment processes. For example, in an effort to foster integration, section 61 of the Act prohibits any land and water board from issuing a licence, permit, authorization, or amendment that does not conform to an approved land-use plan. Furthermore, section 62 ensures that land and water licences, permits, or authorizations comply with all provisions for impact assessment articulated in Part 5 of the Act, including any conditions imposed on the licence as a result of the assessment process (e.g., for mitigation, monitoring). Land and water boards thus play a crucial role in environmental impact assessment, as they act as the principal preliminary screeners of proposed development activities.

MVRMA, Part 5: Mackenzie Valley Environmental Impact Review Board

Part 5 of the MVRMA replaced the Canadian Environmental Assessment Act (CEAA), 1992, as the primary process for environmental impact assessment in the Mackenzie Valley;[4] as well, it provided for the establishment of a three-stage EIA process involving (1) preliminary screening, (2) environmental assessment, and (3) environmental impact review (Figure 11.2). The preliminary screening stage is undertaken largely by the land and water boards—the Mackenzie Valley Land and

Water Board and the two regional panels, the Gwich'in Land and Water Board and the Sahtu Land and Water Board—although a number of other agencies or departments can also serve as the preliminary screeners of a proposed development. The latter two stages—environmental assessment and environmental impact review—are the responsibility of the Mackenzie Valley Environmental Impact Review Board (MVEIRB). Definitions of *development* and *impact on the environment* as outlined in the Act are sufficiently broad and encompassing to ensure that most development activities undergo some level of review prior to the approval of any water licence or land-use permit. In section 111 of the Act, development is defined as follows:

> 'development' means any undertaking, or any part of an undertaking, that is carried out on land or water and, except where the context otherwise indicates, wholly within the Mackenzie Valley, and includes measures carried out by a department or agency of government leading to the establishment of a national park subject to the *National Parks Act* and an acquisition of lands pursuant to the *Historic Sites and Monuments Act*.

Section 111 defines an *impact on the environment* as 'any effect on land, water or air or any other component of environment, as well as on wildlife harvesting, and includes any effect on the social and cultural environment or on heritage resources.'

Proposed projects need not go through all stages of the environmental assessment and review process to receive approval and obtain necessary authorizations and permits. As the MVEIRB's 'Guidelines for Environmental Impact Assessment in the Mackenzie Valley' stipulate, *preliminary screening* is the 'initial examination by a preliminary screener of a development's potential for impact on the environment, and the potential for public concern' (MVEIRB 2001, 18). Preliminary screening, therefore, suggests a limited level of effort and a quick preview of a development proposal to determine if more detailed environmental assessment is required. As outlined in section 125(1) and (2), the thresholds (or 'tests') established to facilitate this decision-making process are fairly sensitive. Preliminary screening bodies need only determine that a 'development *might* have a significant adverse impact on the environment or *might* be a cause of public concern' or, in the case where a proposed development is wholly based within the boundaries of a local government, that the 'development is *likely* to have a significant adverse impact on air, water or renewable resources or *might* be a cause of public concern.' In contrast, *environmental assessment* under the MVRMA is defined as the examination by the MVEIRB of a development's potential for impact on the environment, including social, economic, and cultural environments, and its potential for public concern. In this regard, environmental assessment is considered to entail a more detailed evaluation of potential impacts and mitigation requirements, as well as of the changes in the environment resulting from a development, and an evaluation of the costs and benefits prior to the determination of whether the development should be allowed to proceed (MVEIRB 2001). The *environmental impact review* is the last stage in the process and can be carried out by the MVEIRB if a development appears likely to be a cause of significant public concern. Typically, an impact review is conducted by a panel comprised of MVEIRB members and other appointed experts, and it may trigger a federal environmental assessment review depending on the decision ultimately made by the responsible minister. The impact review may include a public hearing in the affected communities, and at the completion of the process, the review board will report the findings to the final decision-making authority—the minister of Indian and northern affairs Canada (see Figure 11.2).[5]

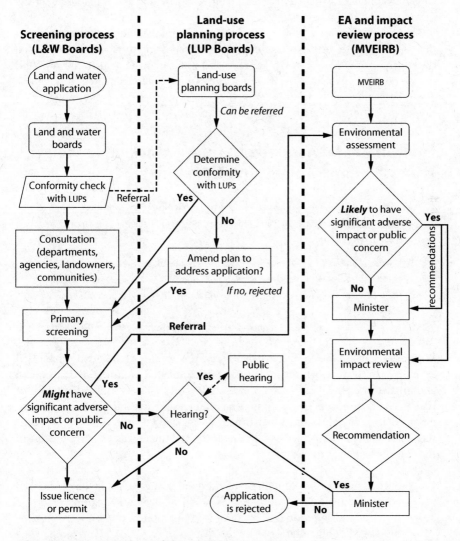

Figure 11.2   Overview of the MVRMA system

Source: Adapted from MVEIRB 2001.

MVRMA, Part 6: Environmental Monitoring and Audit

Part 6 of the MVRMA establishes a system for monitoring and assessment, including the implementation of periodic and independent audits of the land and water management regime. Importantly, the MVRMA also calls for the development of a cumulative impact monitoring framework for the entire Mackenzie Valley. Section 146 of the Act stipulates: 'The responsible authority shall, subject to the regulations, analyze data collected by it, scientific data, traditional knowledge and other pertinent information for the purpose of monitoring the cumulative impact on the environment of concurrent and sequential uses of land and water and deposits of waste in the Mackenzie Valley.'

According to Donihee (2001), this provision is unique and integrates independent auditing with a state-of-the-environment reporting provision to determine the efficacy of the Mackenzie Valley land and water management framework and the EIA process. The provision also makes an explicit call for the integration of scientific data and traditional knowledge when in the development and implementation of the cumulative effects assessment framework.

Environmental Assessment in the Mackenzie Valley: Issues, Opportunities, and Constraints

WHEREAS the Agreements require that those boards be established as institutions of public government within an integrated and coordinated system of land and water management in the Mackenzie Valley;

WHEREAS the intent of the Agreements as acknowledged by the parties is to establish those boards for the purpose of regulating all land and water uses, including deposits of waste, in the settlement areas for which they are established or in the Mackenzie Valley, as the case may be.

—Preamble to the Mackenzie Valley Resource Management Act, 1998

With deep roots in the comprehensive claims process, the MVRMA has established an innovative framework for integrative environmental planning, management, and impact assessment (Fitzpatrick et al. 2008). Nevertheless, implementation of the MVRMA is not occurring without uncertainty, conflict, and difficulty. A number of socio-political, institutional, organizational, and technical issues, opportunities, and constraints influence the evolving environmental assessment process in the Mackenzie Valley, including

1. the administrative complexity of the new framework and the associated politics of institutional and organizational change in the region;
2. issues of coordination and communication among a diverse set of institutional and public interests;
3. challenges associated with the development of common technical approaches to assessment, particularly with respect to cumulative effects assessment;

4. barriers to participation and the challenge of integrating traditional and scientific knowledge in assessment practice; and
5. the capacity of diverse stakeholder groups to participate fully in the environmental assessment process.

While these issues, opportunities, and constraints are specific to the Mackenzie Valley experience, they do serve to illustrate the complexity and uncertainty associated with the construction of an environmental assessment process. The lessons that emerge from the Mackenzie Valley experience, therefore, may prove valuable in other contexts, both in northern Canada and elsewhere.

Administrative Complexity and the Politics of Change

The establishment of the MVRMA framework created a number of uncertainties for the new boards, government departments, industries, and communities in the region. Some of these uncertainties are resolved in due course as roles, responsibilities, and working relationships are clarified. Other uncertainties suggest more difficult issues that must be successfully dealt with to ensure the efficacy of the environmental assessment process. For example, the three-stage preliminary screening and EA process itself was identified by practitioners as a source of uncertainty for organizations, agencies, and project proponents in the Mackenzie Valley. As previously noted, the subjective 'might' or 'likely' tests in section 125 of the Act made the interpretation of the scope and significance of the preliminary screening process, and its point of transition to environmental assessment, problematic at times. The land and water boards, for instance, may consider these tests to be 'overly sensitive triggers' that do not fully account for the decisions that preliminary screeners (e.g., land and water boards, federal and territorial government departments) are required to make (see Terriplan/IER 2001). Land and water boards have a dual mandate that involves initial impact screening as well as regulation of activities following project approval (i.e., providing water licences and land-use permits). As a result, some of the responsibilities of land and water boards are not clearly reflected in the section 125 threshold, such as providing opportunities for community consultation and ensuring that communities possess the level of information they need to participate in the regulatory process.

Interpretations of the MVRMA (notably section 125) that remove decision-making authority from those regional boards (i.e., land and water boards) situated most closely to the communities they serve can undermine the intention of the collaborative framework and hinder opportunities for more constructive participation. What initially appears to be a simple mechanism to facilitate EA practice has on occasion been transformed into a much broader conflict about the mandate and authority of the different boards, capacity issues, different perceptions about how best to implement the Act, and the need of all participants in the assessment process to be empowered (Terriplan/IER 2001). As a result, the roles and responsibilities of different organizations as outlined in the MVRMA are not always reflected in the interpretations of those roles and responsibilities by the staff and members of the various boards.

The challenge of differentiating between screening and assessment, however, is not unique to the Mackenzie Valley. Several jurisdictions across Canada have, or are in the process of developing, criteria or regulations that seek to help determine the level of effort required at screening and assessment stages (Table 11.2).[6] Several ways to address this challenge in the Mackenzie Valley context were discussed at a practitioner workshop (Terriplan/IER 2001). The range of options was seen to include

1. considering development of an 'inclusion list' to establish groupings of small and/or similar projects (e.g., class screenings);
2. reviewing and modifying the criteria-based tables in Appendix F of the revised MVEIRB guidelines (2001) to make them more specific, defensible, and of value to screening practitioners;
3. developing a 'type list' of the activities that might be referred directly to the environmental assessment stage; and
4. exploring the potential of developing a regulation to identify those projects that would be subject to a screening or environmental assessment.

The consensus was, however, that the creation of new guidelines, criteria, and regulations could further reduce flexibility for preliminary screeners and enhance the already significant administrative complexity (perceived and real) of the MVRMA process. Although resolution of this issue emerges through experience, the importance of allowing screeners discretion and the opportunity to utilize their significant professional judgment was identified as a critical aspect of the collaborative environmental assessment process (Terriplan/IER 2001).

Other sections in the Act have also created a degree of administrative complexity and uncertainty. As stipulated in section 126(2) of the MVRMA, notwithstanding any determination of a preliminary screening undertaken by a land and water board, the MVEIRB is required to conduct an environmental assessment of proposed developments referred to it—for example, by the Gwich'in and Sahtu First Nations—when development activities may be carried out on settlement lands. The MVEIRB must also undertake an environmental assessment of proposed developments referred to it by a local government in situations where activities will be undertaken within local administrative boundaries *or* by an agent of the federal or territorial government. Once again, the threshold for referral in such situations, where there *might* be an adverse effect, is not supported by specific guidelines. There are, as well, no clear timelines associated with the section 126(2) clause, and conceivably, proponents may receive permits from a land and water board to begin development activities yet still face a period of uncertainty because of the risk of a future referral to the MVEIRB from another organization or interest that may wish to delay a proposed project (Donihee et al. 2000; Terriplan/IER 2001). In the past, practitioners have raised concerns about the timeliness and efficiency of the process (e.g., about time spent developing terms of reference, time frames available for screeners to make review comments) (see MVEIRB 2000), although the evolving roles and responsibilities of participants in the process are recognized as the

Table 11.2 Frameworks for environmental impact assessment in selected jurisdictions

	Canada	Northwest Territories	Inuvialuit Settlement Region	Nunavut	Yukon
Overview of process	• Three stages: (1) screening; (2) comprehensive study; (3) public review (panel review/mediation)	• Three stages: (1) preliminary screening; (2) environmental assessment; (3) environmental impact review	• Two-step process: (1) environmental impact screening; (2) environmental review	• Two-step process: (1) screening; (2) review (by the EIRB or a federal review panel under CEA Agency)	• Five steps or processes: (1) designated office screening; (2) designated office review; (3) executive committee screening; (4) panel review; (5) final decision by decision body
Implementing organizations and responsible authorities	• Canadian Environmental Assessment Agency (CEA Agency) • Responsible authorities (federal government agencies)	• Mackenzie Valley Environmental Impact Review Board • Mackenzie Valley, Sahtu, and Gwich'in Land and Water Boards	• Environmental Impact Screening Committee • Environmental Impact Review Board	• Nunavut Impact Review Board (NIRB) • Nunavut Planning Commission (NPC)	• Yukon Environmental and Socio-Economic Assessment Board (YESEAB) • Designated offices • Executive committee • Decision bodies (regulators)
Legislative driver	• Canadian Environmental Assessment Act (1992)[2]	• Mackenzie Valley Resource Management Act (1998)	• Inuvialuit Final Agreement (1984)[1]	• Article 12 of the Nunavut Land Claims Agreement (1993)	• Proposed/draft Yukon Environmental and Socio-Economic Assessment Act (YESSEA)
Useful links	• CEA Agency (www.ceaa-acee.gc.ca)	• MVEIRB (www.mveirb.nt.ca) • MVLWB (www.mvlwb.com) • GLWB (www.glwb.com) • SLWB (www.slwb.com)	• Inuvialuit Joint Secretariat (www.bmmda.nt.ca) • Impact review process (www.jointsec.nt.ca)	• NIRB (www.polarnet.ca/nirb) • NPC (www.npc.nunavut.ca) • PLANNER (www.planner.nunavut.ca)	• YESEA (www.dapyukon.yk.net)

1. Implemented through the Western Arctic (Inuvialuit) Claims Settlement Act.
2. Following a five-year review process, Bill C–9, 'An Act to amend the CEAA,' was given royal assent 11 June 2003.

Source: Adapted from Terriplan/IER 2001.

basis for such issues (Terriplan/IER 2001). Efforts to streamline the process, such as revising the MVEIRB Guidelines for Environmental Impact Assessment, reduced some concerns related to timing and efficiency. There was also an effort to address issues of administrative complexity in advance of the major EA activities expected with the proposed Mackenzie Valley pipeline project (see Northern Pipeline EIA and Regulatory Chairs' Pipeline Committee Working Group 2002).

There are, finally, more fundamental challenges. Not all groups were supportive of the new legislative framework. The MVRMA results in a significant redistribution of decision-making authority from the federal government (especially Indian and Northern Affairs Canada) to the boards. This transferring of roles and responsibilities has not always been smooth, as different individuals and interests have in some cases sought to retain their roles and influence. First Nations groups with unsettled claims, moreover, have at times expressed concerns about the imposition of the MVRMA environmental regime. Opposition from certain First Nations groups highlights the challenges of developing a collaborative environmental assessment process throughout the Mackenzie Valley region. As other claims negotiations are completed, it is likely that the environmental regime will be modified accordingly and that new boards will be established. Yet regardless of the changes made to the MVRMA environmental regime, the wording and intent of the comprehensive claims will take precedence over existing interpretations of the law (see Donihee 2001). The challenges that will emerge from these negotiations and the likely restructuring of the MVRMA framework will be largely political, not technical. Already, the politics of comprehensive claims negotiations and the processes of regulation have created conflict and additional uncertainty in the region. As previously noted, citizens in the settlement regions may view the boards as representing their interests rather than the public good. If the EA process is to function effectively, however, political interests must not be reflected in the decisions that boards make; nor should they conflict with the principles and intent of the collaborative EA framework envisioned in the comprehensive claims agreements.

Coordination and Communication

Environmental assessment in a co-management regime is dependent upon effective coordination and communication. As highlighted at the practitioners' workshop (Terriplan/IER 2001), five issues constrain effective coordination and communication among the boards, government agencies, communities, and proponents. Several of these issues are closely connected to related concerns about administrative complexity, capacity issues, and the politics of change that influence this evolving EA framework.

First, the nature of many working relationships among practitioners can be linked to individual personalities—an often positive attribute of northern working conditions—yet, given that the membership of the boards and other organizations are often in flux, mechanisms to develop 'institutionalized' relationships are necessary.

Second, efforts to enhance coordination and collaboration require the sharing of information, to a significant degree, among land and water boards, the MVEIRB, federal and territorial government departments, and industry. In part, the challenge of information sharing and accessibility can be linked to the capacity of boards and other key actors (see below). However, issues related to the emergence of new roles and responsibilities and to the devolution of power have also been identified as complicating factors in efforts to improve the accessibility and sharing of data (Terriplan/IER 2001).

Third, practitioners characterized a number of the early assessment processes under the MVRMA regime as fairly rigid and hierarchical. Typically, the MVEIRB provided an EA work plan and schedule that mapped out the timing and scope for input from screeners and other stakeholder groups—often with limited input or consultation (MVEIRB 2000). Moreover, an implicit hierarchy of roles and responsibilities among the different boards emerged in which the MVEIRB tended to play a more dominant position. This implicit hierarchy among the various boards, however, is not universally accepted or envisioned in the MVRMA, nor is it consistent with the philosophy and intent of the Gwich'in and Sahtu land claims agreements. The existence of hierarchical arrangements between the current boards will not be acceptable if or when new boards or decision-making processes are established following completion of other claims negotiations in the Mackenzie Valley.

Fourth, the perceived and/or actual quasi-judicial status of the MVEIRB has influenced opportunities for achieving the type of coordination and communication (often informal) among boards, government agencies, and proponents that facilitates effective EA practice. With the establishment of the MVRMA process, there has been a reduction in the number of ad hoc meetings, inter-agency working groups, and informal communications among practitioners for fear of compromising the integrity of the process and creating scope for legal challenges to the decisions made and the terms and conditions set. The process may also have resulted in an increasing reliance on formal written communications for even minor issues, which may have slowed down the assessment process (see Fitzpatrick et al. 2008).

Fifth, and finally, at the core of coordination and communication challenges are the value sets or 'cultures' that characterize the different boards and government agencies. Each claim-mandated board, for instance, is an amalgamation of the chosen board members and staff whose values are entwined with their roles and responsibilities and the manner in which they perform their functions. Thus, in combination with dealing with the larger issue of board representation (i.e., claimants versus the public good), board members have to face significant challenges in achieving the levels of coordination and communication central to effective EA practice.

Practitioners have identified a number of specific mechanisms that enhance coordination and collaboration in the environmental assessment process. Of central importance was a renewed commitment by the MVEIRB to review and modify the EIA guidelines collaboratively, a process that took place in 2001. As well, practitioners identified the value of organizing multi-stakeholder post-mortems or project-specific review teams to identify process issues and learn from past experiences; they have therefore recognized the importance of formal and informal opportunities in which board members, agency staff, and other interested parties can address general and project-specific issues (MVEIRB 2000; Terriplan/IER 2001). The federal government (Indian and Northern Affairs Canada), in cooperation with the MVEIRB and the land and water boards, also established a 'Board Relations Unit' to identify and correct structural (e.g., funding, capacity) and procedural (e.g., communication, information sharing) issues that have emerged in the co-management regime. Although the creation of the Board Relations Unit was a unique response to the challenge of EA practice in the Mackenzie Valley, the unit's role as a vehicle to foster coordination and collaboration among diverse interests and mandates may be applicable in other circumstances.

Common Approaches and Cumulative Effects Assessment

As evident in the Mackenzie Valley proposal, simply legislating a new approach to environmental assessment is not necessarily a catalyst for the greater integration or development of common approaches. As assessment practitioners in the Mackenzie Valley have noted (Terriplan/IER 2001), there are several areas of policy and practice that require more consistent application in the Mackenzie Valley, including guidelines on how to incorporate social, cultural, and economic information and analysis into the screening and assessment process. There has also been a lack of common technical standards and criteria regarding air/water quality, habitat protection, and wildlife harvesting and protection. Despite these challenges, advances in the development of common standards and approaches have been made. The development and implementation of the land-use planning process mandated in the Gwich'in and Sahtu land claims, for example, are important components of the overall framework within which the environmental assessment and other resource management activities take place. The land-use planning process contributes to the development of a broader vision for certain areas of the Mackenzie Valley, to corresponding goals and objectives, and to greater sensitivity to ecosystem-based management frameworks. However, because of the challenges inherent in their development, the land-use plans have not always been central to assessment and management decision making.

Efforts to establish a framework for cumulative effects assessment provide a further example of the advances made to foster consistency and commonality in EA practice across the Mackenzie Valley (chapter 8). As previously outlined, Part 6 of the MVRMA provides for an audit process and the assessment of cumulative effects in the Mackenzie Valley. The creation of consistent guidelines to address cumulative effects in the screening and assessment stages was identified in the region as a critical need, despite the many technical and jurisdictional challenges (see Axys 2000; IEMA 2001; NWT Cumulative Effects Assessment and Management Framework Steering Committee 2001). For example, the requirements for cumulative effects assessment are not compatible with the short time frames allowed for project screening. The urgency and complexity of project-specific analyses limit the ability of practitioners to address cumulative effects assessment goals and methodologies. This is an outcome of the current assessment process in the Mackenzie Valley, which runs counter to the integrated management philosophy of the MVRMA. Yet preliminary screeners in the Mackenzie Valley are well positioned to explore the potential for cumulative effects by focusing on the type of project (e.g., forestry, tourism), its areal extent and geographic location, and its proposed time frame. Land and water boards in particular should be able to assess the potential cumulative effects of development proposals, especially where there are many discrete initiatives in a particular region that may be small in scale (e.g., linear developments) but that have the potential for significant cumulative impacts (Terriplan/ IER 2001). The time and resource requirements associated with cumulative effects assessment, however, are significant and cannot be accomplished at sub-regional scales. Federal and territorial government leadership has been necessary to ensure that cumulative effects assessment is regional in scope and neither proponent driven nor project-specific (see Kennett 2000). A significant development for the environmental assessment process in the Mackenzie Valley, therefore, has been the establishment of the Mackenzie Valley Cumulative Impact Monitoring Program (CIMP) and the regional Cumulative Effects Assessment and Management Strategy and Framework (CEAMF).[7] These initiatives provide a unique and consistent framework in which to develop common approaches and standards for cumulative effects assessment in the region.

Participation and Traditional Environmental Knowledge

Environmental assessment practitioners in the Mackenzie Valley have had to face the challenge of ensuring that widely dispersed First Nations communities participate in meaningful public consultation processes (Wismer 1996; NRTEE 2001; Fitzpatrick et al. 2008). As stipulated in the MVRMA, several requirements are associated with participation and consultation; for example, sections 63 and 64 of the Act require that all landowners, communities, First Nations agencies, and other potentially affected groups, as well as the appropriate government departments, be notified of all applications for water licences and land-use permits. Increasing development pressures associated with mineral extraction and oil and gas exploration, however, require technically complex, time-consuming, multi-million-dollar environmental assessment processes in the Mackenzie Valley (Northern Pipeline EIA and Regulatory Chairs' Pipeline Committee Working Group 2002). Moreover, as a result of these activities, there has been a dramatic increase in the number of land- and water-use applications that need to be screened. As all boards are required to ensure an opportunity for the public to comment on development applications, the logistical challenges of meaningful participation are increasing. To address these challenges, boards have turned to a technological solution—they disseminate and post notices of applications, materials for review, and requests for comments in electronic form. Communication is largely written and formalized, a trend with both positive and negative implications (Fitzpatrick et al. 2008). For instance, the verbatim transcripts of the DeBeers Snap Lake diamond mine review were readily accessible through the MVEIRB website, as are all EA documents and notices through the public registry.[8] Land and water boards have similar registries. A few years ago, this level of access to information would not have been possible. On the other hand, many First Nations communities in the North still do not have ready access to this form of communication technology. Moreover, First Nations communities have maintained strong traditions of oral communication, with elders in particular preferring opportunities for personal contact.

Given the volume of development applications in the Mackenzie Valley and the trend towards electronic communication, the capacity of small communities to be properly engaged in all stages of the environmental assessment process is far from certain. Boards simply do not have the full-time staff necessary to work directly with, or engage, First Nations communities, even though the MVRMA has established a more decentralized institutional and organizational framework. This challenge has been recognized and strategies to deal with it implemented. During the Diavik diamond mine review, for example, participant funding was provided to communities and other organizations so that they could better participate in the process. The amount of funding and its timing, however, have not always been considered adequate. The high-profile nature of the Diavik diamond mine review has not necessarily been a feature of other development applications or assessment processes, and in these others, the necessary support to ensure community participation may not have been as forthcoming.

The development of a 'Cooperation Plan' represented an attempt to deal with the magnitude and scope of the proposed pipeline development in the Mackenzie Valley (Northern Pipeline EIA and Regulatory Chairs' Pipeline Committee Working Group 2002); the Cooperation Plan requires the development of a plan that would entail public involvement and the likely provision of participant funding. Meaningful participation of diverse Aboriginal communities in the region requires significant time and expenditure on the part of industry, governments, and the boards.

Related to the challenge of meaningful participation (see chapter 4) is the importance of integrating traditional environmental knowledge (TEK) into the assessment and decision-making process (see Stevenson 1996; Usher 2000). The territorial government has a formal policy in support of the use of traditional ecological knowledge in planning, management, and assessment (GNWT 1993), and strategies to include traditional knowledge in regional cumulative effects assessment activities have been tested in the Mackenzie Valley (see Barnaby 2000; Waehdoo Naowo Ko 2000; Sly et al. 2001; Thorpe and Hakongak 2001). However, actual success in utilizing traditional knowledge in the determination of impact significance or in the setting of mitigation strategies has been limited. In part, this is because the primary mechanism for inclusion of traditional environmental knowledge in assessment activities—the proponent-driven process in the context of a specific development application—limits how, when, and by whom TEK can (and should) be utilized. Traditional environmental knowledge is often transmitted (and filtered) through the third-party consultants working for proponents, governments, or First Nations groups, despite the preference of some groups for greater interaction with decision makers. The creation of the boards and the application of the requirements for public consultation set forth in the MVRMA have enhanced Aboriginal representation in key decision-making activities. Nevertheless, the actual structures and processes of the boards are largely Euro-Canadian in their orientation (see White 2000) and have at times been characterized by an adversarial orientation, an emphasis on formalized rules and procedures, and a focus on Western science in the assessment process. At the same time, it is worth noting that several boards, including the MVEIRB, have become interested in the development of guidelines on the use of traditional knowledge in environmental screening and assessment (MVEIRB 2005). For example, based in part on a traditional environmental knowledge workshop it sponsored, the Sahtu Land and Water Board drafted preliminary guidelines on the role and use of TEK in the land-use permitting and water licensing processes (SLWB 1998, 2002). This was an important step forward, as failure to address this issue could weaken the legitimacy of the MVRMA environmental assessment process among Aboriginal communities (see Wismer 1996). Unless proponents, regulators, intervenors, and communities have greater clarity on this issue, a hard-fought battle to foster more inclusive environmental assessment and decision making in the Mackenzie Valley may be undermined (see Howard and Widdowson 1996; Stevenson 1997).

Capacity Challenges

Capacity can be defined simply as the ability of individuals, organizations, and/or organizational units to perform functions effectively, efficiently, and sustainably. The capacity of the Mackenzie Valley boards is central to the successful implementation of the planning, management, and assessment regime. However, implementation of the MVRMA and of the structural and procedural changes required by this legislation occurred quite quickly considering the dynamic socio-political, institutional, and economic context. As a result, the boards confront a number of technical, financial, and institutional capacity challenges. For example, the staff of the land and water boards face the challenge of processing the high volume of development applications, while the MVEIRB also faces a significant workload. A historically high rate of staff turnover within the boards is a further complicating factor. Yet, since the boards are not considered federal government agencies, there are limited opportunities for crossover of staff or supplemental staffing from government departments. Such human resource solutions would, at any rate, be viewed with some skepticism given the desire of the boards to maintain their independence. As a point of comparison, the MVEIRB has a handful of staff members to assess the many applications it may review, and it thus relies on technical input from external reviewers and consultants to help it determine its recommendations and the terms and conditions it imposes following the assessment process (MVEIRB 2000). In contrast, the National Energy Board (NEB), a partial operational model for the MVEIRB, has several hundred staff, including a full cadre of the specialists required to make informed decisions.

While the pressures on the MVEIRB and the land and water boards are significant, the capacity of government departments (federal and territorial) to provide information or respond to requests for comprehensive reviews in the context of short time frames and limited data requires recognition as well. Other key capacity issues include technical constraints associated with the adequacy of baseline data, a major concern among many practitioners given the emerging requirements for cumulative effects assessment. Gaps in data sets, incompatibility of data sets and information management systems, the willingness to share the information (see IER/Terriplan 2001), and an expectation that industry proponents will adequately fill these information gaps (see Kennett 2000) are all important capacity issues. Finally, perhaps the most significant capacity issue associated with the implementation of the Mackenzie Valley EA process relates to the individuals involved in the decision-making process. As already illustrated, the challenges are significant and the pressures at times very intense. The integrity of the screening and assessment processes depends, therefore, on the ability of board members and staff to make wise decisions in the broader public interest. In turn, boards must be provided with the resources and support they require from other partners in the assessment process, especially those government agencies responsible for implementing impact assessment prior to the proclamation of the MVRMA.

Conclusions and Recommendations

Practitioners, proponents, and communities in the Mackenzie Valley must address many issues associated with environmental impact assessment, including new structures and processes; administrative complexity; coordination, communication, and capacity issues; and the need for meaningful participation. As with any complex process and decision-making framework, solutions to many of these challenges emerge with the evolution of the assessment regime and the clarification of roles, responsibilities, and mandates. Nevertheless, identifying the constraints, opportunities, and interactions that influence the evolution of environmental assessment in the Mackenzie Valley is an important step. Having identified a number of constraints to effective practice, practitioners and researchers can design strategies to overcome those constraints.

This overview has provided a basis upon which to elaborate a set of emerging or working principles for environmental assessment policy and practice in the Northwest Territories context (Table 11.3). Of central importance to the EA process, for example, is the formulation of a vision that encompasses the world views, aspirations, goals, and objectives of Aboriginal and non-Aboriginal interests at regional scales. Expressing this vision through a regional land-use planning process is one tangible method of linking project-specific environmental assessment decisions to diverse value sets regarding resource development and ecological and socio-cultural sustainability. An additional principle highlighted by the experience with environmental assessment in the Mackenzie Valley is the importance of building capacity for decision making, not only in the public boards but in communities as well, where impacts are most directly felt. Integrating traditional knowledge in planning, management, and assessment and establishing innovative mechanisms to facilitate collaboration are likewise important principles for assessment practice (Table 11.3).

Table 11.3   Principles for environmental assessment based on the Mackenzie Valley experience

1. ***Develop a collaborative vision.*** In a co-management context, some form of vision needs to be expressed that captures the world views, aspirations, goals, and objectives of Aboriginal and non-Aboriginal interests. However a collaborative vision is established (e.g., regional land-use planning process, policy statements), its importance to environmental assessment should not be undervalued. Project-specific analyses, particularly cumulative effects assessments, must be linked to broader goals and objectives that balance resource development and economic interests with ecological and socio-cultural sustainability.

2. ***Ensure meaningful consultation with First Nations communities.*** Standard consultation processes (e.g., open houses) do not provide an adequate basis upon which to ensure participation of First Nations communities in project review, impact identification, or determination of acceptable mitigation measures. Difficult language and other cultural barriers exist that must be dealt with in a sensitive manner and that will require significant investments in time and money.

3. ***Build common approaches, strategies, and standards among practitioner groups.*** With the transition to an assessment regime, the roles and approaches of certain organizations (notably government) change, while new organizations with new responsibilities emerge. Finding commonality among different approaches to assessment in an evolving organizational context is essential. Common standards and guidelines need to be developed, such as those related to cumulative effects assessment or to the use of traditional ecological knowledge, that draw on the technical and institutional capacity of all organizations and interests.

4. ***Establish innovative mechanisms that facilitate collaboration and promote adaptation.*** The transition to a collaborative assessment regime is unlikely to be smooth. Sensitivity to changing roles, responsibilities, and mandates associated with the EA process is required. Innovative mechanisms to help alleviate the legitimate concerns of individuals and organizations will facilitate this transition and foster more effective EA practice. The establishment of a 'Board Relations Unit' in the Mackenzie Valley context provided an interesting model aimed at dealing with transition issues. However, less formal mechanisms are also valuable, including the development of preliminary screening groups or project-specific review teams that bring practitioners together to proactively address bottlenecks and that improve relationships in future proceedings.

5. ***Guarantee the development of institutional capacity of key stakeholder groups in the environmental assessment regime.*** In the Mackenzie Valley, the capacity development of the claims-mandated boards responsible for environmental screening and assessment has been a critical factor in the implementation of the new regime. There are, however, ongoing challenges associated with financial support, staffing, and technical capacity (e.g., information management). Just as important, the capacity development within the predominantly First Nations communities in the region is also required to enhance environmental decision making.

Continued

Table 11.3   Continued

6. ***Integrate traditional knowledge into the assessment process and development of mitigation measures.*** Because it is typically based on experiential knowledge that is orally communicated across generations, traditional ecological knowledge does not easily lend itself to EA processes premised on notions of empiricism, risk assessment, and analytical certainty. Nevertheless, the validity of traditional knowledge systems, and the world views and cultural integrity they support, is well documented. Efforts to ensure the incorporation of traditional knowledge into impact assessment and the identification of mitigation measures (through policy development and supporting guidelines) are critical in a collaborative regime.

7. ***Ensure that the assessment framework remains consistent with philosophy and intent of the collaborative model.*** The resource management framework established through the settled land claims in the Mackenzie Valley is based on a philosophy of decentralization, collaboration, and a central role for Aboriginal communities in environmental decision making. With the evolution of new structures and processes for screening and assessment, however, the philosophy and intent of the new regime risk being subsumed by the everyday challenges of practice. In particular, administrative complexity, a legal orientation to the assessment proceedings, and a Euro-Canadian decision-making approach are not always consistent with the intent of the land claims, nor with efforts to ensure that First Nations communities are full participants in the decision-making process.

As a response to the Gwich'in and Sahtu comprehensive land claims agreements, the MVRMA environmental regime represents a reframing of EIA practice in a complex socio-cultural, political, institutional, and economic context. Translating the lessons learned in the Mackenzie Valley into a set of working principles may therefore facilitate the development of EA theory and practice. This is true not only for regions where land claims and co-management governance structures are emerging, but also for those contexts where new partnership models and environmental decision-making structures are being forged among Aboriginal groups, governments, industry, and civil society. Environmental assessment in the Mackenzie Valley provides an innovative framework for policy and practice that can be used by those seeking to set new standards for integration and collaborative decision making. While the challenges in implementing this innovative regime are significant, the emerging lessons and principles for environmental assessment policy and practice should be considered relevant by students, researchers, and practitioners in diverse contexts.

Acknowledgements

I acknowledge the assistance and helpful comments of the anonymous reviewers, of Scott Slocombe and Kevin Hanna, as well as of the practitioners and officials who offered their extensive insight and knowledge about environmental assessment in the Mackenzie Valley. I also thank Pam Schaus for preparing the map.

Finally, I gratefully acknowledge the support provided through a Social Science and Humanities Research Council Post-doctoral Fellowship.

Notes

1. Comprehensive claims agreements establish the rights and benefits of Aboriginal groups in a defined territory (settlement area). Rights and benefits that are often incorporated into comprehensive claims agreements include financial transfers, full ownership of lands, rights to harvest wildlife, and resource sharing from mineral developments.

2. Co-management in this instance refers to arrangements in which governments and Aboriginal groups, primarily, enter into formal agreements that specify their respective rights, powers, and obligations regarding the management and allocation of resources. Members of the EISC and EIRB are jointly appointed by the federal/territorial governments and the Inuvialuit.

3. The regulatory powers of the new boards relate to surface land and water use; the Northwest Territories Waters Act and the Territorial Lands Act may still apply in the land and water management regime (see Donihee et al. 2000).

4. As outlined in section 116 of the MVRMA, the CEAA may still apply in limited and specific circumstances that require public review processes, such as (1) development proposals located entirely within the Mackenzie Valley but considered to be in the 'national interest' and (2) transboundary development proposals located both within and outside the Mackenzie Valley.

5. In addition to the MVRMA, three supporting regulations have been promulgated: the Mackenzie Valley Land Use Regulations (the Northwest Territories Water Regulations still apply in circumstances involving water use and disposal of waste in waters); the Preliminary Screening Requirement Regulations, which include a list of the licences, permits, and authorizations associated with territorial and federal legislation; and the Exemption List Regulation, which outlines projects that do not require a screening assessment.

6. Other jurisdictions and mechanisms that seek to address this common challenge include the Canadian Environmental Assessment Act's Comprehensive Study List; Appendix D of the Inuvialuit Settlement Region's Environmental Impact Steering Committee Operating Procedures and Guidelines; and the draft Project List Regulation in the Yukon Development Assessment Process.

7. See http://www.ceamf.ca/ (Northwest Territories Cumulative Effects Assessment and Management Strategy and Framework).

8. See http://www.mveirb.nt.ca/.

References

Armitage, D. 2005. Collaborative environmental assessment in the Northwest Territories, Canada. *Environmental Impact Assessment Review* 25:239–58.

Axys Environmental Consulting. 2000. Regional approaches to managing cumulative effects in Canada's North. Unpublished report submitted to Environment Canada, Yellowknife, NWT.

Barnaby, J. 2000. Traditional knowledge in cumulative effects monitoring. Unpublished report prepared for the Dogrib Treaty 11 Council Waehdoo Naowo Ko Program.

Berger, T. 1977. *Northern frontier, northern homeland: The report of the Mackenzie Valley Pipeline Inquiry*. Ottawa: Environment Canada.

Byrd, C. 2006. *Diamonds: Still shining brightly for Canada's North*. Canadian Trade Review Analytical Paper. Statistics Canada (65-507-MIE-No. 007). Ottawa.

Canada, Government of. 1998. *Mackenzie Valley Resource Management Act*. Assented to 18 June 1998. First Session, Thirty-Sixth Parliament, Ottawa.

Chase, S. 2000. Yukon, NWT battle for pipeline route. *Globe and Mail*, 11 April, section B2.

Donihee, J. 2001. *Implementing co-management legislation in the Mackenzie Valley*. Calgary: Canadian Institute of Resources Law, University of Calgary.

Donihee, J., J. Gilmour, and D. Burch. 2000. *Resource development and the Mackenzie Valley Resource Management Act: The new regime*. Report prepared for a conference convened by the Canadian Institute of Resources Law, University of Calgary, 17–18 June 1999.

Fitzpatrick, P., J. Sinclair, and B. Mitchell. 2008. Environmental impact assessment under the Mackenzie Valley Resource Management Act: Deliberative democracy in Canada's North? *Environmental Management* 42:1–18.

Gibson, R. 1993. Environmental assessment design: Lessons from the Canadian experience. *Environmental Professional* 15 (1): 12–24.

———. 2001. Specification of sustainability-based environmental assessment decision criteria and implications for determining 'significance' in environmental assessment. Unpublished report prepared for the Canadian Environmental Assessment Agency Research and Development Programme, Ottawa.

GNWT (Government of the Northwest Territories). 1993. *Traditional Knowledge Policy*. Yellowknife, NWT: Government of the Northwest Territories.

Green, N., and R. Binder. 1995. Environmental impact assessment under the Western Arctic (Inuvialuit) Land Claim. In J.A. Bissonette and P.A. Krausman (Eds), *Integrating people and wildlife for a sustainable future*, 343–5. Proceedings of the First International Wildlife Management Congress. Bethesda, MD: Wildlife Society.

Howard, A., and F. Widdowson. 1996. Traditional knowledge threatens environmental assessment. *Policy Options* 17 (9): 34–6.

IEMA (Independent Environmental Monitoring Agency). 2001. *Towards improved environmental management in the North*. Yellowknife, NWT: IEMA.

IER/Terriplan. 2001. *Information Management Workshop Supporting CEAMF and MVCIMP*. Workshop report prepared for Indian and Northern Affairs Canada, Yellowknife, NWT.

Kennett, S. 2000. The future for cumulative effects assessment: Beyond the environmental assessment paradigm. *Resources: Newsletter of the Canadian Institute of Resources Law* 69:1–8.

Lindsay, K.M., and D.W. Smith. 2001. *Review of environmental assessment processes*. Northern Resources Research Centre and Canadian Circumpolar Institute. Occasional Paper No. 50.

MVEIRB (Mackenzie Valley Environmental Impact Review Board). 2000. *Report of Lessons Learned from the Environmental Assessment of the Ranger et al. Fort Liard Pipeline Development Proposal*. Yellowknife, NWT: MVEIRB.

———. 2001. *Guidelines for environmental impact assessment in the Mackenzie Valley*. Yellowknife, NWT: MVEIRB.

———. 2005. Guidelines for incorporating traditional knowledge in environmental assessment. Mackenzie Valley Environmental Impact Review Board, Yellowknife, NWT.

Northern Pipeline EIA and Regulatory Chairs' Pipeline Committee Working Group. 2002. *Cooperation plan*. Yellowknife, NWT: MVEIRB.

NRTEE (National Roundtable on the Environment and the Economy). 2001. *Aboriginal communities and non-renewable resource development*. Ottawa: NRTEE.

NWT Cumulative Effects Assessment and Management Framework Steering Committee. 2001. Current context, 'lessons learned', gaps and challenges. Working draft discussion paper prepared by IER Planning Research and Management Services and Terriplan Consultants Ltd, Yellowknife and Toronto.

Peters, E. 1999. Native people and the environmental regime in the James Bay and Northern Quebec Agreement. *Arctic* 52 (4): 395–410.

Reed, M. 1990. *Environmental assessment and Aboriginal claims: Implementation of the Inuvi-aluit Final Agreement.* Ottawa: Canadian Environmental Assessment Research Council.

SLWB (Sahtu Land and Water Board). 1998. *Summary of traditional environmental knowledge workshop.* Deline, NWT.

———. 2002. Traditional environmental knowledge (draft). Fort Good Hope, NWT: SLWB.

Sly, P.G., L. Little, R. Freeman, and J. McCullum. 2001. *Updated state of knowledge report of the West Kitikmeot and Slave Geological Province.* Report prepared for the West Kitikmeot/Slave Study Society, Yellowknife, NWT.

Stea, D., and B. Wisner. 1984. Introduction. *Antipode* 16 (2): 3–13.

Stevenson, M. 1996. Indigenous knowledge in environmental assessment. *Arctic* 49 (3): 278–91.

———. 1997. Ignorance and prejudice threaten environmental assessment. *Policy Options* 18:25–8.

Terriplan/IER. 2001. *Preliminary screening and environmental assessment under the Mackenzie Valley Resource Management Act: Developing a collaborative approach for the Northwest Terri-tories.* Report prepared for the Mackenzie Valley Environmental Impact Review Board, Mackenzie Valley Land and Water Board and Indian and Northern Affairs Canada, Yellowknife, NWT.

Thorpe, N., and G. Hakongak. 2001. Cumulative effects and the Tuktu and Nogak project in the Slave Geological Province. Presentation to the West Kitikmeot/Slave Study Society Research-ers Workshop, 15 October, Yellowknife, NWT.

Usher, P. 2000. Traditional ecological knowledge in environmental assessment and manage-ment. *Arctic* 53 (2): 183–93.

Waehdoo Naowo Ko. 2000. *Developing a plan to include indigenous knowledge in the NWT Cumulative Effects Assessment and Management Framework.* Dogrib Treaty 11 Council Report submitted to the NWT Cumulative Effects Assessment and Management Program, Yellow-knife, NWT.

White, G. 2000. And now for something completely northern: Institutions of governance in the territorial North. *Journal of Canadian Studies* 35 (4): 80–99.

Wismer, S. 1996. The nasty game: How environmental assessment is failing Aboriginal com-munities in Canada's North. *Alternatives* 22 (4): 10–17.

Wood, C. 1995. *Environmental impact assessment: A comparative review.* Harlow, UK: Longman Scientific and Technical.

World Bank. 1991. *Environmental assessment sourcebook.* Washington, DC: World Bank, Envi-ronment, Social Development and Rural Development Unit.

Chapter 12

Environmental Assessment and Land Claims, Devolution, and Co-management: Evolving Challenges and Opportunities in Yukon

D. Scott Slocombe, Lyn Hartley, and Meagan Noonan

Introduction

Environmental assessment (EA) in Canada's northern territories has long been of broad interest. On the one hand, as they are territories rather than provinces, their lands and resources were long administered under federal regulations and laws. On the other, their unique environments and cultures have called for and created unique resource and environmental institutions and policies. Recognition and development of this can be traced back 30 years and more to the Berger Inquiry into a Mackenzie Valley pipeline (Berger 1977) and the Territorial Water Boards originally established under the 1970 Northern Inland Waters Act. More recently still, from the late 1980s to the present, unique local institutions and regimes have been developed through processes of comprehensive land claims, devolution from federal to territorial governments, and generally increased resource co-management.

In the territories, long-standing questions of how land claims would affect resource and environmental management are beginning to be answered (Cassidy and Dale 1988; Notzke 1994). Canada's territories are vast and encompass a tremendous diversity of environments, peoples, and cultures with regional commonalities. Many of the recent innovations in resource and environmental management, paralleling land claims, focus on particular regions, largely reflecting territorial boundaries in Yukon and Nunavut, but much less so in the Northwest Territories (see Bone 2002 for an overview). Also, in most of the North these new institutions are just that—new—and it is too early for many conclusions. In chapter 11 of this volume, Derek

Armitage looks at the Mackenzie Valley Resource Management Act and its provisions, which have now been in effect since late 1998. The corresponding, though quite different, acts and regulations for Yukon are still quite recent, and evolving, so here we focus more on a review of Yukon/federal history and outline the new processes and recent experiences with them, for a sense of how different current regimes can be and to illustrate the range of new, northern EA processes.

Yukon is the ninth-largest province or territory in Canada, with 483,450 km², or 4.8 per cent of Canada's total land area. However, the population is relatively small, about 33,000 in June 2008, with about 74 per cent of the population non-Aboriginal (Yukon, Bureau of Statistics 2008). Yukon has a diverse terrain with extensive mountains and valleys and tundra plains in the north. Euro-Canadian exploration began in the late nineteenth century, first for furs and then for minerals. The Klondike Gold Rush began in 1896, and mining activity has been a key economic activity in the territory ever since. The Alaska (Alcan) Highway was built in 1942–3 and has been upgraded repeatedly ever since. It and other transportation infrastructure development represent important activities in Yukon. Tourism in Yukon grew steadily in the decades up to the early 1990s, to the point where it had a major role in the economy. Since then, tourism has fluctuated but has remained roughly stable and one of the strongest elements in the economy. Forestry activities have become politically significant in the last 10 years, primarily in southeast Yukon, and more recently have started to be economically significant.

Early territorial government began in the Klondike days, lapsed somewhat in the ensuing decades, and grew steadily from the 1960s onward. In the last 25 years, responsibility for a range of resources and government services has devolved to the territory from the federal government. Paralleling this development has been the increasing economic significance of federal, territorial, and First Nations government activity. There are 14 First Nations in Yukon, and since the 1970s they have taken steps to expand their governance responsibilities through land claims, self-government, and program and service transfer agreements. The Inuvialuit Final Agreement (IFA) for the peoples of the Beaufort Sea/Mackenzie Delta region was completed in 1984 and covers the Yukon North Slope region. Environmental impact assessment (EIA) provisions under that agreement include an Environmental Screening Committee and Environmental Impact Review Board. These were operational by the late 1980s, have evolved considerably in role and influence, and remain in operation today. They have been extensively studied and written about (Keith and Mulvihill 1995; Reed 1990; Green and Binder 1995), are more closely linked to other NWT claims and processes than to those in Yukon, and will not be discussed further here. The NWT/Yukon Gwich'in Comprehensive Land Claim Agreement, signed in 1992, established a development assessment process for Gwich'in lands, of which 1,554 km² are in northeast Yukon. That process, too, continues, and its influence is evident in more recent changes in the rest of Yukon, discussed below.

The Umbrella Final Agreement (UFA) for the rest of Yukon was concluded in 1993; individual First Nations have been negotiating their specific and self-government agreements since then. All but the White River First Nation and the Kaska First

Nations in southeast Yukon have now completed or are in the process of finalizing their agreements. Henderson (1992) provides a useful comparison of the IFA, UFA, and Canadian Environmental Assessment Act (CEAA) environmental assessment provisions.

This chapter was developed via a review of academic and grey literature and a series of interviews with impact assessment practitioners in Yukon. In the following sections, we outline the development and evolution of environmental assessment in Yukon through its recent devolution to the territory, and discuss several key issues for Yukon EA (federal/territorial relations and devolution; First Nations land claims and traditional environmental knowledge [TEK]; mining; and cumulative effects and land use planning). We aim to outline the continuity and change in Yukon environmental assessment over the last 30 years and highlight challenges and opportunities for attention and comparison with northern and other jurisdictions.

Development and Evolution of Yukon Environment Assessment

Resource development in northern Canada, including Yukon, prior to the 1960s and early 1970s was relatively unregulated, which did nothing to reduce the negative consequences of projects. The vast northern environment was viewed as a source of resources needed in the south and of employment needed locally.

For the greater part of Yukon's history, most of the responsibility for resource management in the territory rested with the federal government. During the 1950s and 1960s, the Department of Indian Affairs and Northern Development (DIAND) emerged as a province-like administrative entity for Yukon, according to the Yukon Act and other federal statutes (Reed 1990). Through the DIAND Act, the minister of Indian affairs and northern development was the lead federal minister in the North. The minister's responsibilities included social and economic development, environmental management, and, more recently, the promotion of economic and sustainable development of northern resources for northern communities. Up until recently, DIAND had extensive jurisdiction, with responsibilities for land, water, minerals, and forestry, and over the years the department played a key role in the development of environmental assessment in Yukon. Formal EA began in the early 1970s with the establishment of the federal Environmental Assessment and Review Process (EARP). Land- and water-use permitting processes have also had some assessment role, although their relationships to EA processes have not always been clear (see Figure 12.1).

The Federal Environmental Assessment and Review Process

The Canadian federal government established the Environmental Assessment and Review Process in December of 1973. EARP was amended in February 1977, and an internal review took place in 1979. The formal assessment process at this time was minimal except in regard to major projects. In Yukon in this period, the only project to reach a panel review was the Alaska Highway Gas Pipeline (Alaska High-

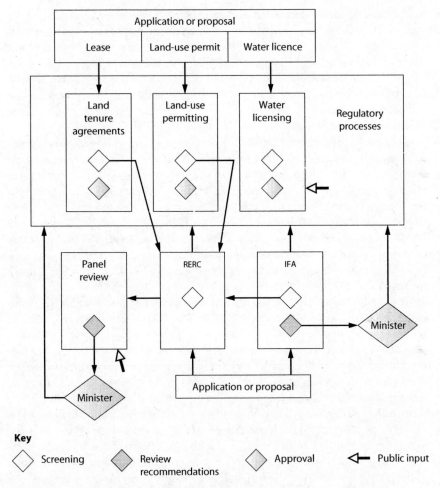

Figure 12.1   Schematic diagram of main relationships among Yukon environmental assessment processes before initiation of the Development Assessment Process (DAP)

Source: After Everitt et al. 1988.

way Gas Pipeline Environmental Assessment Panel 1979). In the early 1980s, the Beaufort Sea Environmental Assessment and Review Process took place, touching on the Yukon North Slope. This very large and contentious review contributed significantly to later reforms (see Rees 1983; Wallace 1986; Sadler 1990; Mulvihill and Baker 2001; Koivurova 2002).

Throughout the late 1970s and early 1980s, EARP was debated and reviewed throughout Canada (Gibson 1984; Rees 1980). Not the least of the issues was the limited and uneven application and implementation of EARP within the federal government; this was of particular significance to Yukon given the federal control of land, resources, and water. At this time, the Government Organization Act granted the federal Department of the Environment the power to establish environmental guidelines for all federal authorities, and thus, on 21 June 1984, the department established the Environmental Assessment Review Process Guidelines Order (EARPGO). EARPGO created a self-assessment process that applied to all projects with federal interest or participation, and it made each federal department responsible for screening the projects, programs, or activities that it initiated or sponsored (Reed 1990).

In the early 1980s, DIAND started using environmental assessment as a planning tool for large and more complex projects, particularly in the area of mining. As DIAND, because of its land and water responsibilities, was usually the lead government department in Yukon for environmental assessments, it was important for the department to consult with other departments. During this time, the Yukon Regional Environmental Review Committee (RERC) was created. The RERC was formed to provide cross-governmental input to an EARP secretariat within DIAND that would facilitate technical review of major development proposals. Representatives from the federal and territorial governments and, later, from the Council for Yukon Indians made up the membership.

Throughout this period, it was court judgments and cases that largely influenced the EA process and content. In Yukon, this meant that government had to respond quickly to changes in the EA process as it negotiated with proponents. Not only did these court judgments impact the major projects, but they also had significant ramifications for smaller projects. Once court decisions directed federal departments to conduct environmental reviews under EARPGO, the Yukon region was required to conduct assessments on all projects, not just the major ones.

By the late 1980s DIAND had developed an administrative process to distinguish between the types of projects. Level I projects involved simple and small-scale screenings by DIAND resource managers, with input from advisory groups. Generally, these projects met the regulatory requirements and their environmental impacts were considered to be insignificant. In such cases, the resource manager would acknowledge insignificant or mitigable environmental impacts and approve the project with the appropriate mitigation. Generally, Level I projects were routine and involved the issuing of permits according to pre-established standards and implications.

Significant environmental impacts, public concern, or insufficient information would lead to a Level II screening. The more complex projects at this level required more comprehensive studies. The Regional Environmental Review Committee, chaired by DIAND's environment director and including representatives of other federal and territorial departments, provided advice on the design of and mitigation for Level II projects. The RERC made recommendations to DIAND, and these were included in the screening report, which was generally made available for public comment. The Council for Yukon Indians (now Council of Yukon First Nations) would participate if the project was of direct concern to those it represented. The representation and participation of public interest groups remained a concern (Prystupa 1994).

Especially large, complex, controversial, and/or potentially harmful projects would be referred to a Level III screening. Level III screening reviews were coordinated by the Federal Environmental Assessment Review Office (FEARO) in Ottawa and involved a full environmental impact statement (EIS), the creation of a review panel, and public hearings (as described in chapter 14). Projects could be referred to Level III either directly, by the minister of DIAND, OR after an initial Level II review.

Canadian Environmental Assessment Act

Owing to a number of disputed projects, notably the Rafferty-Alameda and Oldman River dams, there were a range of pressures to create a statutory basis for federal EA in order to maintain the consistency and integrity of assessments. On 19 January 1995, the Canadian Environmental Assessment Act was proclaimed.

Although the CEAA replaced EARPGO, the environmental assessment process in Yukon was essentially the same as it had been under EARPGO, but with a few significant differences. First, there was a much broader definition of environmental and related socio-economic effects; it now included archaeological heritage and traditional use of lands and resources. Second, the Environment Directorate of DIAND and the Northern Affairs Program became jointly responsible for EA in Yukon. Third, some differences were introduced to the screening and assessment processes (outlined in DIAND 2000). The majority of projects were assessed by Level I screening, and several comprehensive studies were conducted over the years.

The CEAA included four additional regulations: the Law List, the Inclusion List, the Comprehensive Study List, and the Exclusion List (DIAND 2000). The inclusion of the Comprehensive Study List regulation made a difference in Level II assessments. The regulation provided a list of projects that could potentially have adverse environmental effects or were likely to raise public concern and as a result would receive a more comprehensive environmental assessment. Adding to the screening procedures of EARPGO, the CEAA allowed for comprehensive study and mediation (albeit never used) methods in environmental assessment, each of which had additionally to consider alternative means, the need for follow-up programs, and the capacity of renewable resources affected by the project. Under the CEAA, screening had to consider additional factors, including cumulative effects and effects of accidents and malfunctions. Lastly, a public registry of assessments needed to be established and maintained for public access. Approximately 100 to 150 projects were screened by DIAND in a year under the CEAA. Major reviews in the 1990s included the Shakwak Highway Redevelopment and the Aishihik River Dam relicensing.

DAP and YESAA

Chapter 12 of the Umbrella Final Agreement, concluded in 1993 between the governments of Canada and Yukon and the Council for Yukon Indians, provided for the creation of the Development Assessment Process (DAP) to assess the environmental and socio-economic impacts of proposed developments (Government of Canada et al. 1993). The purpose of DAP was to help decision makers ensure that proposed development activities would not harm Yukon's environment, people, or communities. The DAP was to recognize and guarantee First Nations participation in assessing projects on public lands and in considering social, economic, environmental, and cultural impacts. A range of other co-management boards were also called for and have been established: Renewable Resource Councils, Yukon Land Use Planning Council, and Yukon Salmon Council; in addition, other, existing boards have been regularized, such as the Yukon Water Board.

The new assessment process was to apply to all lands in Yukon, including federal, territorial, First Nation, and private. A single process that was to apply to all levels of government was intended to provide consistency throughout the territory, and clear guidelines and timelines were to facilitate planning for proponents. DAP was to establish a clearly public process to guarantee access to all relevant assessment information and ensure that the public had meaningful input in project assessments. It was also hoped that DAP would provide some progress on the long-standing issue of the relationship between EA and non-EA permitting processes in Yukon, notably land-use and water board permitting (see Everitt et al. 1988).

Through the late 1990s and into the 2000s, the Government of Canada, the Yukon government, and Yukon First Nations engaged in a complex, public, and not-so-public process to develop draft DAP legislation, now called the Yukon Environmental and Socio-economic Assessment Act (YESAA). The process began with two workshops and discussions with a variety of people; then, in 1998, draft DAP legislation was released for public review. The detailed comments received led to substantial changes in the proposed process, as many felt that the draft did not live up to the UFA/DAP principles (DIAND et al. 1999 and, for example, YCEE 1999). The process became very long, partly because of consultation results and partly because of other issues among the parties (see Government of Canada et al. 2000). At this time, the agencies involved in EA were consulting among themselves about technical and process issues, including information guidelines, duplication, interpretations of the CEAA, adequacy of application, integrated screening, and the need to bring processes together and clarify roles and responsibilities (Christiane Boisjoly and Associates 1999).

In August 2001, a new draft was released to reflect the comments from the 1998 consultations. The main changes made as a result of the public consultations included the following:

- Clearer opportunities for public involvement in the process
- A more streamlined process
- Protection of confidential information
- Clarity on the enforcement of decisions
- More certainty for project proponents
- Greater consideration of the positive impacts of development

A final round of consultations took place in November 2001. Many stakeholders felt that there was a 'lack of meaningful consultation', that the 'consultation process had been inadequate and that their views had not been given sufficient consideration' (Hébert and Hilling 2002, 42). Nevertheless on 3 October 2002, the minister of Indian affairs and northern development introduced the proposed Bill C-2 in the House of Commons. The YESAA received royal assent on 13 May 2003 despite concerns voiced to the House of Commons Standing Committee by several First Nations groups. The White River First Nation, the Kwanlin Dun First Nation, and the Liard First Nation felt that land claims should be settled under the Umbrella Final Agreement before an additional EIA process was initiated. The commentary attached to Bill C-2 cites perceived shortcomings that include an unclear interface with the CEAA, duplication of assessment on the Yukon North Slope owing to the Inuvialuit Final Agreement, a lack of timelines, inconsistencies between designated offices, and failure to place municipalities in a decision-making position. The new process was to be commonly known as the Development Assessment Process, although the Act was named to reflect both its intended purpose and how proposed development activities might affect the environment, economies, health, lifestyles, and heritage of Yukoners.

The YESAA is composed of three parts:

- Part 1: Yukon Environmental and Socio-economic Assessment Board and Designated Offices (clauses 8–39)
- Part 2: Assessment Process and Decision Documents (clauses 40–123)
- Part 3: Transitional Provisions, Consequential and Coordinating Amendments and Coming into Force (clauses 124–34)

Royal assent initiated Part 1 of the Act, establishing the Yukon Environmental and Socio-economic Assessment Board (YESAB). The board consists of up to seven members. Three members form the executive committee (one nominated by the Yukon First Nations, one nominated by the Government of Canada, and the third member and chair to be appointed by the minister of DIAND in consultation with the committee). The remaining four members include two Yukon First Nations nominees, one Yukon territorial government nominee, and one minister of DIAND nominee (DIAND 2001). Part 1 also includes the creation of designated offices across the territory. The six communities for these offices, announced in October 2004, are Dawson City, Haines Junction, Mayo, Teslin, Watson Lake, and Whitehorse. The offices were intended to be accessible and to involve those most affected by a project (DIAND 2001). In addition, the new Act was introduced over an extended implementation period. During this period, assessments continued to be administered under the Yukon Environmental Assessment Act (YEAA).

Meanwhile, the federal government devolved responsibility for environmental assessment, and for most lands, waters, and resources in Yukon, to the Yukon government on 1 April 2003. As noted above, until the YESAA came fully into force, EA in the Yukon was implemented under the YEAA and through the Yukon Environmental Assessment Unit, with many of the former DIAND staff having transferred to the territorial government (some chose to pursue other opportunities). Planning for this transition began well in advance. On 22 November 1999, the Yukon government tabled the Yukon Environmental Assessment Act to be passed by the Yukon legislature. Bill 90, YEAA, was passed in March 2003. The YEAA is 'mirror' legislation, developed to closely reflect the CEAA. It applied to all lands where the Yukon government was the 'proponent of a project, funds a project, licenses or authorizes a project or disposes of land for a project.' Where projects were being assessed by DIAND under the CEAA at the time of transfer, the Yukon government would continue the assessment under the YEAA. The YEAA operated in conjunction with existing federal EA legislation until both were replaced by the new development assessment legislation, that is, by the YESAA (see Yukon, Executive Council Office 2003, 2004).

The YESAA establishes an assessment process that is independent of government. Regional offices assess the majority of projects, and the executive committee of a new board guides the assessments of complex or larger projects. At the end of a project assessment, recommendations are issued to existing decision makers, who make the final decision on whether a project should go ahead or not. Full implementation of the YESAA depends on development of the Yukon Activity and Project Regulations; on Decision Body Time Periods and Coordination Regulations, which define what activities will be assessed under the Act; on the timelines for assessment processes; and on which administrative bodies hold final decision authority. The decision bodies include federal, territorial, and First Nations governments. The first project assessments were to have been conducted under the new legislation when Part 2 came into effect on 13 November 2004. However, this was delayed owing to the continuing development of the regulations, and full implementation did not begin until November 2005 (Yukon, Executive Council Office, DAP Branch 2005).

Since the lengthy implementation period, assessments have been administered under the new legislation, the YESAA. There are four different types of assessment under the YESAA. The first is a *designated-office screening,* which is the entry point for the majority of projects. The designated office recommends that the 'project proceed, proceed with terms and conditions or not proceed' (DIAND 2001, 2) or that the project requires further assessment by a designated-office review. A *designated-office review,* the second type of assessment, is required if the project may have 'adverse impacts, raise public concerns, or require more information to do a thorough assessment' (DIAND 2001, 2). After the review, the designated office recommends that the 'project proceed, proceed with terms and conditions or not proceed.' The third type of assessment is an *executive committee screening.* The designated offices may refer a project to the committee, or larger projects may enter this level of the assessment process immediately. The executive committee recommends that the 'project proceed, proceed with terms and conditions or not proceed' (DIAND 2001, 2). Should the committee determine that a 'project might have a significant adverse impact, raise significant public concerns or it involves untested technology' (DIAND 2001, 2), a panel review occurs. A panel review, the fourth type of assessment, permits a more detailed assessment. The panel will recommend that a 'project proceed, proceed with terms and conditions or not proceed' (DIAND 2001, 2).

The Umbrella Final Agreement and the YESAA Implementation Plan require a Five-Year Review. That process is currently underway. Terms of reference were released in early 2008. The review is intended to address all aspects of the DAP process, including

- YESAA and its regulations;
- the implementation, assessment and decision-making processes;
- the implementation plan;
- funding;
- opportunities for public participation in the process;
- phases and timelines;
- process performance expectations;
- process documents such as rules, guides, forms;
- the responsibilities, duties and functions of Decision Bodies, YESAB and other participants and their timelines and supporting documentation.
  (YTG et al. 2008)

Information packages were made available in June 2008, and consultant selection took place in the fall of 2008. Opportunities for public and First Nations input in the review process were required by the Statement of Work for the consultants to run the project. Details have not been released as of the time of writing (October 2008). Given the issues that have arisen in the last four years (see below), it is likely that there will be some revisions to YESAA following the review.

Key Issues

Clearly, environmental assessment in most of Yukon is now in a state of flux. It has been evolving over the last 35 years, and the last several years in particular have been a period of significant change. Several themes, or key issues, can be identified in EA evolution in Yukon, not all of which were touched upon in the brief administrative history above. The following sections will discuss some of these issues in the context of the processes discussed above.

Federal/Territorial Relations and Devolution

On 1 April 2003, responsibility for the Department of Indian Affairs and Northern Development's Northern Affairs Program (land and resource management) was transferred to the Government of Yukon. Through devolution, approximately 275 positions from the Northern Affairs Program and roughly $34 million/year of funding were transferred to the Yukon government so that it might administer programs and services. The transfer of land and resource management was intended to increase Yukoners' control over land and resources and to reflect regional values and priorities in land and resource management.

With the devolution of the Northern Affairs Program, Yukon was required to pass legislation that mirrored the CEAA, and in March 2003 the Yukon's Environmental Assessment Act was passed. As noted above, the YEAA was an interim measure to facilitate the transition to the Development Assessment Process. The YEAA applied to all lands under Yukon government jurisdiction after 1 April 2003. The YEAA specifically applied where the Yukon government was the proponent of a project, funded a project, licensed or authorized a project, or disposed of land for a project. While the YEAA applied to the Yukon government, federal departments were still required to conduct assessments under the CEAA. A harmonization agreement was created to coordinate the different assessments when both federal and territorial governments were triggered by the same project.

The DIAND Environment Directorate remained responsible for coordinating the assessments of major projects and was devolved to the Yukon Executive Council Office (ECO), Development Assessment Process Branch. In June 2003, the Regional Environmental Review Committee's name was changed to the Yukon Environmental Review Committee (YERC), although the committee largely operates with functions and roles similar to those it had under the CEAA. The manager of ECO's Environmental Assessment Unit chairs the YERC, and the committee continues to facilitate the assessment of major projects, disseminate EA information, and serve as a forum to review and participate in national initiatives.

The most significant change in environmental assessment under the YEAA has been that more assessments are being conducted owing to the requirement that the Yukon government conduct assessments whenever there are triggers. When the YESAA came into effect, environmental assessments started under the YEAA will continue through to completion under the YEAA rather than be transferred to the new regime. While most EA in Yukon will now be undertaken under territorial rules and guidance, major projects may still trigger CEAA reviews and there will be a need (as in the provinces) for a means to coordinate these.

Yukon First Nations, Land Claims, and TEK

The UFA framework allows the First Nations to negotiate individual agreements. As of 2008, most Yukon First Nations have signed and ratified final agreements, the exceptions being Ross River, Liard, and White River First Nations. These individual final agreements reflect the provisions of the broader UFA while recognizing and respecting each First Nation's power and responsibility. At the same time, self-government agreements have been, and are being, negotiated with First Nations. These agreements create new government-to-government relationships between the First Nations and the federal and Yukon governments. They also give the First Nations explicit law-making powers over their citizens and settlement land. Claim agreements formally acknowledge Yukon First Nations as legally constituted governing bodies that work collaboratively with other levels of government. The successful settlement of land claims increases certainty about Aboriginal interests on settlement lands, as it leads to clear definitions of ownership and jurisdiction over land. Both final and self-government agreements seek to promote economic self-sufficiency for First Nations.

A range of resource and environmental management initiatives stem directly from the Umbrella Final Agreement, among them the Yukon Development Assessment Process and the Yukon Land Use Planning Council. The implementation of claims-derived processes is still in an early phase, with the exception of the Inuvialuit Settlement Region. In 1984 the Western Arctic (Inuvialuit) Claim Settlement Act was passed, becoming one of the first comprehensive land claim settlements to empower First Nations formally in resource and environmental management. The Inuvialuit Final Agreement outlined a joint government-Inuvialuit environmental impact screening and review process. The introduction of the YESAA raises issues of duplication, as an environmental assessment must occur under both the IFA and the YESAA on any proposed project on the North Slope (see Henderson 1992 for an early review of the issue).

Reed (1990) among others has argued strongly for the inclusion of Aboriginal peoples in environmental assessment exercises, from which they have historically been excluded; she sees their inclusion as important because of their dependence on land and resources for their livelihood and because of the convergence of EIA and claims structures in co-management for sustainability. Involvement of First Nations in EA is critical owing to the continued significance of the environment to them and their strong ethical and legal claims to the right to participate. There are continuing challenges in implementing true First Nations involvement in resource and environmental management (Natcher and Davis 2007), and the YESAA experience reflects this (see below).

Another common rationale for First Nations' participation is their possession of indigenous knowledge, or traditional environmental (ecological) knowledge. Many authors have called for the inclusion of TEK in environmental assessment (e.g., Inglis 1993), and substantial work has been done on that within co-management processes created by the IFA (e.g., WMAC, NS, and AHTC 2003) and elsewhere in Canada. There remain, however, many challenges in integrating or, perhaps better, in learning from TEK in EA and other decision-making processes (see Usher 2000; and on inclusion of TEK in southwest Yukon wildlife co-management, see Nadasdy 2003). While it is a more holistic approach to include TEK (see Fischer and Keith 1977) and improved communication and understanding, as well as shared decision-making power, are cited as benefits to using traditional knowledge, there are few formal methods of collecting and interpreting TEK and the process is often limited by a lack of financial resources (Frayne 1998). Nevertheless, innovative approaches are being developed (e.g., Wolfe et al. 2007).

Mining

While very large mining projects have received full CEAA attention, environmental stakeholders in Yukon have long been concerned that smaller- and medium-scale placer and hard rock mines have been inadequately assessed. This issue relates to old law and practice dating back to the Klondike Gold Rush and legislation passed then (especially the Yukon Quartz and Placer Mining Act and the Yukon Placer Authorization). For many years, such activities were not regulated, and there was only a little change from then through the 1980s. Beginning in the 1990s, Yukon environmental non-governmental organizations (ENGOs) made a concerted effort to introduce fuller regulation and assessment of Yukon placer mining activities (for an EA perspective, see YCS 2001). While it remains a contentious issue, progress has been made, although neither the industry nor the territorial government has been consistently supportive of stronger environmental protection. The opposing viewpoint is that there is too much regulation from too many different agencies and acts, and that this inhibits mineral development. In late 2002, it was announced that the 1993 Yukon Placer Authorization under the federal Fisheries Act would be phased out by 2007 and replaced by a new regime jointly determined by the governments of Canada, Yukon, and the Council of Yukon First Nations. An agreement and implementation steering committee was established in May 2003 (see Yukon, EMR 2003), and took effect in April 2008.

Under the CEAA, one of the most significant changes to the environmental assessment process was informally known as the 'blue book'. In response to the mining industry's complaints, a steering committee was created to clarify and streamline the administration of environmental assessment processes. Representatives from the Yukon Territorial Water Board, the Yukon Chamber of Mines, the Yukon government, and the Department of Indian Affairs and Northern Development worked to address industry concerns about the federal permitting process. The final draft also benefited from the public review of the committee report, and industry endorsed the proposed administrative changes.

In October 2000, the blue book, titled *Administrative Procedures for Environmental Assessment of Major Mining Projects in the Yukon,* was released (see Environment Directorate 2001). Although the majority of the processes identified in the blue book applied to major mining projects only, key aspects include

- the process for Level I/Level II assessments (project description, environmental assessment report and environmental screening report);
- terms of reference for the Regional Environmental Review Committee;
- generic guidelines for the preparation of project descriptions and EA reports;
- timeline estimates;
- project agreements for management and environmental assessment; and
- the Yukon Major Projects Advisory Committee.

Additionally, DIAND and the Yukon Territorial Water Board developed a protocol to clarify the relationship between environmental assessment and regulatory approval. The Yukon Water Board is an adjudicative board that acts as an impartial tribunal and is responsible for issuing water licences for water use and waste deposit in water. These licences include conditions that are intended to mitigate potential adverse environmental effects. The protocol includes several key changes:

- It extends consultation on draft environmental screening reports until the water licence public hearing is completed.
- It ensures that intervenors have time to review the draft screening report prior to preparing interventions.
- It provides for technical edits of draft water licences prior to approval.

It is not yet clear how these developments will relate to the YEAA and YESAA. Currently, the Yukon Water Board cannot issue a water use license, or set terms of a license, that contradict a decision under YESAA.

Cumulative Environmental Effects and Land-Use Planning

Under the CEAA and successor legislation, the cumulative effects of projects have had to be considered, and this is an area where Yukon has placed considerable focus recently. The mid-1990s screening of the Kluane National Park Reserve Management Plan was an early study and part of the development of guidelines for cumulative environmental effects within the CEAA (Hegmann 1995). Other work on early northern cumulative effects took place in the Western Arctic (see Glaholt 1994). In 1997 the Yukon region of DIAND developed a user's guide to give guidance on screening cumulative effects. From 2001 to 2003, the DIAND Environment Directorate coordinated workshops to explore cumulative effects assessment and management strategies and developed guidelines for cumulative effects thresholds, largely in a forestry and wildlife context (e.g., Axys 2001). In the first two workshops there was recognition of the link between cumulative effects strategies and regional land use plans.

Regional land-use planning is, ideally, a participatory process that seeks to balance a wide range of interests within a specific geographic area. Both environmental assessment and land-use planning require cooperation and communication within similar sets of diverse interests and agencies, and they should take a regional perspective. There is renewed recognition academically (e.g., Kennett 1999), as well as in regional land-use planning and among EA practitioners, that EA and land-use planning activities should be integrated and that regional land-use planning might be one of the best tools to manage cumulative effects. It is worth noting that the Council for Yukon Indians recognized this more than 25 years ago (CYI 1982).

Land-use planning is outlined in chapter 11 of the Yukon First Nation Umbrella Final Agreement as the process for determining regional land-use priorities in Yukon. The Yukon Land Use Planning Council (YLUPC) makes recommendations to government and affected First Nations on land-use planning, the identification of planning regions, the priorities for land-use plans, and terms of reference for the Regional Land Use Planning Commissions in Yukon (YLUPC 2004). The council's role is increasingly coordinative. Several planning commissions are now in operation in Yukon, and the North Yukon Draft Plan was released in early 2008. The council held two workshops on cumulative effects and land-use planning in 2003 (with proceedings and background documents available online). It is to be hoped that EA and land-use planning can be linked in Yukon; as plans are being released, this is increasingly important. Some efforts in this direction can be found in the Mackenzie Valley Resource Management Act regime (see Lindsay and Smith 2001; Armitage, chapter 11 of this volume).

YESAA

Although the five-year review of YESAA is just getting underway, a number of concerns have been raised by numerous actors in the process. While there is clear support for a Yukon-based assessment process, the online registry, and the regional offices, there are some problematic areas. At a detail level, it is certainly the case that many more, and smaller, activities have required assessment than previously. Not all would agree all of these truly required the full process. More formal provisions for strategic environmental assessment might help to address this issue for some classes of activities. The timelines for response by agencies and intervenors are perceived as too short by many, although likely not by industry. It is also not clear what happens to stakeholder input when it is submitted. There are difficulties in recruiting experienced, qualified professional staff for the designated offices, and the diversity of the projects going through those offices is a serious challenge to the expertise of any single member, or even a few, of the assessment staff. There may be a need for more opportunities to acquire external, independent advice for even lower-level assessments. High thresholds for the assessment of some activities (e.g., the cutting of fuel wood) mean that cumulative effects are not considered because activities above the threshold are exempted.

Larger issues include the overturning of YESAB recommendations by territorial government decision bodies and no requirements for written reasons in the decisions made by decision makers. Some people feel that there have been several examples of weak YESAB recommendations (at least from a conservation perspective; see YCS 2008). There are concerns that narrow scoping may undermine EA due to rigid requirements for a specific, non-government proponent for an activity before it can be assessed. There has been some feeling that socio-economic factors have been inadequately considered in assessments. Then there is the fact that First Nations are finding that they do not always have the capacity to engage meaningfully in such an elaborate and bureaucratic process. They are also not happy with their limited ability to influence assessment outcomes within their traditional territories but outside their settlement lands. There are government-to-government relationship issues, as well as some debate over whether there can be more than one decision body in some cases (CBC News 2008). The need for a land-use planning framework to help guide environmental assessment is also increasingly recognized.

Discussion and Conclusions

Environmental assessment in Yukon has seen considerable change and evolution in the last 45 years. Much of that change was driven, given the federal control of land and resources, by changes in federal EA policy, in turn driven by legal, political, and environmental concerns far from Yukon. These common changes include the switch from Cabinet order to legislation, the broadening of the definition of *environmental*, the initial moves to tighten up self-regulation and assessment, the initial attention given to cumulative effects, and the early efforts to foster public participation in at least major EA processes.

Other changes have been driven by concerns within Yukon, most obviously and importantly by land claims and devolution, but also by environmental campaigns and litigation within Yukon. The Yukon region of DIAND was challenged in court on the assessment of the Faro Vangorda project in Faro, and this resulted in the department developing detailed policy and procedures for environmental assessment. Another challenge during this period involved incorporating mitigation that exceeded the DIAND minister's responsibilities. The department worked on ways to incorporate these 'super-added responsibilities' into their assessment and regulatory processes. DIAND implemented these responsibilities through regulatory authorizations, environmental agreements, compensation agreements, and impact-benefit agreements.

Yukon remains a large territory with a small population and much actual and potential development activity. This makes environmental assessment in the territory contentious, characterized by a range of substantive and process-oriented issues common to northern EA processes (Mulvihill and Baker 2001). Devolution not only of EA responsibility but also of most land, water, and resource management responsibility, at the same time as land claims–mandated co-management processes are being developed, offers the unprecedented opportunity to implement collaborative, participatory management, build on indigenous knowledge, and integrate and link different environmental management activities, such as EA and land-use planning, which are usually separate. Outside the IFA region, all this is quite new to Yukon, although it has had several years of experience with regional Renewable Resource Councils, the Yukon Land Use Planning Council, and now YESAA. Recent experience in Yukon suggests that YESAA has likely reflected local concerns and realities better than earlier regimes have, and has improved certainty about planning, assessment, and jurisdictional and ownership issues. But there is room for improvement, and a well-defined review process has been put in motion to identify needed change, which governments will hopefully implement. These processes have not necessarily made resource and environmental management and assessment processes simpler or faster. But they have improved local relevance and control, and by balancing development and conservation and incorporating traditional and scientific knowledge, they should foster more effective co-management.

Acknowledgements

We are grateful to the numerous people who gave us interviews and to the Social Sciences and Humanities Research Council of Canada and the DIAND Northern Scientific Training Program for support.

References

Alaska Highway Gas Pipeline Environmental Assessment Panel. 1979. *Alaska Highway Gas Pipeline, Yukon Hearings (March–April 1979): Report of the Environmental Assessment Panel.* Ottawa: FEARO.

Axys Environmental Consulting Ltd. 2001. Thresholds for addressing cumulative effects on terrestrial and avian wildlife in the Yukon Territory. Calgary: Axys Environmental Consulting Ltd.

Berger, T. 1977. *Northern frontier, northern homeland: The report of the Mackenzie Valley Pipeline Inquiry.* Ottawa: Supply and Services Canada.

Bone, R.M. 2002. *The Geography of the Canadian North.* 2nd ed. Toronto: Oxford University Press.

Cassidy, F., and N. Dale. 1988. *After Native claims? The implications of comprehensive claims settlements for natural resources in British Columbia.* Halifax: Institute for Research on Public Policy.

CBC News. 2008. Dawson First Nation claims right to license Yukon Queen. Available at http://www.cbc.ca/technology/story/2008/09/24/yukon-queen.html. Accessed 12 October 2008.

Christiane Boisjoly and Associates Inc. 1999. Report on one day meeting to develop a working partnership between the agencies involved in the environmental assessment. Whitehorse, YT, May.

CYI (Council of Yukon Indians). 1982. Land use planning, environmental assessment and land ownership in Yukon: A discussion paper. Whitehorse, YT: CYI, August.

DIAND (Department of Indian Affairs and Northern Development). 2000. The Northern Affairs Program, DIAND, Yukon Region and the Canadian Environmental Assessment Act. Ottawa: Indian and Northern Affairs Canada.

———. 2001. Yukon Environmental and Socio-economic Assessment Act and Regulations. Ottawa: Indian Affairs and Northern Development.

DIAND, YTG (Yukon Territorial Government), and CYFN (Council for Yukon First Nations). 1999. Development assessment process legislation—Comments on consultation. Whitehorse, YT: DIAND, March.

Environment Directorate, DIAND. 2001. Guidelines for the environmental assessment of major mining projects in the Yukon. Whitehorse, YT: DIAND.

Everitt, R.R., J.H. Andrew, B. Sadler, and D. Loeks. 1988. Environmental assessment processes in the Yukon. Essa Ltd (Vancouver) report to Biological Resources Division, Indian and Northern Affairs Canada, Whitehorse, YT, August.

Fischer, D.W., and R.F. Keith. 1977. Assessing the development decision-making process: A holistic framework. *American Journal of Economics and Sociology* 36 (1): 1–17.

Frayne, T.L. 1998. An examination of the development assessment process, Yukon. Master's thesis, Department of Geography, University of Guelph, Guelph, ON.

Gibson, R.B. 1984. Environmental assessment in Canada. *Environmental Education and Information* 3 (3): 30–52.

———. 1993. Environmental assessment design: Lessons from the Canadian experience. *Environmental Professional* 15:12–24.

Glaholt, R.D. 1994. Cumulative effects management in the Canadian Western Arctic. In A.J. Kennedy (Ed.), *Cumulative effects assessment in Canada: From concept to practice,* 265–74. Edmonton: Alberta Society of Professional Biologists.

Government of Canada, CAFN, CYFN, Na-Cho-Nyak Dun, VGFN, YTG, and TTC. 2000. Five-year review of the Umbrella Final Agreement Implementation Plan and Yukon First Nation Final Agreement Implementation Plans for the first four Yukon First Nations. Ottawa: DIAND.

Government of Canada, Council for Yukon Indians, and Government of Yukon. 1993. Umbrella Final Agreement between the Government of Canada, the Council for Yukon Indians and the Government of the Yukon. Ottawa: DIAND.

Green, N., and R.M. Binder. 1995. Environmental impact assessment under the Western Arctic (Inuvialuit) Land Claim. In J.A. Bissonette and P.R. Krausman (Eds), *Integrating people and wildlife for a sustainable future*, 343–5. Proceedings of the first International Wildlife Management Congress. Bethesda, MD: Wildlife Society.

Hébert, M., and C. Hilling. 2002. Bill C-2: The Yukon Environmental and Economic Assessment Act. Legislative Summary 441E. Parliamentary Research Branch, Ottawa. Available at http://www.parl.gc.ca/common/bills_ls.asp?Parl=37&Ses=2&ls=c2.

Hegmann, G. 1995. A cumulative effects assessment of proposed projects in Kluane National Park Reserve, Yukon Territory. Report to Kluane National Park Reserve, Department of Canadian Heritage. Haines Junction, YT, September.

Henderson, L.A. 1992. Environmental impact assessment in the North Yukon: Options for harmonization. Whitehorse, YT: Tuak Environmental Services for Yukon Department of Renewable Resources.

Inglis, J. (Ed.). 1993. *Traditional ecological knowledge: Concepts and cases.* Ottawa: International Program on Traditional Ecological Knowledge/International Development Research Centre.

Keith, R., and P. Mulvihill. 1995. Organizational development and environmental assessment in Canada's North. *Environments* 23 (1): 71–81.

Kennett, S.A. 1999. Towards a new paradigm for cumulative effects management. Canadian Institute of Resources Law, Occasional Paper 8, University of Calgary, December.

Koivurova, T. 2002. *Environmental impact assessment in the Arctic: A study of international legal norms.* Aldershot, UK: Ashgate Publishing.

Lindsay, K.M., and D.W. Smith. 2001. *Review of environmental assessment processes.* Occasional Publication 50. Edmonton: Canadian Circumpolar Institute.

Mulvihill, P.R., and Baker, D.C. 2001. Ambitious and retrospective scoping: Case studies from Northern Canada. *EIA Review* 21 (4): 363–84.

Nadasdy, P. 2003. Reevaluating the co-management success story. *Arctic* 56 (4): 367–80.

Natcher, D.C., and S. Davis. 2007. Rethinking devolution: Challenges for Aboriginal resource management in the Yukon Territory. *Society and Natural Resources* 20:271–79.

Notzke, C. 1994. *Aboriginal peoples and natural resources in Canada.* Concord, ON: Captus Press.

Prystupa, M.V. 1994. Evaluation of Yukon Environmental Assessment Interest Representation. Ph.D. thesis, Department of Geography, University of Western Ontario, London.

Reed, M. 1990. *Environmental assessment and Aboriginal claims: Implementation of the Inuvialuit Final Agreement.* Ottawa: Canadian Environmental Assessment Research Council.

Rees, W.E. 1980. EARP at the crossroads: Environmental assessment in Canada. *EIA Review* 1 (4): 355–77.

———. 1983. Environmental assessment of hydrocarbon production from the Canadian Beaufort Sea. *EIA Review* 4 (3/4): 539–55.

Sadler, B. 1990. An evaluation of the Beaufort Sea Environmental Assessment Panel Review. Minister of Supply and Services, Ottawa.

Usher, P. 2000. Traditional ecological knowledge in environmental assessment and management. *Arctic* 53 (2): 183–93.

Wallace, R.R. 1986. Assessing the assessors: An examination of the impact of the federal environmental assessment and review process on federal decision making. *Arctic* 39 (3): 240–6.

WMAC, NS, and AHTC (Wildlife Management Advisory Council [North Slope] and the Aklavik Hunters and Trappers Committee). 2003. Aklavik Inuvialuit describe the status of certain birds and animals on the Yukon North Slope, March 2003. Final report. Whitehorse, YT: WMAC (NS).

Wolfe, B.B., D. Armitage, S. Wesche, B.E. Brock, M.A. Sokal, K.P. Clogg-Wright, C.P. Mongeon, M.E. Adam, R.I. Hall, and T.W.D. Edwards. 2007. From isotopes to TK interviews: Towards interdisciplinary research in Fort Resolution and the Slave River Delta, Northwest Territories. *Arctic* 60 (1): 75–87.

YCEE (Yukon Council on the Environment and the Economy). 1999. Building an effective Yukon Development Assessment Process: Report on the Draft Yukon Development Assessment Act. Whitehorse, YT, February.

YCS (Yukon Conservation Society). 2001. Environmental assessment reform in the Yukon. Whitehorse, YT, June.

————. 2008. YESAA. Whitehorse, YT, June. Available at http://www.yukonconservation.org/. Accessed 10 October 2008.

YLUPC (Yukon Land Use Planning Council). 2004. YLUPC website, newsletter. Available at http://www.planyukon.ca/index.php. Accessed 4 February 2004.

YTG (Yukon Territorial Government), CYFN (Council of Yukon First Nations), and Government of Canada. 2008. The YESAA Five-Year Review. Available at http://www.yesaareview.ca/en/review/.

Yukon, Bureau of Statistics. 2008. Yukon Factsheet. Available at http://www.eco.gov.yk.ca/stats/yukonfactsheet.html. Accessed 10 October 2008.

Yukon, Department of Energy, Mines and Resources (EMR). 2003. New placer mining regime. Available at http://www.emr.gov.yk.ca/Mining/PlacerMining/new_placer_regime.htm. Accessed 4 February 2004.

Yukon, Executive Council Office. 2003. Yukon's Environment Assessment Act information sheet. Available at http://www.gov.yk.ca/depts/eco/dap/information.html. Accessed 6 February 2004.

————. 2004. Environmental assessments in the Yukon government. Available at http://www.gov.yk.ca/depts/eco/dap/Transition%20info%20sheet.pdf. Accessed 31 January 2005.

Yukon, Executive Council Office, DAP Branch. January 2005. Yukon environmental assessment newsletter. Whitehorse, YT. Available at www.eco.gov.yk.ca/dap/dab_news.html.

Chapter 13

Impact Assessment in Nunavut

J. Jeffrey Rusk, Sophia C.R. Granchinho, and Ryan W. Barry

Introduction

Impact assessment in Canada's three territories demonstrates an array of similarities and differences. As the governance structure and regulatory regimes of these jurisdictions continue to develop, the impact assessment process and procedures also evolve. This evolution has likely been most precipitous in Canada's newest territory, Nunavut. Impact assessment in Nunavut is currently experiencing the effects of a wide range of changes—political, institutional, economic, environmental, and cultural. Previously included in the Northwest Territories (NWT) regime, Nunavut-specific impact assessment was not a reality until the implementation of the Agreement Between the Inuit of the Nunavut Settlement Area and Her Majesty the Queen in Right of Canada (Government of Canada and Tungavik Federation of Nunavut 1993). This document, typically referred to as the Nunavut Land Claims Agreement (NLCA), represents the settlement of land claims with the Inuit of Nunavut as well as the framework for the creation of the Territory of Nunavut. However, the territory was not formally created until April 1999.

Since the signing of the NLCA in 1993, impact assessment in Nunavut has been affected by many changes, including the incorporation of the territory itself, the creation of 'institutions of public government' (IPGs) under the NLCA, intense development pressure from the mineral and mining sector, cultural changes in the societies of Inuit and other northern residents, the recognition of Inuktitut and Inuinnaqtun as official languages territorially, and a number of environmental changes, such as climate change and species decline. Changes of similar magnitude are still occurring today and are having a profound effect on the impact assessment process.

While in the past the federal impact assessment system (see chapter 14) under the Canadian Environmental Assessment Act (CEAA) was used in the Nunavut Settlement Area, it is no longer applied in Nunavut. Impact assessment in Nunavut is currently almost the exclusive purview of the Nunavut Impact Review Board (NIRB), a relatively new IPG. Although the NIRB has been conducting screening, review, and monitoring tasks since 1996 (prior to 1996 the NIRB was structured as a 'transition team'), it currently lacks the enacting federal legislation necessary for it to provide implementation direction to supplement its mandate described in the NLCA. Any discussion of the impact assessment process in Nunavut is necessarily focused on the process as defined by the NIRB. The NIRB is officially designated an IPG under the NLCA, and in its general structure it bears a strong resemblance to the co-management body described in Plummer and Armitage 2007.

As a creation of the NLCA, the Nunavut Impact Review Board is meant to ensure that Inuit have an opportunity to be formally involved in—and even to direct—impact assessment in Nunavut. However, as a public body established and funded by the federal government, it does inherit a number of the bureaucratic features that have been identified by White (2006) in similar institutions (such as the Nunavut Wildlife Management Board) as being in conflict with the traditional Inuit understanding of resources and their management. Although the NIRB has a collaborative role with other IPGs in the land and resource management regime in Nunavut, it exists primarily to conduct impact assessment activities in accordance with the NLCA, specifically screening, review, and monitoring operations. It does so primarily at the level of the individual project proposal, although it does have a limited ability to provide recommendations related to regional issues. For example, the NIRB can play a specified advisory role in marine management as a member of the Nunavut Marine Council (Mulrennan and Scott 2000). Although the focus of the NIRB is on impact assessment in the holistic sense, this chapter necessarily focuses on the large subset of this domain referred to as environmental assessment (EA); the two terms are often used interchangeably in Nunavut.

The purpose in this chapter is to analyse what can be learned about impact assessment generally through the Nunavut experience and also what can be learned about the conduct of EA in northern cross-cultural environments. Practitioners and researchers alike can capitalize on these experiences and further improve impact assessment in Nunavut and in other, similar jurisdictions. This chapter is based on extensive experience in conducting impact assessments in Nunavut, as well as on a review of the relatively sparse body of literature dealing with matters relating to Nunavut-specific impact assessment. It begins by outlining the basis of impact assessment in Nunavut as defined by the NLCA and then proceeds to describe, with appropriate cases, the screening, review, and monitoring practices that comprise the overall EA process. A number of opportunities and challenges become evident in this review and are detailed in a later section. Following the discussion of these opportunities and challenges, recommendations and conclusions are presented. While there exists a body of literature pertaining to EA in Canada's North, most of it involves only the NWT and Yukon. Very little work has covered development projects in Nunavut, and there is virtually no literature that critically evaluates how EA is conducted within Nunavut's unique framework. This chapter serves as a novel contribution in this regard.

The Nunavut Land Claim Agreement

The Nunavut Impact Review Board was established, along with other IPGs, under Article 10 of the NLCA. However, the board's role is largely defined, in detail, within Article 12. The NIRB is relegated to conducting EA for project proposals within the Nunavut Settlement Area and the Outer Land Fast Ice Zone on the East Baffin coast (Figure 13.1). It has jurisdiction to fulfil its mandate on all lands in Nunavut, in-

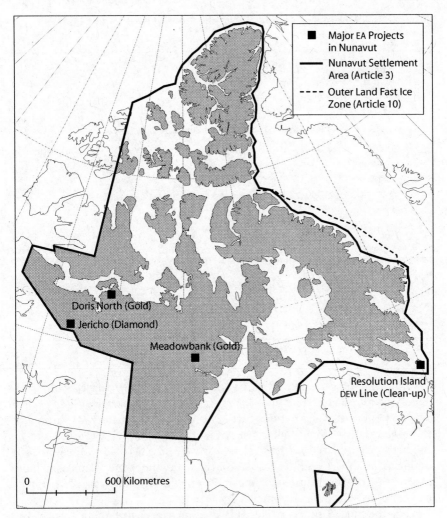

Figure 13.1   Map of Nunavut Settlement Area, Outer Land Fast Ice Zone, and major EA projects completed by the Nunavut Impact Review Board

cluding Crown lands (including national parks and other protected areas), commissioner's lands (territorial), Inuit-owned lands, and other private lands. While the Mackenzie Valley Resource Management Act (MVRMA) effectively removed the application of the CEAA in the NWT (Valiela 2006), initially the federal interpretation of the NLCA did not remove the role of the CEAA in Nunavut. The NLCA has since been amended to formally remove the CEAA from the territory (INAC 2008), leaving the NIRB as the sole agency responsible for conducting environmental assessments, with the exception of EAs for projects sent to a federal panel for a Part 6 review, under Article 12 of the NLCA. Article 12 assigns and defines the responsibilities of the NIRB in conducting EA in Nunavut. The sections of this article are listed in Table 13.1.

Table 13.1   Sections of Article 12 within the Nunavut Land Claims Agreement

Section	Theme
1	Definitions
2	Nunavut Impact Review Board
3	Relationship to land-use planning provisions
4	Screening of project proposals
5	Review of project proposals by NIRB
6	Review by a federal environmental assessment panel
7	Monitoring
8	Flexibility in relation to certificates
9	Implementation
10	Enforcement
11	Transboundary impacts
12	Application

Although it has been noted that the NIRB lacks federal enacting legislation, it is important to recognize that the NLCA gives, for the most part, clear and explicit direction to the NIRB and to those parties the board may interact with during the EA process. The sections guiding the screening (Part 4), review (Parts 5 and 6), and monitoring (Part 7) of project proposals are very detailed and require little clarification in most cases. The EA tasks assigned to the NIRB are similar to those conducted by other institutions in jurisdictions throughout Canada, although a number of modifications are unique to Nunavut.

While the NIRB is an IPG created in the co-management fashion described in the NLCA, it is essential to recognize that the federal government retains an overriding authority over EA conducted on most of the land in Nunavut (there is very little territorial commissioner's land in Nunavut, most falling within hamlet boundaries)—even on Inuit-owned lands. The NIRB's screening decisions serve only as recommendations to an authorizing minister, and even with the stronger project certificates associated with a Part 5 review, the minister has a number of options regarding how the NIRB's decision is implemented. The Nunavut process is similar to that found in the NWT, where the board reports its findings from the EA to the final decision-making authority—the minister of Indian and northern affairs (see chapter 11). However, it differs from the process found in Yukon, where the final decision-making authority may include federal, territorial, and/or First Nations governments (see chapter 12). The overall process for EA in Nunavut is illustrated in Figure 13.2.

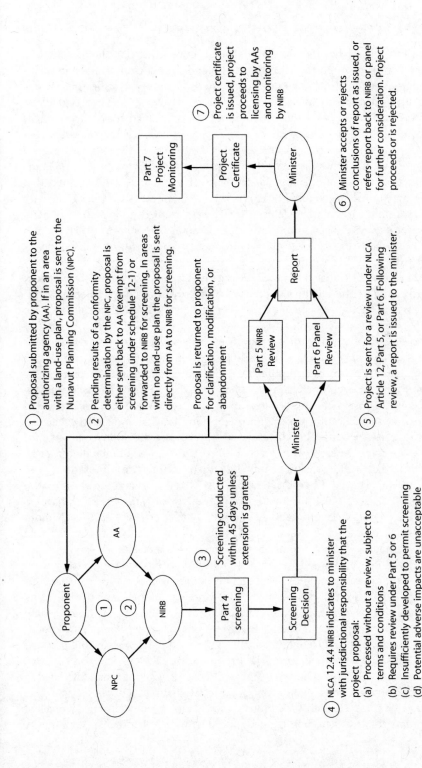

Figure 13.2 Environmental Assessment Process in Nunavut

Part 4 Screening of Project Proposals

Part 4 of Article 12 of the NLCA outlines the process the NIRB should follow in screening project proposals. The project's potential for environmental and socio-economic impacts should be documented, and measures for mitigating the adverse effects developed. Alternatively, the project may be deemed unacceptable as proposed and the proposal returned for modification, or further assessment might be required through a review by the NIRB or a federal environmental assessment panel as determined by the minister. As defined in Article 1 of the NLCA, a "'project proposal'" means a physical work that a proponent proposes to construct, operate, modify, decommission, abandon or otherwise carry out, or a physical activity that a proponent proposes to undertake or otherwise carry out, such work or activity being within the Nunavut Settlement Area, except as provided in section 12.11.1" (Government of Canada and Tungavik Federation of Nunavut 1993). Section 12.11.1 of the NLCA speaks to the environmental assessment of transboundary impacts by the NIRB, making it possible for the NIRB to review projects physically located outside of Nunavut but with the potential for impacts within the Nunavut Settlement Area. This section also directs the government(s) (both federal and territorial) and the NIRB to negotiate with other jurisdictions to achieve collaboration in the review of project proposals.

The NIRB may receive a project proposal for screening in one of two ways, either from the Nunavut Planning Commission (NPC) or directly from an authorizing agency (AA) with which a proponent has filed an application (see Figure 13.2). In areas with an established land-use plan (LUP), the NPC makes a determination regarding the proposal's conformity with the regional LUP. Should the proposal conform but not meet the criteria for exemption from screening (NLCA Schedule 12-1), the NPC forwards both the proposal and the determination to the NIRB for Part 4 screening. In areas with no established LUP, the AA submits the proposal directly to the NIRB for screening. At present, there are only two approved land-use plans in Nunavut, the Keewatin Regional LUP (NPC 2000a) and the North Baffin Regional LUP (NPC 2000b). In the remaining regions of Nunavut, which have no approved plans, project proposals are submitted directly to the EA process without any formal consideration of land-use planning issues, which means that the NIRB commonly addresses issues somewhat unrelated to its mandate.

Once the NIRB acknowledges receipt of a project proposal, it conducts an internal check for completeness to ensure that the necessary information requirements are met. The board corresponds with the proponent and relevant AAs regarding any information deficiencies and attempts to resolve these prior to proceeding with the screening process. Prior to being sent out for public comment, all documents pertaining to a screening are made available for public viewing online at the NIRB's ftp site (http://ftp.nirb.ca). The ftp site mirrors the public registry located in Cambridge Bay, Nunavut, and is updated regularly with information related to NIRB screenings, reviews, monitoring, and other initiatives.

Once the completeness of the project proposal has been determined, copies of the proposal are sent to those named on a distribution list, such as community representatives, designated Inuit organizations, hunters' and trappers' organizations (HTOs), co-management boards, community councils, federal and territorial government departments, relevant wildlife management boards, and other potentially interested agencies or individuals. Members of the distribution list are asked to comment on the project proposal from the perspective of their knowledge area, expertise, and mandate within a stated time frame, usually between one and three weeks. The commenting period of the Part 4 screening process allows for meaningful public input into EA projects in Nunavut and serves as one method of gauging the potential for significant public concern. Comments may include, but are not limited to, an indication of support for or against the project proposal; a summary of the regulatory role and/or mandate of the commenting party, where applicable; an indication of approval authority; a request for additional information; an expression of concern for potential impacts; and recommendations for terms and conditions, including monitoring and mitigation.

During the public commenting period, parties such as HTOs and community hamlet offices commonly offer traditional knowledge about harvesting activities or archaeological sites in a proposed project area, and may communicate their desire that the proponent continue to consult with them throughout all stages of the project. Depending on the issues raised, further information or commitments may be required from the proponent before the NIRB can make a determination. Concerns related to the protection of caribou, for example, are often raised by commenting parties, especially when exploration programs are proposed within the traditional caribou calving or post-calving grounds of Nunavut's various resident barren ground caribou herds. During the screening of a uranium exploration program proposed by Cameco Corporation (NIRB file no. 08EN015), revised wildlife mitigation and monitoring plans were required, and the terms and conditions recommended by the NIRB detailed extensive adaptive mitigation measures for the protection of caribou during sensitive periods. In addition to the immediate protection that these terms and conditions offer, the reporting requirements allow for improved information sharing with the governments and co-management boards responsible for managing the caribou herds in Nunavut.

Once comments and any additional requested information have been received, the NIRB determines whether the project proposal has such significant impact potential or is of such significant public concern as to require review under Part 5 or 6 of Article 12 of the NLCA. In the screening assessment, the NIRB considers the following: the completeness of the project proposal; further information requests from the distribution list; ecosystemic impacts and specific environmental impacts; whether impacts can be mitigated with terms and conditions; and monitoring requirements. Four possible determinations are available to the NIRB at the conclusion of the screening assessment, as detailed in section 12.4.4 of the NLCA:

a)   The proposal may be processed without a review under Part 5 or 6; NIRB may recommend that specific terms and conditions be attached to any approval, reflecting the primary objectives set out in section 12.2.5.

b) The proposal requires review under Part 5 or 6; NIRB shall identify particular issues or concerns that should be considered in such a review.

c) The proposal is insufficiently developed to permit proper screening and should be returned to the proponent for clarification.

d) The potential adverse impacts of the proposal are so unacceptable that it should be modified or abandoned.

The NIRB issues its determination in a screening decision report to the minister with jurisdiction over the relevant authorizing agency, most commonly the minister of Indian and northern affairs Canada for land-use permits on Crown land. Although the NIRB recommends an appropriate course of action, it is the minister who is responsible for the final decision. Where the NIRB determines that a review is not required (section 12.4.4(a)), the screening decision report will include recommended project-specific terms and conditions for incorporation into the relevant authorizations. The AA responsible may elect to use some or all of the terms and conditions as recommended, either because of their limited legislated mandate or for other reasons. This limits the power of the NIRB, as the recommendations are not legally enforceable unless they are incorporated into a licence or permit. In practice, the NIRB works closely with the various AAs to ensure that properly worded and implementable terms and conditions are issued in its screening decision reports and subsequently adopted into the project authorizations. In cases where the NIRB determines, and the minister agrees, that a public review is necessary, the minister has the authority to send project proposals for a review under either Part 5 or Part 6 of Article 12 of the NLCA.

The NIRB has a responsibility under section 12.4.5 of the NLCA to complete its screenings within 45 days or, where an AA has a legal requirement to make a decision, within a shorter time frame. An important characteristic of Nunavut's EA process is the absence of an intermediary step between screenings and reviews, such as the comprehensive study option under the Canadian Environmental Assessment Act. Nunavut is a vast territory and the project proposals referred to the NIRB for environmental assessment each year vary greatly in scale and impact potential. Although the amount of effort required to screen a scientific research permit may differ considerably from that required to screen a proposal for a bulk sampling program for mineral exploration, the 45-day rule applies to both screenings equally. When necessary, the NIRB may apply in writing to the minister for an extension to the period of time allowed for screening. For example, an extension might be requested when the proponent needs additional time to address information deficiencies identified during the check-for-completeness stage of the screening process or perhaps to address an information request from a member of the distribution list.

Part 5 and Part 6 Reviews of Project Proposals

Article 12 of the NLCA provides for two different review processes for the environmental assessment of project proposals; these are commonly referred to as Part

5 reviews and Part 6 reviews. A Part 5 review is conducted by the NIRB, whereas a Part 6 review is conducted by a federal environmental assessment panel. The review process is initiated following the screening of the project proposal when it has been determined that the project may have significant adverse effects on the ecosystem and/or significant adverse socio-economic effects, or that there is significant public concern. The NIRB issues a screening decision report to the minister with jurisdiction over the AA, recommending a review under Part 5 or Part 6 and detailing the reasons for the recommendation. It is the minister who determines the type of public review there will be (i.e., Part 5 or Part 6). Section 12.4.7 of the NLCA directs the minister to take into account any relevant law, as well as the national and regional interests, when making this decision. Since the establishment of the NIRB in 1996 under the NLCA, all projects referred to the minister by the NIRB for review have undergone a Part 5 review (see Table 13.2). The following sections describe in detail the NIRB's Part 5 and Part 6 review processes. A detailed look at the NIRB's review of the Meadowbank gold mine project shows the flexibility inherent in the Part 5 review process (Table 13.3).

Part 5: Review of Project Proposals by NIRB

The review of a project proposal involves a detailed evaluation of the project's potential environmental impacts and an examination of the measures proposed to mitigate these impacts. The Part 5 review generally follows a typical procedure:

- Scoping of the project proposal
- Development and issuance of environmental impact statement (EIS) guidelines
- Submission of an EIS by the proponent
- Technical review of the EIS
- Technical meetings and hearings
- Submission of the assessment report to the minister

In sending a project proposal to the NIRB for review, the minister may identify particular issues or concerns that the NIRB must consider. This authority, however, does not limit the NIRB from reviewing any matter within its mandate as outlined in section 12.5.5 of the NLCA (see Table 13.3).

Following the minister's referral of a project for Part 5 review, the first step in the NIRB's review process is to determine the scope of the project. Scoping seeks to establish the potential environmental impacts; to consider alternative means of carrying out the project (including technical and technological alternatives); to identify the potential effects on the sustainability of resources in the project area; and to clarify the mitigation measures that will be analysed in the EA process. Scoping literally sets the stage for the entire environmental assessment and review process (Mulvihill and Baker 2001), and thus it is important that the scoping process take a broad view of the issues to be dealt with in the assessment. One important feature of EA in Nunavut is the early and full involvement of Inuit and

Table 13.2   Timelines for reviews completed under Article 12, Part 5, of the Nunavut Land Claims Agreement (1996–2006)

Project Name (NIRB File #)	Project Proposal Filed with the NIRB	NIRB Screening Decision Report Issued	Minister's Referral to Part 5 Review Issued	NIRB Pre-hearing Conference Report Issued	NIRB Final Hearing Report Submitted to Minister	Minister's Decision on Final Hearing Report	Project Certificate Issued	Time Taken to Complete Review
The Clean-up of PCBs at Resolution Island (98D01N074)	08-May-98	29-June-98	02-July-98	26-Sep-00	10-Nov-00	13-June-01	08-July-02	~ 4 Years
Jericho Diamond Project (00MN059)	07-May-99	07-Feb-01	14-Mar-01	17-July-01	03-Feb-04	07-June-04	20-July-04	~ 3.4 Years
Doris Hinge Project (02MN134)	01-Mar-02	05-June-02	27-Aug-02	12-June-03	13-Aug-04	06-Dec-04	Project not approved—insufficient information	n/a
Doris North Gold Mine Project (05MN047)	14-Feb-05	07-Mar-05	22-Apr-05	13-Sept-05	06-Mar-06	01-Aug-06	13-Sept-06	~ 1.4 Years
Meadowbank Gold Mine Project (03MN017)	31-Mar-03	23-Sept-03	03-Dec-03	14-July-05	30-Aug-06	14-Sept-06	30-Dec-06	~ 3.1 Years

Notes

1. 'Minister' refers to the minister of Indian and northern affairs Canada.

2. 'Time taken to complete review' refers to the duration of time between the minister's referral of the project proposal to a Part 5 review and the NIRB's issuance of a project certificate.

3. n/a = not available.

Table 13.3   Flexibility within the Part 5 review process: A closer look at the NIRB's review of the Meadowbank project proposal (NIRB File No. 03MN107)

Review Stage	Event
Minister's referral	Minister of Indian and northern affairs highlighted a particular issue for the NIRB to consider during its review (INAC 2005), the environmental and socio-economic effects of the construction of a 102 km all-weather road, the first of its kind in Nunavut.
Draft EIS conformity review	The draft EIS generally conformed to the EIS guidelines issued by the NIRB but did not capture all of the requirements. The proponent was required to provide clarification on many of the items within its submission, as well as provide supplementary information to address deficiencies identified during the draft EIS conformity review.
Final EIS compliance review	The internal review of the final EIS submission (Cumberland Resources 2005) found that it did not comply with the directions contained within the PHC decision report, and an addendum submission was required from the proponent prior to commencement of a technical review.
Final hearings	Final hearings were held in three communities in the Kivalliq region. Following the hearings, the NIRB decided that additional information was required to permit an adequate analysis of the impacts and to reach a decision to report to the minister. Final hearings were completed by way of written submissions, obtaining the required information from the proponent, and providing an opportunity for interested parties to comment on the additional information.
Exceptions from the review	Field investigation work for the proposed all-weather road was authorized during the review after the NIRB granted an exception from the review process under NCLA section 12.10.2 (NIRB 2005a).

other northern residents, particularly those with unique knowledge of the area (traditional knowledge). A successful scoping process is essentially a design exercise that influences the prospect of a successful EA (Mulvihill and Baker 2001). In 1997, guidelines for EA in the Arctic were developed by the Arctic Council (Arctic Environment Projection Strategy 1997), of which Canada is a member. These guidelines focus on the circumstances and issues of special importance in the Arctic, and have influenced the scoping procedures implemented by the NIRB.

Taking into account the issues raised during the scoping period, the NIRB issues project-specific guidelines for the preparation of a *draft* EIS. An EIS is a detailed document prepared by the proponent in accordance with these guidelines; it identifies, predicts, evaluates, and communicates information about the ecosystemic and socio-economic impacts of a proposed project. An EIS identifies the mitigation measures that can be developed to control, reduce, or eliminate the potentially adverse impacts of an activity or project. Once the proponent completes and submits a draft EIS, the NIRB will conduct an internal review of the material to determine whether the submission addresses the provisions of the guidelines. The guideline conformity review is focused on determining whether any information requirements of the NIRB's project-specific guidelines or any of the NIRB's 'Minimum EIS Requirements' (NIRB 2006) have been overlooked or omitted from the EIS. Guideline conformity review is strictly a presence-or-absence analysis; it is not intended to evaluate the quality of the information presented, although the NIRB may point out significant deficiencies encountered. Should any omissions be identified, the proponent may have to submit supplementary information or may be required to revise and resubmit the draft EIS (see Table 13.3).

Once the draft EIS is deemed satisfactory, the submission is released to the public for a minimum 60-day technical review period, during which the parties may request additional technical information to complete their review. A technical review is a detailed assessment of the potential project-specific, cumulative, and ecosystemic-level impacts and the proposed mitigation measures. Once the technical review of the document is complete, a technical meeting may be held to facilitate discussions on technical matters related to the draft EIS. Kept as informal as possible, technical meetings provide an opportunity for experts to resolve various technical issues prior to the preliminary hearing conference (PHC). During the technical meeting, the proponent may make commitments based on the discussion of the proposed project. These commitments are compiled as a list and carried forward to the PHC for incorporation in the PHC decision report.

The technical meeting and the PHC also provide a public forum for the discussion of the proposed project. Parties and intervenors, including those communities, organizations, and individuals that are potentially affected by a proposal, are given the opportunity to voice their concerns and present information to the NIRB. Both the PHC and the final hearing are quasi-judicial processes, open to the public, with sworn testimonies and the filing of evidence for the public record. Through section 12.2.25 of the NLCA, the NIRB is given the power to subpoena witnesses and documents during its proceedings. Discussions at the PHC help to establish a timeline for the submission of the proponent's *final* EIS, as well as the dates, locations, and other logistics of future meetings and the final hearing. Following the PHC, the NIRB prepares a report that gives the proponent direction regarding the requirements for the final EIS and the procedures for the review of the final EIS and final hearing. It is the responsibility of the proponent to prepare the final EIS in accordance with the PHC decision report and the list of commitments formulated at the technical meeting and approved by the NIRB.

Once the final EIS is complete, the proponent is responsible for circulating electronic and hard copies of it to all parties involved in the review. Prior to its being formally accepted, the final EIS submission is reviewed internally by the NIRB for compliance with the PHC decision report. As with the draft EIS review, this internal review is a presence-or-absence analysis; it is not intended to evaluate the quality of the information presented. Should any omissions be identified, the proponent is responsible for submitting supplementary information; if the final EIS is found to be significantly non-compliant with the PHC decision, it may be returned to the proponent (see Table 13.3).

Once the NIRB completes the compliance review and any shortcomings are adequately resolved, the submission is distributed to the public for a minimum 60-day technical review period. The intent of this technical review is to allow the analysis of the quality of the new and/or revised information, as well as of all the pre-existing information and the overall project in light of the information in the final EIS. Once the technical review is completed, the NIRB holds a final hearing to further solicit views and opinions from the public on the merits of the proposal. Significantly, the final hearing gives due regard and weight to the opinions of Inuit elders and community members and to the tradition of Inuit oral communication and decision making. In exceptional circumstances, final hearing proceedings can be completed by written submissions (see Table 13.3).

Following the final hearing, the NIRB prepares a report containing its conclusions and recommendations for distribution to the responsible minister (typically the minister of Indian and northern affairs), to the proponent and to members of the public on the distribution list. Although the NIRB makes a determination on projects under review, it is the minister who makes the final decision on whether or not the project should proceed based on the NIRB's determination as per section 12.5.7 of the NLCA. Once the review process is complete and it has been determined that a project should proceed, the NIRB issues a project certificate with the terms and conditions that have been accepted or varied by the minister.

Public participation is an important component of the NIRB Part 5 review process. Public involvement starts with screening, occurs throughout the review process, and continues during the NIRB Part 7 monitoring of the project. The public is informed when a project proposal is first submitted to the NIRB, and it is given information on how to participate in the scoping process and take advantage of the ample opportunity to review and comment on each phase of the EA. A variety of methods are used to involve the public, including letters, media announcements, information sessions and meetings held within the communities, and public hearings. Additionally, general information about the NIRB review process, the project undergoing the review, and any relevant correspondence is placed on the public registry and thus made available to the public. Moreover, participant funding may be made available to individuals and not-for-profit organizations interested in participating in the review of a project proposal. Participant funding programs are established and administered by Indian and Northern Affairs Canada (INAC) on a project-specific basis, helping to ensure effective public participation during the hearing or review processes.

Section 12.10.1 of the NLCA prohibits the issuance of the licences or approvals required for a proposed project to proceed until after a Part 5 or Part 6 review has been completed and a NIRB project certificate has been issued. However, in some instances, specific components of the project proposal (i.e., approvals or licences for exploration or development activities) may proceed without a review as outlined in section 12.10.2 of the NLCA (see Table 13.3).

Part 6: Review by a Federal Environmental Assessment Panel

As previously mentioned, upon receipt of a NIRB screening decision recommending that a project proposal receive a review, the responsible minister may deem it necessary to refer the proposal to the minister of the environment for a public review by a federal assessment panel in accordance with Part 6 of the NLCA. In making the decision for a Part 6 review, the responsible minister will consider the following questions: (1) does the project proposal involve matters of important national interest and thus might it best be reviewed under Part 6? and (2) is the project proposal to be carried out partly within and partly outside the Nunavut Settlement Area? Like a Part 5 review, a Part 6 review includes a review of both socio-economic and ecosystemic impacts. To date, no project proposals in Nunavut have been referred to a Part 6 review by a minister. Therefore, the example given here is by necessity a summary of a panel review process as outlined in section 12.6 of the NLCA.

The members of a federal assessment panel are appointed by the minister of the environment in accordance with the minister's general practice, with the exception that at least one-quarter of the panel members must be appointed from a list of nominees provided by the designated Inuit organization and a further quarter of the panel members must be appointed from a list of nominees provided by the appropriate territorial government minister. Nothing prevents the designated Inuit organization or the territorial government minister from nominating candidates who are already members of the NIRB. Panel members must be unbiased and free from conflict of interest, and have special knowledge and experience relevant to the potential environmental effects of the project proposal under review.

Once the panel members have been appointed, the panel develops a set of guidelines for the preparation of an EIS, based on scoping meetings that are held in the communities potentially affected by the proposed project. The NIRB reviews the guidelines and provides input into their development. After the EIS document is submitted to the panel, it is released to the public for review and to allow an opportunity for comments on its adequacy. The NIRB, too, is given adequate opportunity to review the EIS prior to the commencement of any public hearings, and the panel must take into account any recommendations or concerns that the NIRB identifies. Once a review of the EIS is completed, the panel reviews the comments from all parties and schedules a public hearing to receive additional views and opinions on the merits of the proposal. Following the completion of hearings, the panel will prepare a report containing its conclusions and recommendations to both the minister of the environment and the responsible minister. The minister will make the report available to the public and forward a copy to the NIRB for review. The NIRB has 60 days to review the report and forward to the minister its findings and conclusions with respect to deficiencies in the report and to potential ecosystemic and socio-economic impacts in the Nunavut Settlement Area. The NIRB may also identify additional terms and conditions, data requirements, and anything else deemed pertinent, including whether or not the project proposal should proceed. The minister considers the recommendations in the panel's report and the recommendations from the NIRB, and subsequently makes a decision on whether the project proposal should proceed with the proposed terms and conditions as they apply to the Nunavut Settlement Area. Once the proposal is accepted by the minister, the NIRB prepares and issues a project certificate that includes all the terms and conditions that have been accepted or varied by the minister.

Part 7 Monitoring of Projects

Under section 12.7.1 of the NLCA, the terms and conditions contained in a NIRB project certificate, screening decision report, or Nunavut Water Board licence may indicate the need for a project-specific monitoring program with specific responsibilities for the proponent, the NIRB, or other government agencies. The monitoring of projects verifies the accuracy of predictions made during the environmental assessment and determines the effectiveness of measures taken to mitigate any potentially adverse environmental effects. In this way a feedback loop for future decisions can be established (Morrison-Saunders et al. 2007). Monitoring activities can include effects monitoring and compliance monitoring.

A project certificate may require that a proponent develop plans to measure the effects of the project on the ecosystemic and socio-economic environments. For example, the project certificate issued by the NIRB for the Jericho diamond mine project (NIRB file no. 00MN059) required that a post-environmental assessment monitoring program (PEAMP) be developed for the project, in accordance with commitments made in the final EIS. Similar plans may be required by regulatory instruments and would thus be developed in partnership with regulators (e.g., a fisheries authorization from Fisheries and Oceans Canada). However, in order to avoid the unnecessary duplication of processes and to increase the efficiency of monitoring in general, the NIRB ensures that its monitoring programs are coordinated with those of other agencies. In cases where a regulatory instrument does not exist to capture monitoring requirements, the responsibility for ensuring that adequate project-specific monitoring takes place would fall to the NIRB.

Monitoring plans developed by the proponent should be designed to assess the accuracy of predictions contained in its final EIS. The proponent may be asked to submit the results of this monitoring on a regular basis. For the Jericho project, the proponent is required to provide an annual report to the NIRB, summarizing how the project was carried out in compliance with the terms and conditions contained within the project certificate and all relevant authorizations, as well as a summary of all the activities planned for the upcoming year. Additionally, the NIRB may also require annual reports from AAs detailing their compliance monitoring for the project.

Once the NIRB is in receipt of all the annual reports related to a project, a monitoring report is prepared. This report serves as a record of all monitoring activities over the reporting period and summarizes the results of the various monitoring programs in one comprehensive document; recommendations for revisions to the monitoring program or for corrective actions for identified problems are also offered. The monitoring report produced by the NIRB is a public document directed to the proponent and all relevant AAs. Ultimately, the monitoring program provides the NIRB with feedback on whether or not the terms and conditions are achieving their purpose and thus plays an important role in future decision-making processes. However, provisions within the NLCA allow the NIRB to reconsider terms and conditions if it is found that certain terms and conditions are not meeting their purpose as had been anticipated at the time the project certificate was issued, that the circumstances relating to the project or the effect of the terms and conditions are significantly different, or that technological developments or new information may suggest a more efficient method of accomplishing the purpose of the term and conditions.

Opportunities and Challenges

Gaps in the Land-Use Planning Regime

The linkages between land-use planning and environmental assessment have long been recognized in both theory and practice (Gibson 1992; McDonald and Brown 1995). In jurisdictions where land-use planning is a relatively mature process, these linkages may appear intuitive or may even be taken for granted. However, they are readily exposed in areas where the land-use planning process and its institutions are more recently developed, as is the case in Nunavut. In the Nunavut regulatory system, all applications for project proposals must first go to the Nunavut Planning Commission to determine how well the proposal conforms to the land-use plan—where an approved land-use plan exists. In the absence of an approved land-use plan, applications for project proposals go directly to the NIRB for screening. Weakness in the area of land-use plans represents a major gap in the management of the environment in Nunavut, for currently, as noted earlier, there are only two approved land-use plans for the territory, the Keewatin Regional Land Use Plan (NPC 2000a) and the North Baffin Regional Land Use Plan (NPC 2000b). Issued approximately one year after Nunavut's official recognition as a territory but based on community consultations conducted much earlier, both of these plans are now dated. Yet there are large areas of Nunavut entirely without approved land-use plans, where project proposals go directly to the EA process without any formal consideration of land-use planning issues. As may be expected in this scenario, many concerns related to land-use planning are consequently raised during the EA process. Some of these issues are readily addressed during EA, while others may be more difficult to address within the constraints of the EA process.

Without approved and regularly updated land-use plans, important land-use issues—such as the regional monitoring of cumulative effects, harmonization with existing municipal or protected-area management plans, regional caribou protection measures, or even the general goals and objectives of a region—are referred directly to the EA process in a default, or ad hoc, manner. The options available to the NIRB during a Part 4 screening do not include rectifying larger-scale regional issues that extend beyond the regulation of the proposed activity put before the board for consideration. To further exacerbate the situation, several conformity requirements of the two approved land-use plans in Nunavut can be perceived as being contrary to the NLCA owing to how they interact with the EA process. Concerns expressed by the Qikiqtani Inuit Association (QIA) during the Part 4 screening of the Baffinland Iron Mines' Mary River railway project (NIRB file no. 08MN053), a proposed iron mine development in the North Baffin region of Nunavut, serve to illustrate this paradox. In this instance, the QIA had difficulty reconciling the NPC's conformity decision with an outstanding requirement for an amendment to the North Baffin Regional Land Use Plan, necessary because of the transportation corridor that would be created by the proposed railway (QIA 2008). Section 12.3.4 of the NLCA instructs the NIRB not to conduct screening on project proposals that do not conform to land-use plans, except where a variance has been granted by the NPC or where a minister has granted an exemption from land-use plan conformity. The outstanding requirement for a land-use plan amendment related to this project proposal called into question the NIRB's jurisdiction in commencing a Part 4 screening. In its screening decision dated 27 June 2008, the NIRB recommended that the Mary River project proposal be subject to a review under Part 5 or 6, and it sought the minister's advice on the dilemma posed by this issue. This incident and others like it have allowed uncertainty to enter the process by complicating the transition from land-use planning to EA.

Issues raised by the Government of Nunavut (2008) with regard to uranium development in the Keewatin region are further evidence of the uncertainty that can occur. Referring again to the screening of a project proposed by Cameco Corporation (NIRB File 08EN015), commenting parties voiced their concerns about permitting uranium exploration in the region, which centred around the consequent potential cumulative effects and the need to ensure adequate protection of important caribou habitat. Therefore, even though the proposed project was in an area with an established land-use plan (Keewatin Regional Land Use Plan), stringent mitigation measures restricting the permitted activities were necessary. These examples and others offer clear evidence that when land-use plans are not kept current, or when land-use planning does not occur at all, the EA process inevitably struggles with larger regional issues in screenings or reviews at the level of the individual project.

Protection of Marine Areas

There is growing public concern about increased shipping in Nunavut's marine areas, and this has led to a call to ensure that proper mechanisms are in place to address the potential for related environmental and socio-economic impacts. While the NIRB's jurisdiction includes marine areas in the Nunavut Settlement Area and the Outer Land Fast Ice Zone, shipping activities typically do not trigger the Part 4 screening process unless they require an authorization for related land-based activities that meet the definition of a project proposal. For example, cruise ships navigating through Nunavut's waters are not subject to EA by the NIRB unless there is a related authorization application, such as a Parks Canada authorization for access to a national park or a Canadian Wildlife Service authorization for access to a migratory bird sanctuary. Shipping related to normal community resupply or individual ship movements not associated with a project proposal are not subject to screening by the NIRB.

Several proposed mining developments in the territory include the installation of deep water ports and associated shipping, which are subject to both screening and review by the NIRB. As currently proposed, the Baffinland Iron Mines' Mary River railway project includes a substantial shipping component for daily transport of iron ore, year-round, via specially built carriers capable of breaking the ice that covers the surrounding waters most of the year. The Mary River project alone would represent an unprecedented level of shipping in Nunavut, and concern has been expressed by residents who harvest wildlife and travel on the marine ice in the area (CBC 2008). Additionally, the public has expressed considerable concern regarding the shipping-related components of both the proposed Bathurst Inlet Port and Road (BIPR) project and the High Lake Mine project, currently undergoing NIRB Part 5 reviews (Boychuk 2004; Nunatsiaq News 2008).

Section 15.4.1 of the NLCA contains provisions for a Nunavut marine council. The council consists of the four Nunavut IPGs with mandates for land and resource development: the NIRB, the Nunavut Planning Commission, the Nunavut Water Board, and the Nunavut Wildlife Management Board. The intended purpose of this marine council is to advise and make recommendations to other government agencies regarding marine areas. Although the Nunavut Marine Council has been underutilized to date, this body can potentially be an effective tool for developing an integrated management approach, one that would strengthen efforts to protect the marine areas and compensate for shortcomings in the current EA regime.

Capacity and Cooperation among EA Participants

Because of ever-increasing levels of development in Nunavut, each of the IPGs created under the NLCA are constantly challenged in maintaining staffing levels and allocating adequate resources to carry out their duties. Capacity challenges are common to many organizations in Canada, but they are often exacerbated in Northern Canada and Nunavut, given the relatively small population and high rate of staff turnover. Nunavut IPGs and government agencies are also confronted with a number of technical, financial, and institutional capacity challenges. The NIRB processes a high volume of development applications and carries a significant workload of screening, review, and monitoring functions. With approximately 15 staff and one office for the entire territory, the NIRB has to depend, on a regular basis, on technical input and comments from federal and territorial governments and Inuit organizations when developing its recommendations, terms, and conditions. As capacity issues are not limited to the NIRB but are seen throughout the North, it is often difficult for these parties to provide such input within the short time frames required.

Time frames for environmental assessments tend to be longer in the Arctic than in Canada's south, thanks in part to cultural and socio-economic realities, coupled with the physical isolation of towns or communities (Arctic Environment Projection Strategy 1997; Koivurova 2002; McCrank 2008). Nunavut encompasses a vast region of more than 2,000,000 km^2, and its population of approximately 30,000 is contained within just over two dozen communities (Statistics Canada 2006). The communities can only be accessed by air, which further constrains working relationships among the various agencies involved in EA, as face to-face meetings are difficult and costly to arrange. Other factors that may affect the time frames for Part 5 reviews include the time the proponent takes in the preparation of the EIS document, alterations to the project design, and the technical review of the EIS. In the NIRB's review process, the schedule for filing both the draft and final EIS is generally left to the proponent's discretion; during previous reviews, this part of the process contributed to Part 5 reviews taking more than three years to complete (see Table 13.2). Adjustments or changes to the design of the proposed project may also affect the time needed for review of the project, as new components change the scope and require inclusion in the EIS. For example, the addition of an all-weather road to the list of project components significantly changed the scope of the assessment during the Part 5 review of the Meadowbank project, after the draft EIS had already been submitted to the NIRB. In this instance, the NIRB decided that the Part 5 review could not be completed until conformity with the existing land-use plan (Keewatin Regional Land Use Plan) for the new component was determined by the NPC (NIRB 2005b). Finally, the technical review period for each draft and final EIS document can range from 60 to 90 days. However, delays may be encountered if parties require additional time to review the project or if participant funding was not issued in the time required to facilitate inclusion of qualified intervenors during the initial stages of the review process.

A report commissioned by the minister of Indian and northern affairs Canada provides recommendations for improving Northern Canada's regulatory system (McCrank 2008). The report recognizes the current capacity issues within the organizations responsible for conducting EA and recommends simplifying the regulatory system, putting particular emphasis on the NWT. In addition to recognizing the current working relationships among the IPGs created under the NLCA, Article 10 of the NLCA contains provisions for the further consolidation of their processes and for the reallocation of their powers, though the discrete functions of each IPG is to be preserved. Previous discussion has highlighted the connectedness of land-use planning and EA functions; many similarities also exist between the requirements of the EA and those of the water licensing process. The NLCA explicitly encourages additional coordination of efforts between the Nunavut Water Board (NWB) and the NIRB. The majority of projects requiring review under Part 5 or 6 of Article 12 are also subject to the water licensing process, with additional hearings and requirements. To reduce potential duplication of efforts, the two boards maintain the ability to hold joint public hearings during the EA process. Alternatively, the NWB can fulfil its requirements for hearings through participation in a NIRB review.

Traditional Knowledge or Inuit *Qaujimajatuqangit*

The NIRB's mandate is to use both traditional knowledge and recognized scien-. tific methods in an ecosystems analysis to assess and monitor, on a site-specific and regional basis, the environmental, cultural, and socio-economic impacts of those proposals for which it has responsibility (NIRB 2007). The incorporation of traditional knowledge (TK) or Inuit Qaujimajatuqangit (IQ) into the EA process in Nunavut is one of the distinctive features of the board's mandate. The NIRB is uniquely placed in that it is one of the few EA boards in Canada where the majority of stakeholders and residents are Aboriginal.

Generally, IQ has been used interchangeably with TK. Both are difficult to define because of their dynamic nature (Ellis 2005; Hansen 2003; White 2006). IQ literally translates as 'what has always been known'. Thorpe et al. (2001) define it as

> Inuit knowledge, insight and wisdom that is gained through experience, shared through stories and passed from one generation to the next. More than just knowledge, as it is typically termed, IQ includes a finely tuned awareness of the forever-changing relationship between 'nuna',[1] 'hila',[2] wildlife and the spiritual world. . . . IQ is local in scale, changing, aggregating, iterative, adaptive, based on oral tradition, intergenerational, complex and spiritual.

TK has also been used interchangeably with traditional ecological knowledge (TEK). However, both IQ and TK are interpreted not only as including TEK, but also as being more encompassing than TEK is alone (Wenzel 1999).

IQ is an important aspect of EA in Inuit communities, as it makes the process more inclusive, thorough, and meaningful. The territory's land and traditional way of life have been vital to its Inuit inhabitants for thousands of years and continue to play important roles. It is imperative to incorporate this knowledge in the EA process in a manner that leads to sustainable development, both from an economic and environmental perspective. Additionally, IQ is useful in the EA process as a tool for understanding the possible consequences of predicted impacts and reducing associated uncertainties. It is a vital component of research in the ecology and environment of the Arctic and is intended to complement and support scientific and ecological findings (Arctic Environment Protection Strategy 1997).

Inclusion of IQ in the EA process improves the effectiveness of impact studies in the Nunavut Settlement Area and increases the knowledge base of the region. IQ is incorporated into all aspects of the NIRB's EA process, including the screening of project proposals (public comment period), review of project proposals (scoping, creation of guidelines, public meetings, community visits, community round tables, intervenor funding), and the monitoring of approved projects. During the public commenting period for screenings, for example, concerns related to the traditional use of land, Inuit harvesting, effects on wildlife, and so on have been raised by local hunters' and trappers' organizations and community hamlet offices. To illustrate, during the screening of the proposed construction of a breakwater at Resolute Bay by South Camp Enterprises (NIRB file no. 06UN073), the

Resolute Bay HTO had concerns about the traditional use of the area for the harvesting of marine mammals. These concerns were incorporated into the screening decision report issued to the minister of INAC. The Mary River project (NIRB file no. 08MN053) provides another illustration of the importance of traditional knowledge during Part 4 screening. The representatives of several hamlets (Arctic Bay, Pond Inlet, and Igloolik) stated that the project might impact the marine environment and affect Inuit harvesting activities. These potential impacts were highlighted as 'Issues of Concern' in the NIRB's screening decision report to the minister, which recommended a review under Part 5 or 6 of the NLCA.

Challenges do exist, however, in effectively incorporating IQ into the EA process, particularly in ensuring the meaningful translation of scientific concepts into Inuktitut or Inuinnaqtun and in making IQ concepts that are unique to Inuit culture (or other Aboriginal cultures) understandable to an English-speaking audience. Most elders and other TK experts are only comfortable when they can communicate orally in their native language (Ellis 2005); therefore, interpreters must be used if these experts are to participate in the EA process. Interpreters must be well versed in both English and Inuktitut or Inuinnaqtun, and able to understand very technical scientific terms and concepts. These terms and concepts are often difficult, if not impossible, to translate, as Aboriginal languages simply do not have the necessary words. This means that the translations received by TK experts are frequently oversimplified or even incorrect (Ellis 2005). The challenges in effectively incorporating IQ into the EA process are not limited to Nunavut; they have been observed in other organizations across the North.

Recommendations and Conclusions

This chapter has outlined the EA process in the Nunavut Settlement Area, with particular emphasis on the challenges and opportunities involved. The EA process as conducted by the NIRB is consistent with the Nunavut Land Claims Agreement, and it has resulted in effective EA decisions, as reflected in the federal minister's acceptance of the project certificates associated with all the reviews to date. However, some shortcomings are apparent. Because of strong direction from the NLCA, the lack of federal enabling legislation has not affected EA in a negative way thus far in Nunavut; yet this deficiency does represent potential uncertainty for future projects. There are some finer aspects of EA that are not explicitly specified within the NLCA, and these omissions may cause problems for participants in the EA process down the road. One example of such uncertainty is the complete lack of specified timelines for Part 5 or Part 6 reviews. While in principle these procedures remain largely proponent driven, this lack of guidance on time frames could lead to unrealistic expectations on the part of participants and proponents alike. Completion of the legislation governing the NIRB and other IPGs could prevent this from becoming an issue in the future.

Related to the matter of legislation, the collaboration—even integration—of the various IPGs is envisioned in the NLCA. This has the potential to significantly alleviate the capacity issues these organizations are now facing. However, without a legislative framework, such collaboration or integration could only be accomplished in an ad hoc manner. Closer collaboration between land-use planning and EA entities in Nunavut has the potential to ensure that land-use plans provide useful input into the EA process, and the lessons learned in the EA process feed back into land-use planning. Finally, to make this collaboration work within the context of Nunavut's regulatory system, the key role of Inuit Qaujimajatuqangit must be recognized at both the organizational and legislative levels; this would ensure that the environmental assessment reflected the goals and priorities of the Inuit of Nunavut, effectively incorporating their extensive knowledge of the arctic environment.

This chapter has highlighted several important challenges currently facing Nunavut's EA process. These include the need for enabling legislation for the various IPGs, adequate capacity and effective collaboration among the IPGs, and the advancement of Inuit Qaujimajatuqangit at all levels within the regulatory regime. Each of these challenges presents a unique opportunity for additional learning and experience, particularly in the areas of land-use planning, marine policy and protection, and the development of regional cumulative effects strategies. It will be important to continue studying the EA process in Nunavut, especially in relation to the other jurisdictions in Northern Canada, in order to monitor successfully the effectiveness of the NIRB as future development pressures come into play.

Notes

1.   *Nuna* = Inuktitut/Inuinnaqtun for 'land'.
2.   *Hila* = Inuinnaqtun for 'air', 'atmosphere', 'climate', or 'outside'.

References

Arctic Environment Protection Strategy. 1997. *Guidelines for environmental impact assessment (EIA) in the Arctic: Sustainable development and utilization.* Finland: Finnish Ministry of Environment.

Boychuk R. 2004. The road from Bathurst Inlet. *Canadian Geographic,* March/April, 39–56.

CBC (Canadian Broadcasting Corporation). 2008. Baffin Island residents resist proposed iron mine plans. Available at http://www.cbc.ca/canada/north/story/2008/04/07/mine-walrus. html. Accessed 10 August 2008.

Cumberland Resources Ltd. 2004. *Meadowbank Gold Project Draft Environmental Impact Statement.* Available at http://ftp.nirb.ca/REVIEWS/PREVIOUS_REVIEWS/03MN107-MEADOWBANK/02-REVIEW/05-DRAFT_EIS/DEIS/. Accessed 10 August 2008.

———. 2005. *Meadowbank Gold Project Final Environmental Impact Statement.* Available at http://ftp.nirb.ca/REVIEWS/PREVIOUS_REVIEWS/03MN107-MEADOWBANK/02-REVIEW/08-FINAL_EIS/174._051108-CRL-FEIS-MB-ITAE/. Accessed 10 August 2008.

Ellis, S.C. 2005. Meaningful consideration? A review of traditional knowledge in environmental decision making. *Arctic* 58 (1): 66–77.

Government of Canada, Tungavik Federation of Nunavut. 1993. *Agreement Between the Inuit of the Nunavut Settlement Area and Her Majesty the Queen in Right of Canada.* Ottawa: Indian and Northern Affairs Canada.

Government of Nunavut. 2008. Letter to the Nunavut Impact Review Board re: NIRB file no. 08EN037. Available at http://ftp.nirb.ca/SCREENINGS/COMPLETED%20SCREENINGS/

2008_SCREENINGS/08EN037-Uravan%20Minerals%20Inc/1-SCREENING/02-
DISTRIBUTION/COMMENTS/080424-08EN037-GN%20DOE%20Comments-IT4E.pdf.
Accessed 18 June 2008.

Gibson, R.B. 1992. Environmental assessment design: Lessons from the Canadian experience. *Environmental and Resource Economics* 2 (6): 12–24.

Hansen, S., and J.W. VanFleet. 2003. Traditional knowledge & intellectual property: A handbook on issues and options for traditional knowledge holders in protecting their intellectual property and maintaining biological diversity. *American Association for the Advance of Science.* Available at http://shr.aaas.org/tek/handbook. Accessed 10 August 2008.

INAC (Indian and Northern Affairs Canada). 2005. INAC letter to the Nunavut Impact Review Board re. NIRB file no. 03MN017. Available at http://ftp.nirb.ca/REVIEWS/PREVIOUS_REVIEWS/03MN107-MEADOWBANK/02-REVIEW/07-TECH_MTG_PHC/158._050914-Minister_INAC-MB-Ltr-Pre_Hearing_Decision-IEAE.pdf. Accessed 10 August 2008.

———. 2008. Government moves ahead on improvements to northern regulatory regime. Available at http://www.ainc-inac.gc.ca/nr/prs/m-a2008/2-3053-eng.asp. Accessed 10 August 2008.

Koivurova, T. 2002. *Environmental Impact Assessment in the Arctic: A study of international legal norms.* Great Britain: Ashgate Publishing.

McCrank, N. 2008. Road to improvement: The review of the regulatory systems across the North. Available at http://www.ainc-inac.gc.ca/nr/prs/m-a2008/ri08-eng.asp. Accessed 10 August 2008.

McDonald, G.T., and L. Brown. 1995. Going beyond environmental impact assessment: Environmental input to planning and design. *Environmental Impact Assessment Review* 15 (6): 483–95.

Morrison-Saunders A., R. Marshall, and J. Arts. 2007. EIA *Follow-up International Best Practice Principles.* Special Publication Series No. 6. Fargo, ND: International Association for Impact Assessment.

Mulrennan, M.E., and C.H. Scott. 2000. Mare Nullius: Indigenous rights in saltwater environments. *Development and Change* 31 (3): 681–708.

Mulvihill, P., and D. Baker. 2001. Ambitious and restrictive scoping: Case studies for Northern Canada. *Environmental Impact Assessment Review* 21 (2001): 363–84.

NIRB (Nunavut Impact Review Board). 2005a. NIRB letter to Brad Thiele, Cumberland Resources Ltd. re: NIRB file no. 03MN017. Available at http://ftp.nirb.ca/REVIEWS/PREVIOUS_REVIEWS/03MN107-MEADOWBANK/02-REVIEW/14-EXEMPTIONS/112._050425-NIRB-MB-F-Ltr-CRL_AccessRd_exempt-OTAE.pdf. Accessed 10 August 2008.

———. 2005b. NIRB letter to Craig Goodins, Cumberland Resources Ltd. re: NIRB file no. 03MN017. Available at http://ftp.nirb.ca/REVIEWS/PREVIOUS_REVIEWS/03MN107-MEADOWBANK/02-REVIEW/07-TECH_MTG_PHC/144b._050714-NIRB_Final_CRL_Ltr_PHDecision-OHCE.pdf. Accessed 10 August 2008.

———. 2006. Guide 7: Guide to the preparation of Environmental Impact Statements. Available at http://ftp.nirb.ca/GUIDES/NIRB-F-Guide%207-the%20Preparation%20of%20Environmental%20Impact%20Statements-OT3E.pdf. Accessed 11 August 2008.

———. 2007. Guide 1: Guide to the Nunavut Impact Review Board. Available at http://ftp.nirb.ca/GUIDES/NIRB-F-Guide%201-The%20NIRN-OT3E.pdf. Accessed 11 August 2008.

NPC (Nunavut Planning Commission). 2000a. Keewatin Regional Land Use Plan. Taloyoak, Nunavut.

———. 2000b. North Baffin Regional Land Use Plan. Taloyoak, Nunavut.

Nunatsiaq News. 2008. But does it make sense? Editorial dated 1 August 2008. Available at http://www.nunatsiaq.com/opinionEditorial/editorial.html. Accessed 10 August 2008.

Plummer, R., and D. Armitage. 2007. Crossing boundaries, crossing scales: The evolution of environment and resource co-management. *Geography Compass* 1 (4): 834–49.

QIA (Qikiqtani Inuit Association). 2008. Letter to the Nunavut Impact Review Board re: NIRB file no. 08MN053. Available at http://ftp.nirb.ca/SCREENINGS/COMPLETED%20SCREENINGS/2008_SCREENINGS/08MN053-Baffinland%20Iron%20Mines%20

Corporation/1-SCREENING/02-DISTRIBUTION/COMMENTS/080603-08MN053-QIA%20Comments%202-IPG%20Process-IT4E.pdf. Accessed 18 June 2008.

Statistics Canada. 2006. Community highlights for Nunavut. Available at http://www.statcan.com. Accessed 7 August 2008.

Thorpe, N.L., S. Eyegetok, N. Hakongak, and Qitirmiut Elders. 2001. *Tuktu and Nogak project: A caribou chronicle.* Final report to the West Kitikmeot Slave/Study Society. Ikaluktuuttiak, NT.

Valiela, D. 2006. *Effects of land claims agreements on environmental impact assessment.* Available at http://www.lawsonlundell.com/resources/EffectsofLandClaimsAgreements.pdf. Accessed 1 June 2008.

Wenzel, G.W. 1999. Traditional ecological knowledge and Inuit: Reflections on TEK research and ethics. *Arctic* 52 (2): 113–24.

White, G. 2006. Cultures in collision: Traditional knowledge and Euro-Canadian governance processes in Northern land-claim boards. *Arctic* 59 (4): 401–14.

Chapter 14

The Canadian Federal EIA System

R. Jamie Herring

Introduction

Environmental assessment, or EA, is a planning and decision-making process that is used to determine and respond to the environmental effects of proposed projects. It is a critical tool intended to protect and maintain a healthy, sustainable environment. Environmental assessment is also intended to help decision makers and citizens understand how proposed development projects could potentially harm the natural environment, and to prompt them to seek alternative means and mitigation measures to avoid these impacts.

At the federal level, EA is given legislative powers through the Canadian Environmental Assessment Act (CEAA). When the Act is applied to projects that are initiated, funded, or permitted to occur by a federal department or agency, its purpose is to mitigate adverse environmental effects resulting from the development of these projects. Considering environmental effects at the early stages of a project can improve planning and help avoid expensive and sometimes controversial corrective action. In this way, the CEAA is an attempt to bridge the potentially conflicting interests between economic development and environmental sustainability.

Despite many successes over the past several years, several criticisms have been raised about the CEAA. Government officials, citizen groups, environmental activists, and business owners alike have critiqued the Act for a variety of different reasons. However, this chapter will argue that achieving environmentally sustainable development is contingent upon a legislative foundation such as the one provided for by the CEAA. Therefore, informed debate surrounding the federal EA process should be aimed at strengthening the Act and moving it towards a paradigm based on environmental sustainability as its core value. To achieve a better appreciation for these debates, this chapter will provide an overview of the CEAA process, highlight the strengths and weaknesses of the Act, and discuss how amendments resulting from the implementation of Bill C-9 have changed the original CEAA.

History

The CEAA is a product of over 20 years of federal-level EA experience. Its origins can be found in a 1972 task force initiated by the federal government to review environmental assessments. In response to the report of the task force, Cabinet determined that all new federal projects initiated by the federal government, as well as those that were classified under its jurisdiction, should be screened for 'potential pollution effects'. To accomplish this, the environmental assessment and review process (EARP) was initiated, and the Federal Environmental Assessment Review Office (FEARO) was established to implement the review process.

In 1984 EARP was further modified and was formally registered as a 'Guidelines Order' under the Government Organization Act of 1979. However, it became progressively evident towards the end of the decade that the new Guidelines Order was not securing greater commitment to environmental assessment by federal authorities. This changed in 1992 when Mr Justice Cullen of the Federal Court of Canada ruled that the Guidelines Order was legally binding in the Rafferty-Alameda water management undertaking in southern Saskatchewan (Hazell 1998).

In 1990 the federal government introduced a bill to establish the Canadian Environmental Assessment Act. The bill received legislative approval in 1992, and after a second round of consultations in 1995, the CEAA was proclaimed to be in force, along with a set of regulations governing its application.

Coverage

The CEAA introduced vital changes to the federal government's decision-making process. For the first time, the requirement that federal departments and agencies conduct environmental assessments was enshrined in legislation. The Act established a set of requirements and factors that need to be considered in its application, and established sustainable development as a primary reason for applying the assessment process. It also offered formal opportunities for the public to contribute to the federal EAs of activities that could affect their lives.

The CEAA establishes the responsibilities and procedures for completing project-level environmental assessments that involve the federal government. Under the Act, federal departments are required to conduct EAs for prescribed projects and certain activities before federal approval or support is possible. The stated purposes of the Act are

(a)   to ensure that the environmental effects of projects receive careful consideration before responsible authorities take actions in connection with them;

(b)   to encourage responsible authorities to take actions that promote sustainable development and thereby achieve or maintain a healthy environment and economy;

(c)   to ensure that projects that are carried out in Canada or on federal lands do not cause significant adverse environmental effects outside the jurisdiction in which the projects are carried out; and

(d)   to ensure that there be an opportunity for public participation in the environmental assessment process. (CEAA, s. 4)

The CEAA applies when a federal authority exercises a power or performs a duty that enables a project to proceed. An environmental assessment is legally mandatory when

(a) any federal authority is the proponent of a project,
(b) makes or authorizes payments or provides a guarantee for a loan for the purposes of enabling a project to be carried out,
(c) has the administration of federal lands and sells, leases or otherwise disposes of those lands for the purposes of enabling a project to be carried out,
(d) and/or issues a permit or license for the purposes of enabling a project to be carried out. (CEAA, s. 5)

In cases where the Act applies, the responsible authority may not provide federal support for the project before the environmental assessment is completed or if it is concluded that the implementation of the project proposal could result in significant adverse environmental effects. However, a recent proposal advanced by the Minister of the Environment would see a two year moratorium on impact assessment for certain projects. This is intended to speed up the implementation of the federal government's 'stimulus package' which was introduced to help alleviate the impacts of the recession.

A *project* is defined as a proposed physical work and any related construction, operation, modification, decommissioning, or abandonment, or where prescribed by regulations known as the Inclusion List Regulations. The definition of *project* is limited to physical changes and not to plans or transactions (Northey and Tilleman 1998, 201). Under section 7 of the CEAA, provision is made for the exemption of projects from assessment if they fit into one of three categories provided for within the Exclusion List Regulations. These include projects where national security is deemed vital, where the federal role in authorizing them is minimal, or where their potential environmental effects are known to be minimal.

The complexity of the CEAA's coverage is heightened for projects that require assessments at both the provincial and federal levels of government. For this reason, the Act enables the minister of the environment to enter into agreements with other jurisdictions to harmonize the EA process. In January 1998, the Canadian Council of Ministers of the Environment (CCME), together with the environment ministers of all provinces and territories except Quebec, signed the Canada-Wide Accord on Environmental Harmonization and the Sub-agreement on Environmental Assessment. The stated purpose for harmonization is to permit coordination in EA activities and to prevent overlapping activities and inter-jurisdictional disputes (CCME 1998a, 1). In order to accomplish this, these agreements are aimed at creating a 'one job, one level of government' approach (Hazell 1998, 98). The sub-agreement was created 'to achieve greater efficiency and the most effective use of public and private resources, where assessment processes involving more than one jurisdiction are required by law, through a single environmental assessment and review process for each proposed project' (CCME 1998b, s.1.1.2). To achieve this harmonization, each administration involved for each proposed project must identify a one-window contact for that assessment. This contact window is determined as follows:

> 5.6.1.   The federal government will be the lead Party for proposed projects on federal
> lands where federal approval(s) apply to a proposed project.
> 5.6.2.   The provincial government will be the lead Party for proposed projects on lands
> within its provincial boundary not covered under 5.6.1 where provincial approval(s) apply
> to a proposed project. (Sub-agreement on Environmental Assessment, s. 5.6)

In this way, the federal authority is restricted to projects proposed for federal and Aboriginal lands and is limited to a supporting role in the conduct of assessments outside these jurisdictions (Hazell 1998, 100). Although the stated objectives of both the sub-agreement and the accord are there to ensure that environmental effects are carefully considered in the pursuit of greater efficiency, they are likely to have substantial consequences in the area of environmental assessment. The long-term environmental benefits of such an approach is debatable, a point we will return to in our discussion of the strengths and weaknesses of the CEAA.

In response to growing concerns about the transport of pollutants across national and international boundaries, the CEAA also aims to reduce or minimize adverse transboundary environmental effects through its transboundary provisions. In 1998 the minister of the environment announced that Canada had ratified the United Nations Economic Commission for Europe (UNECE) Convention on Environmental Impact Assessment in a Transboundary Context. The convention requires that the parties to it ensure that an EA is undertaken prior to the authorization or funding of projects that may have significant environmental transboundary impacts. Specifically, the minister of the environment has the authority to refer a project directly to a mediator or panel if it is deemed that the development of the project could result in significant adverse transboundary environmental effects. Such authority is intended to be exercised in cases where the project would not require an environmental assessment or where no other federal acts or regulations apply.

In addition to international transboundary effects, the CEAA is also intended to apply to international projects. Under section 54.2, international projects can be subjected to an environmental assessment when a federal authority exercises power or performs a duty in relation to that project. The project will not be subjected to an assessment under the CEAA if it is subject to another agreement that provides for the assessment of environmental effects. The original Act, moreover, applied directly to the Canadian International Development Agency (CIDA) because of CIDA's being defined as a federal authority, despite strong concerns about subjecting funding for developing countries to domestic assessment requirements. The inclusion of CIDA as a federal authority has since been changed. In a controversial amendment to the Act, the requirement for a comprehensive study for any CIDA-sponsored project outside Canada was eliminated. However, the precise applications of other CEAA requirements to CIDA remain unclear. Also unclear are the legal obligations of responsible authorities to assess international projects through the Canada Account. The Canada Account is an investment system administered by Export Development Corporation (EDC) and is used to support export transactions that the EDC is unable to support but which the minister for international trade considers to be in Canada's national interest. The controversial sale of CANDU reactors to China and Romania using EDC funds highlights the need for clarity on this issue.

The Process

There are four primary stages to completing an environmental assessment under the CEAA. These are (1) the triggering of the Act and start-up of the assessment, (2) the development of the environmental assessment, (3) the decision by the responsible authority, and (4) post-decision activities, including implementation and follow-up.

As mentioned previously, the CEAA is triggered, or initiated, whenever a federal authority proposes a project; contributes money or any other form of financial assistance to a project; sells, leases, or otherwise transfers control or administration of land to enable a project to be implemented; or issues a permit for a project to proceed. When this occurs, the federal authority becomes known as the responsible authority. The Act may also be triggered by the minister of the environment, who may request that an EA be completed if it is believed that a particular project could result in significant adverse environmental effects across boundaries or between federal and non-federal lands, provincial boundaries, or international boundaries.

At the start-up of an assessment, the responsible authority determines which of four EA tracks need be pursued: screening, comprehensive study, mediation, or panel review (Figure 14.1). Screenings and comprehensive studies are regarded as self-assessments because the responsible authority determines the scope of the environmental assessment and directly manages the EA process in compliance with the requirements of the CEAA. Mediations and panel reviews are considered independent assessments, as they are conducted independently of government. Regardless of what track is followed, a primary objective of environment assessment is to determine whether the project is likely to result in significant adverse environmental effects, taking into consideration the implementation of appropriate mitigation measures. In the case of a project deemed to have unjustifiable significant adverse effects, federal contributions to that project must not be made.

Screening

In the majority of cases, environmental assessments under the CEAA are self-directed screenings. Screenings, the most basic level of environmental assessment, are applied to 99 per cent of all projects. This type of assessment was developed for use in many federal decisions, including the issuing of permits or licences (Northey and Tilleman 1998, 215). Screening is a systematic approach to documenting the environmental effects of a proposed project and determining whether there

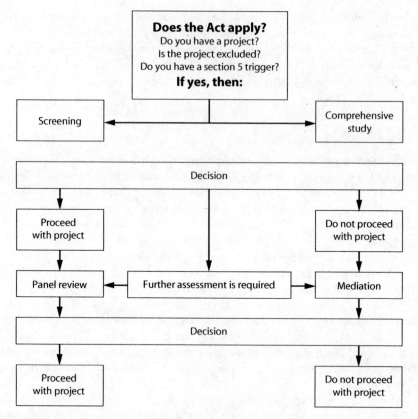

Figure 14.1 The environmental assessment process

is a need to eliminate or minimize these effects, to modify the project plan, or to recommend further assessment through mediation or panel review. A screening must address the environmental effects of the project, focusing on cumulative effects and the effects of possible accidents and malfunctions, the significance of the environmental effects, the measures needed to reduce or eliminate adverse effects, and public comments and concerns.

Once the screening process is completed, the responsible authority must determine whether or not to take action that will enable the project to proceed. If the screening reveals potentially significant adverse effects or if there is significant public concern, the responsible authority should request that the minister of the environment refer the project to mediation or panel review.

In some cases, routine and repetitive projects, such as road maintenance, culvert installations, and building construction, may be assessed through what is known as a *class screening*. A class screening report is generally considered acceptable for projects where the environmental effects and mitigation measures are well known. It is still important, however, that the responsible authority take into consideration site-specific circumstances and conditions. If the assessment of a particular project has been approved by the Canadian Environmental Assessment Agency (CEA Agency) as a class screening, the report can be used in its entirety or partially by a responsible authority in conducting screenings of other projects within the same class.

Comprehensive Study

Certain projects, by nature, require a more thorough review. All such projects are described in the Comprehensive Study List and include projects such as major works for national parks, water management, oil and natural gas development, nuclear power facilities, and large industrial plants. The comprehensive study must address the same factors as the screening as well as other considerations. Importantly, the purpose of the project must be examined, and alternative means of carrying out the project that are technically and economically feasible must also be explored. In addition, the comprehensive study must examine the capacity of those renewable resources likely to be affected by the project and evaluate the need for and requirements of any follow-up program.

An important difference between the screening and the comprehensive study is that prior to making a final decision, the responsible authority must submit the comprehensive study report to the Canadian Environmental Assessment Agency for review. The CEA Agency ensures that the report was completed in compliance with the CEAA and publishes a notice advising that the study is available to the public. Individuals or groups are invited to submit comments to the CEA Agency relating to any aspect of the report. Following this period, the minister of the environment must determine the next step of the project, taking into consideration the comprehensive study, the CEA Agency's review, and the public's comments on it. If the project is unlikely to cause significant environmental effects and if there was no significant public concern, the minister will then refer the project back to the responsible authority. If the project satisfies the requirements of the CEAA, then the responsible authority may allow the project to proceed. After the assessment is completed, the responsible authority must ensure that prescribed mitigation measures are implemented. If the project is likely to cause significant environmental effects or if there is significant public concern, the responsible authority should then recommend that the assessment move to mediation or public review.

Mediation

Mediation, the first type of independent EA, is a voluntary process of negotiation in which an independent mediator attempts to help various parties resolve their differences. The mediator is appointed by the minister of the environment after the minister has consulted with the responsible authorities and the various interested parties. Mediation is the best option when the interested parties are willing to participate and a consensus on the issue appears to be attainable. It is particularly effective in resolving disputes where there are only a few interested parties and the issues surrounding the project are limited.

Mediation sessions are not generally open to the public, as it involves mediating between various stakeholders with direct interests in the project. However, information is made available to the general public. If mediation does not appear to be resolving the various disputes, then the mediator can refer the environmental assessment to the minister of the environment for a panel review.

The Panel Review

The panel review, the strongest of the four assessment tracks, allows the public the opportunity to participate in the EA process. The minister of the environment can request a public review at any time during a screening, comprehensive study, or mediation. A project may also be directed to a panel review if it is deemed to have significant adverse transboundary effects. These boundaries include those between non-federal and federal lands and between provinces and/or territories, as well as international lines for projects that would otherwise not fall under the CEAA or not be subject to any other federal statute or regulation.

The panel review is often requested or ordered in cases where there is uncertainty over the project's potential environmental effects and/or where public concern warrants further investigation. Only the minister of the environment can order a panel review. However, the responsible authority may recommend a panel review at any point during lesser assessment tests. A primary difference between mediation and panel review is that the mediation process is voluntary and is viewed as a negotiation between interested parties, whereas the panel review must be requested by the minister of the environment and usually involves parties whose interests are perceived to be at odds. An independent panel completes the panel review. The review provides all parties and the public the opportunity to present evidence, concerns, and recommendations. Under the CEAA, a panel also retains the legal authority to summon any person to give evidence and produce materials or documents that the panel considers necessary for completing the review.

The panel is composed of a group of experts (usually three) selected from a roster of candidates established by the minister. Panel members must be unbiased and free from conflicts of interest. Furthermore, panel members must have knowledge or expertise with regard to the potential environmental effects of the project.

The Participant Funding Program strengthens the panel review's focus on public participation. By providing money to concerned citizens and organizations, this program gives them the opportunity to participate in panel reviews. These funds are also accessible for the mediation process. Funds are available to any member of the public to help him or her prepare for and participate in background scoping meetings, which identify the factors that a project's proponent must address. Furthermore, the funds enable the public to review the proponent's environmental assessment and to prepare for and participate in mediations and panel hearings.

The panel review generally follows a typical procedure. First, the terms of reference are established. The minister of the environment then appoints a panel that develops and releases operating procedures. Once established, the panel holds informal scoping meetings to learn about issues of concern to the public. Taking into account the presentations in these scoping meetings, the panel then drafts guidelines for the proponent to use in the preparation of an environmental impact statement (EIS). Once the proponent completes the EIS, it is released to the public for a minimum 60-day review period to allow for comments on its adequacy. If the panel determines that the EIS is inadequate, it will issue a statement to this effect to the proponent, who must then fill in the necessary gaps. Once the EIS is complete, the panel holds public hearings to receive views and opinions on the merits of the proposals. Following these hearings, the panel prepares a report containing its conclusions and its recommendations to both the responsible authority and the minister of the environment. Members of the public on the distribution list will receive a copy of the report or a summary of its findings. The minister then publicly comments on the report's findings. At this stage, the responsible authority considers the recommendations of the report and makes a decision on whether it should proceed with the action that will enable the project to go ahead. If the responsible authority deems that the project has unjustified adverse environmental effects, the project will not be given approval. If the project is given approval, however, the responsible authority must consider whether a follow-up program is appropriate.

A follow-up program verifies the accuracy of the environmental assessment and/or determines the effectiveness of any mitigation measures. For example, assessments that indicate that a follow-up program may be warranted are often for projects that use new or unproven technologies or mitigation measures, that are being developed in a new or unfamiliar setting, or that were assessed on the basis of a newly developed EA technique. Follow-up programs can contribute to the improvement of future project development by providing an opportunity for learning and feedback.

The Five-Year Review and Amendments to the Act

As required by the Act, the minister of the environment launched a review of the Act's first five years of application. The five-year review was undertaken through a series of national public consultations held in 19 locations across the country. The completion of the review was further supported by an interactive website that presented citizens with the opportunity to express their concerns. The review process also involved consultations with the provinces, Aboriginal groups, and federal departments and agencies (CEA Agency 2001a, 6). In March 2001, the minister of the environment tabled proposed amendments to the Act in the form of Bill C-9 based on this review. These amendments received royal assent on 11 June 2003 and came into force on 30 October 2003.

A number of important changes to the CEAA have resulted from these amendments. The amendments contain key changes intended to address many of the criticisms and problems that were identified through the five-year review. However, there is considerable debate about the effectiveness of these changes, many of which have become very controversial. Although not exhaustive, a discussion of several key amendments is provided below.

One of the more important changes to the original Act is the addition of a precautionary principle regarding the Act's purpose. This principle is important, as it acknowledges that our scientific understanding of the environmental effects of certain projects that use new technologies and/or untested mitigation strategies may be insufficient. In the original Act, section 4.1(a) stated that the purpose of the Act was 'to ensure that the environmental effects of projects receive careful consideration before responsible authority(ies) take actions in connection with them.' With the proposed changes, the Act now reads that its purpose is 'to ensure that projects are considered in a careful and precautionary manner before federal authorities take action in connection with them.'

Conditions for transboundary effects had a legal foundation in the original Act, but this was rarely used, as there were technical problems with the wording of these provisions (CEA Agency 2001a, 15). The transboundary clause in section 46.1 of the original CEAA stated that

> where no power, duty or function referred to in section 5 or conferred by or under any other Act of Parliament or regulation is to be exercised or performed by a federal authority in relation to a project that is to be carried out in a province and the Minister is of the opinion that the project may cause significant adverse environmental effects in another province, the Minister may refer the project to a mediator or a review panel.

This important section had been inoperable, as the minister's ability to refer a transboundary project to a mediator or panel is limited by the phrase 'or conferred by or under any other Act of Parliament or regulation'. Thus, the offending statement will be removed (CEAA, s. 46.1). It is hoped that by removing this clause, the minister will be better able to utilize public reviews to assess projects that may have significant transboundary effects.

As mentioned above, the application of the Act to international projects has been murky at best. With the new amendments, CIDA's obligations and responsibilities to the Act have been modified. Section 59.1(iii) states that 'providing that, in the case of a project in respect of which an agreement or arrangement entered into by the Canadian International Development Agency in accordance with subsection 54.2 applies, no assessment of environmental effects need to be carried out by that agency.' This controversial clause certainly weakens, if not completely excludes, the application of federal environmental assessments to projects funded by CIDA.

Conversely, the CEAA has been strengthened to include Crown corporations and Indian reserves. As a result of Bill C-9, Crown corporations will now be classified as a 'federal authority'. Subsection 8.1 of the Act has been replaced with the following:

> 8.1 A Crown corporation, within the meaning of the Financial Administration Act, or any corporation controlled by a Crown corporation, shall, if regulations have been made in relation to it under paragraph 59(j) and have come into force, ensure that an assessment of the environmental effects of a project under this section is conducted. (CEAA, s. 8.1)

Similarly, funding on reserve lands will be covered under the CEAA. Although assessments were originally to be fulfilled by band councils with self-government, these assessments were rarely done. The Act now mandates that a commitment to doing an EA be undertaken in relation to any funding.

Changes have also been made in the way mitigation measures are implemented. The original Act (section 20.2) states that if a responsible authority has no direct control over a mitigation measure, then this measure cannot be used in determining the significance of adverse environmental effects. The new section now states that 'mitigation measures that may be taken into account include (a) any mitigation measures whose implementation the responsible authority can ensure; and (b) any other mitigation measures that it is satisfied will be implemented by another person or body' (Bill C-9, s. 20.2). Whether this new understanding of mitigation measures strengthens the environmental protection abilities of the Act is unclear. Prior wording mandated that mitigation measures were under the control of the responsible authority. The responsible authority can now simply be satisfied with the measures put in place, although control of their implementation can be transferred to another body, including the project proponent. Considering previous attempts to get unwilling project proponents to implement rigorous mitigation measures, this new wording may lead to the federal government having even less effective environmental protection measures and controls.

In the five-year review, the Commissioner of the Environment and Sustainable Development (CESD) found follow-up lacking. To strengthen this important part of the EA process, follow-up is now mandatory for a comprehensive study, mediation, or panel review. Follow-up is not mandatory if a project passes the screening assessment track, but the responsible authority must now consider whether follow-up would be suitable.

Finally, several issues surrounding public participation noted during the five-year review have been addressed with amendments, although serious unanswered questions remain. Commendably, participation funding for comprehensive studies, previously not available, will now be established. Thus, the facilitation of public participation at this important level of assessment can now occur. However, funds will not be available for screenings, which, as noted, account for 99 per cent of all assessments. 'In terms of screenings, the CEA Agency estimates that participation currently occurs in only 10–15 per cent of cases' (Sinclair and Fitzpatrick 2002, 173). Many of the new amendments also relegate much of what is new for public participation to guidelines, ministerial direction, and policy. Consequently, the Act signals areas for greater public participation but does nothing to guarantee greater public involvement in legislation (Sinclair and Fitzpatrick 2002, 173–4).

Strengths and Weaknesses

Addressing the various strengths and weakness of the CEAA is difficult considering the various interests that it attempts to bridge. As stated in section 4(b), the intention of the CEAA is to achieve or maintain a healthy environment (sustainability) while also promoting a healthy economy (development). However, as Murphy (1997) points out, these two interests are not necessarily harmonious and in many cases are mutually exclusive. Sachs, for example, argues that sustainable development is a utopian ideal because environmental sustainability and economic development are inherently conflicting ideals. Sustainable development, according to Sachs (1988, 7), implies a need for progress that undermines traditional cultures and 'obstructs the wealth of indigenous alternatives' to maintaining a healthy environment. Others argue that the notion of sustainable development, by placing a premium on economic growth, perpetuates the consumption of natural resources that is at the heart of many environmental problems (Nieto et al. 1995).

Penney argues that two competing paradigms of environmental assessment are the result of this conflict. The first he calls the 'development' paradigm and the second the 'sustainability' paradigm (1994, 245). From the development paradigm, environmental assessment is viewed as a vehicle through which environmental issues can be taken into consideration within the context of a larger development plan: 'While the environment must be taken into consideration, other interests will prevail if they are perceived to be more important in a given situation' (Penney 1994, 245). From the sustainability paradigm, on the other hand, the environment is seen as the core value and main task of EA is to restore ecological integrity on a global scale (Penney 1994, 245).

Depending on the paradigm from which the CEAA is viewed, different strengths and weaknesses emerge. According to Penney, the CEAA can be firmly placed within the development paradigm (1994, 246). From this perspective, the strengths of the Act lie in its ability to provide a decision-making process by which variables can be taken into account, including economic growth. Mitigation measures only need to be put in place where significant adverse environmental effects cannot be justified in the circumstances. Where they are justified for economic or other reasons, it is presumed that environmental impacts will remain of secondary concern.

The bias towards the development paradigm is built into the structure of the Act. The process tends to favour the project proponent, since the proponent is in charge of developing the initial environmental impact statement for panel review. Although the statement can be changed, based on the recommendation of the panel, the fact that the proponent creates the statement sets the tone and dictates the context of the debate from the start of the process. Second, the role of a responsible authority in determining which of the four EA tracks should be pursued when the Act is triggered biases the process towards those federal authorities. Not all federal authorities have as their mandate environmental protection. Indeed, many, such as the Ministry of Finance and Industry Canada, have at the core of their mandate the promotion of economic growth, not environmental protection. Furthermore, the final decision of whether a project can proceed or not rests with the responsible authority. Though decisions to approve projects with adverse environmental effects can only be made if the circumstances justify them, the definition of a justified circumstance is at the discretion of the responsible authorities, and they are not required to explain their justification (Penney 1994, 266). In this way, the Act places a disproportionate weight on project proponents and federal authorities whose motivations may be driven by development rather than sustainability.

This is not to say that criticisms of the Act cannot be laid from this perspective. Indeed, one of the main critiques coming from the five-year review was that the CEAA needed to ensure 'a greater measure of certainty, predictability and timeliness' by providing a 'more efficient and effective process that can save time, money and effort for industry, government and the public' (CEA Agency 2001a, v). To these ends, amendments to the Act have been made to ensure faster turnaround times for assessments; specifically, public comment periods are now limited to 15 days for screenings and to 30 days for comprehensive studies (Benevides 2004). Also, the discretionary power of a federal authority to refer a project to a review panel before the preparation of a comprehensive study report is ordered was eliminated, providing greater certainty for project proponents.

Given that the Act leans towards the development paradigm, many critics from the sustainability camp view the CEAA as deficient (Penney 1994; Gibson 2001; Benevides 2004). The power that responsible authorities have in the assessment process is criticized for several reasons: the authority may have a conflict of interest, the EA process may lack the transparency required for ensuring proper compliance with the Act, and the authority's broad discretionary power in determining justifiable circumstances may be problematic (Benevides 2004).

In the interests of streamlining, harmonizing, and avoiding delays, many in the sustainability camp argue that opportunities for positive environmental contributions are being lost. New amendments to the Act require that the minister of the environment make early and irrevocable rulings on whether comprehensive study cases are to be referred to panel review or mediation. However, environmental advocates expect that fewer panel reviews will occur as a result and less open public deliberation will take place regarding cases where assessment work reveals grounds for concern (Gibson 2001, 90–1). With input limited to only 15 days for screenings and 30 days for comprehensive studies, the public's ability to do meaningful research and provide well-expressed responses is severely restricted.

Similarly, a movement under the label 'harmonization' has been undertaken by federal and provincial governments to reduce overlap of assessments and to increase efficiency. However, as Hazell states,

> an argument can be made that the entire duplication argument is a smokescreen for proponents who do not want their projects to be subjected to appropriate environmental assessment, and for provincial governments who want to ensure, for political reasons, that the federal government is shut out of any environmental assessment that does take place. (1998, 91)

Furthermore, he argues that the one job, one level of government approach to EA is especially dubious given that governments such as Ontario, Quebec, and Alberta are reducing the capacity of their environment ministries (Hazell 1998, 98). In fact, a report issued by the Commissioner of the Environment and Sustainable

Development concluded that 'there is little evidence of federal-provincial duplication' (CESD 1998, 6). In the 25 cases reviewed by both federal and provincial assessments, 'there was more evidence of federal-provincial cooperation than of duplication of effort' (CESD, as quoted by Hazell 1998, 100). Thus, we need to regard the harmonization and efficiency debates critically in order to discern whether they are intended to make EA a stronger protective force or whether they seek to limit the scope of federal EA and apply less rigorous standards (Northey and Tilleman 1998, 233).

From a sustainability perspective, public participation is critical to the overall success and validity of an assessment. Many argue that although public participation may be strong at the panel review level, overall participation is lacking, as the majority (99.99 per cent) of EAs done by the federal government are screenings that have limited room for public input. 'Community and environmental groups have indicated general dissatisfaction with their experience with participating in screenings, citing lack of notification, lack of time to prepare and insufficient opportunities to express their views' (CEA Agency 1999a, 63, as quoted by Sinclair and Fitzpatrick 2002, 162–3). Critics have raised similar concerns about comprehensive studies, citing a lack of funding for public participation during this process (Sinclair and Fitzpatrick 2002, 163) as well as a need for more substantive public support given the scale and complexity of projects undergoing this level of assessment (CEA Agency 2001a, 25). In addition, there are concerns that although most of the projects at the screening level have low environmental impacts, the cumulative effects of thousands of smaller projects have major environmental effects that cannot be considered under the current Act (Benevides 2004).

Clarity with respect to the Act's application to Crown corporations, harbour commissions, and Indian reserves is also lacking. The CEAA specifically exempts Crown corporations and harbour commissions from the definition of federal authority. Projects undertaken by these bodies or projects receiving funding, land, or approval by these bodies are not subject to the general provisions. However, sections 8 and 9 of the CEAA provide instructions for requiring some form of environmental effects assessment where these bodies propose, fund, or provide land for projects. 'What is not clear is whether special regulations must be passed before there is a legal requirement on these bodies to carry out such assessments' (Northey and Tilleman 1998, 205). Similarly, the application of the Act to Indian reserves is unclear on the issue of whether a regulation is required before an EA is needed or whether the Act applies generally and a regulation may simply vary these requirements (Northey and Tilleman 1998, 205).

There is also apprehension about how to scope a project properly. The scope of a project refers to those components of the proposed development that should or must be considered part of the project for the purposes of the environmental assessment. Under section 15.1 of the CEAA, the responsible authority is required to determine the scope of a project. Such an evaluation is extremely important, as it will largely determine the focus of the EA. For example, the logging company Sunpine proposed to build a road and two bridges across the Ram River and Prairie Creek to access certain forest areas in Alberta. Using a narrow scope, the responsible authority treated both bridges as separate projects. A broader scope would have viewed the two bridges, the forestry roads that use the bridges, and the forestry operations as one single project. With the narrower focus, the potential environmental effects were limited to the local effects of the individual bridges. A broader focus would have yielded the potential environmental harm of the entire forest development (Hazell 1998, 115). The ability to scope a project properly is therefore key to being able to address all the potential adverse environmental effects that an entire development project may cause. In the report by the Commissioner of the Environment and Sustainable Development, responsible authorities were criticized for scoping projects in a more narrow fashion than advised by the CEA Agency's 'Responsible Authorities Guide' (Hazell 1998, 116).

A related weakness has to do with mitigation measures. For example, in December 2001 the Federal Court found that approvals issued by the federal government for the purposes of Suncor's Millennium oil sands project in Alberta were invalid. The legal failing identified by the court was the reliance of federal authorities on provincial mitigation requirements in conducting their environmental assessment. The court decided that the reliance of the federal authority on a voluntary provincial process intended to identify and mitigate adverse environmental issues related to the project was invalid. This reliance failed to include a legal assurance that Suncor would in fact follow the mitigation measures in the future. According to section 20.2 of the CEAA, if a responsible authority has no direct control over a mitigation measure, then this measure cannot be used in the decision-making process to determine the significance of adverse environmental effects. However, it is important to include such measures in order to determine their validity in a rigorous manner, to provide complementary environmental protections when required, and to enforce their implementation legally. It is imperative, though, that the inclusion of other mitigation measures not lead to the acceptance of lesser environmental standards, but rather to the adoption of innovative and more effective protection measures.

In general, the above criticism from the sustainability camp stems from a conflict with the development paradigm built into the CEAA. Rather than accepting the simple mitigation of adverse effects as sufficient, the sustainability perspective favours the 'higher test' (Gibson 2000, 39) of ecological integrity, restoration, and the creation of projects that contribute to the overall environmental sustainability of the planet. Gibson states:

> [Environmental assessment] would continue to make business as usual a little less damaging. But this is a world, and a nation, in which business as usual is clearly not sustainable, where it is contributing to growing negative pressures on crucial areas of ecological and community integrity. In this context, making new projects a little less damaging is insufficient. (2001, 97)

From a sustainability point of view, environmental assessment needs to move from simply being a peripheral cost-benefit analysis to becoming something that informs the core design and implementation of a project (Gibson 2000, 2001;

Benevides 2004). It also needs to integrate a more strategic assessment approach such as the one laid out in response to the 1999 Cabinet Directive (Environment Canada 2003). A legislative mandate for strategic environmental assessments (SEAs) would ensure that assessments would apply not simply to projects but to the overarching plans, policies, and decisions that help guide them.

Conclusion

Environmental assessment is a potentially powerful tool in our quest for greater environmental protection and security. The CEAA's contribution to environmental sustainability can be viewed as a legal foundation upon which stronger environmental legislation can be built. The need to achieve and maintain environmental health requires not only strong legal commitments but also a strong learning culture within the EA community, as environmental threats and protection measures are constantly changing. The CEAA provides a concrete point of departure, a basis upon which learning can be fostered and the process of environmental protection can evolve.

In this regard, there is some supporting evidence that the Act has contributed to the evolution of more effective environmental protection. According to Gibson, federal environmental assessment in Canada has evolved towards being earlier in planning, more open and participative, more comprehensive, more mandatory, more closely monitored by courts and government auditors, more widely applied, more integrative, more ambitious, and more humble in recognizing and addressing uncertainties (Gibson 2002, 152–3). However, he states: 'Viewed from a sustainability perspective, this evolution has been generally positive, but insufficient. While further improvements are possible and desirable, the experience of process reform in Canada so far suggests that they will be resisted and gradual' (2002, 153). Indeed, the recent move to weaken impact assessment provisions designed ostensibly to speed implementation of the Conservative government's stimulus package supports this view. It is possible that the current economic downturn may be used as an excuse to weaken the CEAA.

While the CEAA has created a legislative foundation for the assessment of the environmental effects of projects, a move towards a sustainability paradigm would provide greater protection. Considering the incredible ecological damage that development has already created, it is not enough simply to lessen the impact of major projects. Rather, it is crucial that we find ways to reverse the harm already done and restore what we have already destroyed. Improving federal EA to force project proponents to incorporate these considerations into the very design and intent of their undertakings would be a major step towards these ends. While these reforms have yet to be achieved, through public vigilance, informed discourse, and continuing pressure, environmental assessment could become this force in the future.

References

Benevides, H.J. 2004. Real reform deferred: Analysis of recent amendments to the Canadian Environmental Assessment Act. *Journal of Environmental Law and Practice* 13 (March):195–226.

Bill C-9. 2003. Ottawa: House of Commons of Canada.

CCME (Canadian Council of Ministers of the Environment). 1998a. *A Canada-Wide Accord on Environmental Harmonization.* Ottawa: Minister of Public Works and Government Services.
———. 1998b. *Sub-agreement on Environmental Assessment.* Ottawa: Minister of Public Works and Government Services.
CEA Agency (Canadian Environmental Assessment Agency). 1999a. *Review of the Canadian Environmental Assessment Act.* Ottawa: Minister of Public Works and Government Services.
———. 1999b. *Consolidated regulations under the Canadian Environmental Assessment Act.* Ottawa: Minister of Public Works and Government Services.
———. 2001a. *Strengthening environmental assessment for Canadians: Report of the Minister of the Environment to the Parliament of Canada on the review of the Canadian Environmental Assessment Act.* Ottawa: Minister of Public Works and Government Services.
———. 2001b. *Canadian environmental assessment process: A citizen's guide.* Ottawa: Minister of Public Works and Government Services.
Commissioner of the Environment and Sustainable Development. 1998. *Report of the Commissioner of the Environment and Sustainable Development to the House of Commons.* Ottawa: Minister of Public Works and Government Services.
Environment Canada. 2003. *Strategic environmental assessment at Environment Canada.* Ottawa: Minister of Public Works and Government Services.
Gibson, R. 2000. Favoring the higher test: Contribution to sustainability as the central criterion for reviews and decisions under the Canadian Environmental Assessment Act. *Journal of Environmental Law and Practice* 10:39–56.
———. 2001. The major deficiencies remain: A review of the provisions and limitations of Bill C-19, an Act to Amend the Canadian Environmental Assessment Act. *Journal of Environmental Law and Practice* 11:83–103.
———. 2002. From Wreck Cove to Voisey's Bay: The evolution of federal environmental assessment in Canada. *Impact Assessment and Project Appraisal* 20 (3): 151–60.
Government of Canada. 1992. *Canadian Environmental Assessment Act,* S.C. 1992, c. 37.
Hazell, S. 1998. *Canada vs. the environment.* Toronto: Canadian Environmental Defence Fund.
Murphy, R. 1997. *Sociology and nature: Social action in context.* Boulder, CO: Westview Press.
Nieto, C., F. Neotropica, and P. Durbin. 1995. Sustainable development and philosophies of technology. *Philosophy and Technology* 1 (1): 1–11.
Northey, R., and W. Tilleman. 1998. Environmental assessment. In E. Hughes, A. Lucas, and W. Tilleman (Eds), *Environmental law and policy.* Toronto: Emond Montgomery Publications.
Penney, S. 1994. Assessing CEAA: EA theory and the Canadian Environmental Assessment Act. *Journal of Environmental Law and Practice* 4:243–69.
Sachs, W. 1988. *The gospel of global efficiency.* New York: IFDA.
Sinclair, J., and P. Fitzpatrick. 2002. Provisions for more meaningful public participation still elusive in proposed new Canadian EA bill. *Impact Assessment and Project Appraisal* 20 (3): 161–76.

Chapter 15

Impact Assessment under British Columbia's Environmental Impact Assessment Act

Murray B. Rutherford

Introduction

British Columbia is richly endowed with forests, minerals, hydroelectric capacity, and other natural resources. In the past, BC's political leaders struggled to design policies that would encourage exploitation of these resources for economic development while ensuring that benefits flowed to local communities and the province as a whole. In recent years, the provincial government has faced the added challenge of making sure that the negative environmental impacts of new developments are properly assessed and, where possible, avoided or mitigated. The balance struck between promoting development and controlling how and where development occurs has fluctuated with changes in the global markets for the province's resource outputs and with domestic elections that have shifted prevailing political ideologies. The current version of the BC Environmental Assessment Act (EAA) (S.B.C. 2002, c. 43), which came into effect on 30 December 2002, is the product of such an ideological shift, from an emphasis on prescriptive command and control regulation to a focus on limiting the role of government and improving efficiency. As the guide to the environmental assessment (EA) process issued by the BC Environmental Assessment Office (EAO) states, 'The provincial government is committed to more flexible, efficient and timely reviews of proposed major projects to help revitalize the provincial economy. This is why a new, streamlined environmental assessment process was introduced in 2002' (EAO 2003, 1).

Overview of the BC Environmental Assessment Process

As might be expected given its origins, the Environmental Assessment Act and associated regulations are lean documents, providing the basic framework for environmental assessment but leaving much of the detail to policies and practices developed by the EAO. Table 15.1 lists the main regulations, guides, and agreements associated with the Act. In brief, the EAA establishes an assessment process for 'major' projects that might be expected to have substantial impacts. The Reviewable Projects Regulation (B.C. Reg. 370/2002) lists the types of projects and thresholds for size that will trigger the assessment provisions. The BC minister of the environment (MOE) can also designate other projects for assessment, and proponents may ask to opt into the BC assessment process even if their projects would not otherwise be captured by regulation or by ministerial designation. The assessment includes potential environmental, economic, social, heritage, and health effects, and identifies positive as well as negative effects within each category. The EAO is established as an office to administer most assessments, but the Act also provides that the MOE can appoint a hearing panel or commission to conduct an assessment where circumstances warrant.

The EAA and accompanying regulations do not spell out the details of how each assessment is to be conducted. Instead, the 'scope, procedures and methods of each assessment are flexible' and are designed individually for each project so the assessment can 'focus on the issues relevant to whether or not that project should proceed' (EAO 2003, 1). The EAO has issued guidebooks and policy pronouncements to encourage consistency across assessments, but these do not have the force of legislation or regulations. The EAA and regulations do, however, provide for time limits, especially for the government to review and respond to proponents' submissions and also for proponents to respond to requests for documents or information. The outcome of an assessment under the EAA is a report and recommendation to the MOE and the provincial minister responsible for the type of project proposed (for some projects the provincial minister is the MOE as well). The minister or ministers then make a decision about whether or not to issue an environmental assessment certificate—the critical document that grants approval in principle for the project under the EAA and establishes the conditions under which it may proceed.

History of Environmental Assessment in British Columbia

Like many other jurisdictions, British Columbia established its first environmental assessment procedures under policy guidelines rather than specific legislation. Early assessment processes in the province were developed separately for different types of projects, including coal development, metal mine development, and energy projects. The assessment processes for coal and metal mines were merged in 1984 to become the Mine Development Review Process, and this was then replaced in 1991 by the Mine Development Assessment Process (Greenwood 2004, 49). In addition, under the 1981 Environment Management Act (S.B.C. 1981, c. 14), the MOE was given broad discretion with respect to requiring an environmental impact assessment (EIA) for any proposal that could have a detrimental environmental impact (this power still exists under the current Environmental

Table 15.1 Key legislation, regulations, guides, and agreements related to the British Columbia environmental assessment process

Title	Purpose
Environmental Assessment Act	Creates the legal framework and authorization for the EA process, provides for the Environmental Assessment Office, authorizes regulations
Reviewable Projects Regulation	Designates projects as reviewable based on type of activity and thresholds for size
Concurrent Approval Regulation	Establishes the process and conditions for review of other required provincial approvals concurrently with an EA
Prescribed Time Limits Regulation	Establishes time limits for steps in the assessment process
Public Consultation Policy Regulation	Establishes general policy requirements for public notice, information, and consultation
Transition Regulation	Exempts certain types of projects for which approvals were granted before the new EAA came into force
EAO Policy on Public Comments	Establishes EAO policy respecting the treatment of public comments
Guide to the BC Environmental Assessment Process	Comprehensive guide to the BC assessment process
Guidelines for Preparing Project Descriptions	Instructions for proponents on how to prepare the initial project description
Mine Proponent's Guide	Instructions for preparing terms of reference and application for an EA certificate for reviewable mine projects
Proponent Guide to the EA Review Process, Working Draft	Detailed information and instructions on the review process for proponents
CEAA Project Description Guide	Additional information on project description for projects that trigger the CEAA process
Guide to Preparing Terms of Reference	Instructions for proponents on how to prepare terms of reference for review
Canada–British Columbia Agreement on Environmental Assessment Cooperation	Harmonization agreement for coordinated assessment of projects that trigger both the federal and BC assessment processes

Management Act, S.B.C. 2003, c. 53). In 1990 the province introduced the Major Project Review Process, which focused on large industrial projects. With the exception of the MOE's power to designate specific proposals for assessment, all of these assessment processes were replaced in 1995 by the first Environmental Assessment Act (R.S.B.C. 1996, c. 119).

In 2001 the BC Liberal Party came to power with a majority government elected on a platform that emphasized regulatory reform, reduced bureaucracy, and improved government efficiency. As part of a sweeping overhaul of the regulatory system in the province, based on a review of 'core' government responsibilities, the government repealed the prescriptive EA legislation then in place and introduced the new EAA. The 'Guide to the British Columbia Environmental Assessment Process' describes the rationale for the change as follows:

> In particular, it was determined that the legislation should provide for greater procedural
> flexibility in order that assessments could be designed to focus on the specific issues and
> circumstances associated with the individual project.... While retaining many funda-
> mental elements of the previous Act, this legislation enables environmental assessments
> to be tailored to the requirements of each project, allowing for a more streamlined and
> efficient review process. The changes are intended to ensure environmental assessments
> are more focused and cost-effective, while remaining thorough, open and accountable.
> (EAO 2003, 4)

Goals and Principles of Assessment under the EAA

The EAA does not contain a preamble or description of legislative purpose, nor does it refer to sustainability, sustainable development, or other broad goals often associated with environmental assessment. It is clear, however, from the description of the decision-making powers of the executive director of the EAO that the focus of the process is on determining whether a project will have significant adverse environmental, economic, social, heritage, or health effects and, where possible, on preventing such effects or reducing them to an acceptable level. The 'Guide to the British Columbia Environmental Assessment Process' contains a more expansive description of the goals of EA, in a section called 'Purpose of Environmental Assessment':

> Its primary goal is to identify and assess the potential effects that may result from devel-
> opment of a proposed project, and to develop measures for managing those effects. En-
> vironmental assessment is an important means of ensuring that project decision-making
> by governments and proponents is informed. In Canada, all provinces and the federal
> government implement environmental assessment procedures to assist in making deci-
> sions on whether large scale projects should proceed. Environmental assessment provides
> a framework to address a broad range of environmental, health and safety, socioeconomic,
> community and First Nation issues through a single, integrated process, ensuring the is-
> sues and concerns of all interested parties are considered together. Through the process
> of environmental assessment, potential effects of a proposed project are identified and

evaluated early, providing the opportunity for a project to be modified before irreversible project design and construction decisions are made. This results in improved project design and helps to avoid costly mistakes for proponents, governments, local communities and the environment. (EAO 2003, 3)

The guide also specifies a series of fundamental principles of EA (adapted from EAO 2003, 10–11):

Access to Information. Interested parties should have access to information and documentation about the assessment through web-based or hard-copy sources.

Balanced Decision Making. Benefits as well as costs should be included in the assessment, and the ultimate decision should be made by ministers of the provincial Cabinet who are politically accountable for the decision.

Comprehensiveness. Assessments can include environmental, economic, social, health, and heritage effects on-site and off-site and for the life cycle of the project.

Consultation. All potentially affected parties should be consulted and given the opportunity to provide input. Consultation methods and procedures should be appropriate for the individuals concerned with each specific project.

Coordination. Projects that trigger both the federal and BC assessment regimes should be assessed in a single coordinated process. In addition, other provincial permitting and approval requirements should be identified during the BC assessment process so that they can be dealt with concurrently or at least coordinated with the BC process.

Flexibility. Each assessment should be individually tailored to the project being assessed so that the assessment can focus on the issues relevant to a ministerial decision about whether or not the project should proceed.

Integration. The proponent should file one integrated submission covering all relevant issues, which should be reviewed by all participants and dealt with in a single assessment process that leads to a single assessment report.

Neutral Administration. The EAO should operate as a neutral, open, and accountable agency and should manage assessments in accordance with procedures specified in legal and policy documents.

Timeliness. The assessment legislation and regulations should specify time limits for major actions and decision points, and additional time limits may be established for individual projects.

The Typical Assessment Process in British Columbia

There are two main stages in a typical assessment led and managed by the EAO. In the pre-application stage, the proponent, in consultation with the EAO and other agencies and interested parties, prepares and submits a project description, develops terms of reference for the application, and submits these for approval by the EAO; the proponent then prepares and submits to the EAO its formal application for an environmental assessment certificate. The application includes the findings and recommendations from studies conducted by the proponent and its consultants to assess the possible environmental, economic, social, heritage, and health impacts of the project. The process then moves into the application review stage, in which the EAO reviews the proponent's application to ensure that it is complete, provides an opportunity for input from other agencies, First Nations, and other interested parties, and then prepares an assessment report and submits it for a ministerial decision. The final step in the application review stage is the ministerial decision about whether to issue an environmental assessment certificate. The assessment process is proponent driven in that the proponent is responsible for preparing the initial project description, the proponent usually prepares the draft terms of reference for the assessment, and the proponent is responsible for preparing the application for an EA certificate, including all necessary supporting studies and supplementary information to be considered in the assessment review. The steps for a typical assessment are described in more detail below (see also Figure 15.1).

Step 1: Determining Whether the Environmental Assessment Act Applies

There are three ways in which a proposed activity can be designated as a 'reviewable project' and subject to assessment under the EAA. First, section 5(2) of the Act provides that projects can be designated by regulation on the basis of 'size, production or storage capacity, timing, geographical location, potential for adverse effects, type of industry to which the projects are related, type of proponent or on any other basis that the Lieutenant Governor in Council considers appropriate.' The Reviewable Projects Regulation includes the following categories: industrial, mining, energy, water management, waste disposal, food processing, transportation, and tourist destination resort. Each category includes descriptions of the types of projects and thresholds for size that will trigger the assessment process. For example, a new hydroelectric power plant with a capacity of 50 MW or more is deemed to be a reviewable project (Reviewable Projects Regulation, Table 7).

Even if a proposed activity is not captured by the Reviewable Projects Regulation, it may still be designated as reviewable by the MOE under section 6 of the EAA or by the executive director of the EAO under section 7. The minister can make such a designation if satisfied that the activity 'may have a significant adverse environmental, economic, social, heritage or health effect, and that the designation is in the public interest' (EAA, s. 6). The minister's designation must be made before

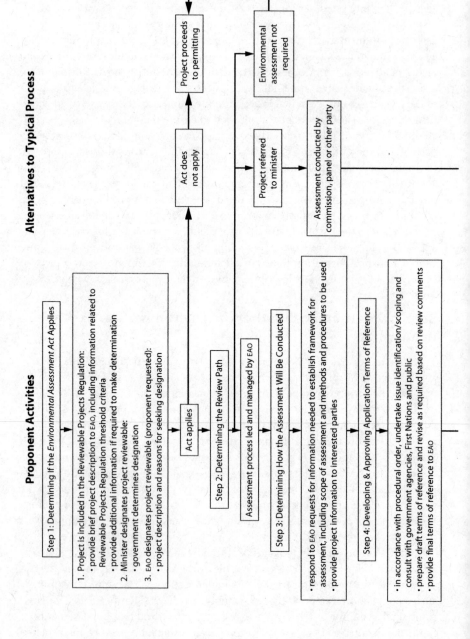

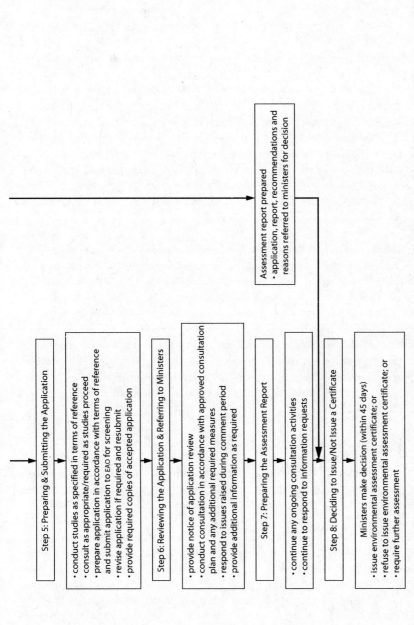

Figure 15.1 Proponent activities in a typical environmental assessment led and managed by the Environmental Assessment Office

Source: EAO 2003, Supplementary Guide to Proponents, 5.

the proposed activity has been substantially started. The executive director may designate a project as reviewable if the proponent requests the designation and provides reasons for the request. In other words, the proponent can ask to 'opt in' to the BC process for a project that would not otherwise require assessment under the EAA (EAA, s. 7).

In practice, the determination of whether a project is reviewable normally begins with the proponent realizing that there is a possibility that an assessment will be required, and then providing a preliminary description of the project to the EAO. The EAO's 'Guidelines for Preparing Project Descriptions' (EAO 2007a) call for information about the proponent; the type, size, location, and purpose of the project; the components of the project; possible effects; ownership of the land; existing land use in the area; consultations that have taken place with interested parties; and the proposed development schedule. The proponent should also provide a list of the permits that may be required for the project. The project description is used by the EAO to determine whether an assessment is required, and copies of the project description are

> provided to First Nations, government agencies, and local governments to enable them to determine whether they are interested in participating in a project review and to initiate discussions on the scope of the assessment. It may also be used by federal government agencies to determine whether the Canadian Environmental Assessment Act is triggered. (EAO 2007a, 1)

The EAO typically establishes an advisory 'Working Group'—with representation from provincial and federal agencies, local government, First Nations, and, where needed, other jurisdictions—to provide technical review and guidance to the proponent and the EAO and to help resolve issues throughout the assessment process (EAO 2005).

Step 2: Determining the Review Path

If an activity is a reviewable project, the executive director of the EAO must choose one of three options for determining the scope, methods, and procedures of the assessment:

1. The executive director may decide that an EA certificate is required for the project, thereby setting in motion a typical EAA assessment process managed by the EAO. This option is available if the executive director considers that the project 'may have a significant adverse environmental, economic, social, heritage or health effect, taking into account practical means of preventing or reducing to an acceptable level any potential adverse effects of the project' (EAA, s. 10(1)(c)).

2. If the executive director considers that the project 'will not have a significant adverse environmental, economic, social, heritage or health effect, taking into account practical means of preventing or reducing to an acceptable

level any potential adverse effects of the project,' the executive director may waive the requirement for an assessment under the EAA (s. 10(1)(b)). The 'Guide to the British Columbia Environmental Assessment Process' indicates that the test for the waiver power is strictly applied and that waivers will only be granted in rare cases (EAO 2003). In the 2007/8 fiscal year, the executive director issued one such waiver, for an extension of the Smithers Regional Airport runway by 762 metres (MOE 2008, 57).

3. The executive director may choose to refer the project to the MOE for a decision about scope, procedures, and methods. The minister's discretion includes the option of ordering an assessment by a commission, hearing panel, or other process. This option is intended for 'special circumstances' (EAO 2003, 33), presumably for particularly large projects or projects with a high level of public interest. Note that the executive director does not have the power to order a hearing panel or commission directly; the decision must be made by the MOE.

The executive director informs the proponent in writing of the review path selected (a 'section 10 order').

Step 3: Determining How the Assessment Will Be Conducted

If the executive director determines that an EAO-managed assessment is required, the EAO then develops and issues a section 11 procedural order setting out the scope, procedures, and methods for the assessment. The EAO considers 'input and advice from government agencies, First Nations, the public and the proponent, as appropriate' (EAO 2003, 16) in preparing the procedural order, and the order itself normally deals with the following issues:

- The facilities and activities that comprise the reviewable project (the project scope);
- The procedures and methods to be used in conducting the assessment;
- The potential effects to be considered in the assessment;
- Information required from the proponent, primarily in its application for an environmental assessment certificate (the order will normally require the proponent to develop terms of reference for the application);
- Information from sources other than the proponent, if any;
- First Nation consultation requirements;
- Public consultation requirements; and
- Time limits for activities in the assessment not otherwise covered by legislated time limits. (EAO 2003, 16)

As part of the public consultation requirements, the procedural order typically requires at least one formal public comment period during the application review stage and, if necessary and practical, an earlier public comment period during the review of the draft terms of reference (EAO 2003, 22). Additional public comment periods may be required if circumstances warrant.

If the executive director has referred the project to the MOE for the determination of scope, methods, and procedure, the MOE may issue a similar procedural order.

Step 4: Developing and Approving Application Terms of Reference

The next step in a typical assessment is for the proponent, in consultation with the EAO, First Nations, other government agencies, and the public, to prepare formal terms of reference for its application for an EA certificate. The terms of reference should 'identify the issues to be addressed and the information to be provided by the proponent in its Application', such as required public consultations, potential effects of the project, measures the proponent will take to minimize or mitigate those effects, and plans for monitoring (EAO 2007b, 5). The proponent submits a draft version of the terms of reference to the EAO, which the EAO releases for public comment. There is a guide to preparing terms of reference (EAO 2007b), and the procedural order for a project may specify additional matters for inclusion and consultations that must be held. Although the proponent may be required to consider alternative designs and methods for completing a project, the BC process does not require consideration of alternatives to the project itself. Once approved by the EAO, the terms of reference establish the standards for the review of the application and for the eventual decision about whether to issue an EA certificate:

> The government decision to accept an Application will be based on whether or not the Application document provides the information required by the approved Terms of Reference.... The government decision to issue an Environmental Assessment Certificate will be based on how effectively the Application addresses the issues identified in the Terms of Reference. (EAO 2007b, 5)

Step 5: Preparing and Submitting the Application for an Environmental Assessment Certificate

After the final terms of reference are approved by the EAO, the proponent undertakes the study and consultation required by the terms of reference and the procedural order, and prepares and submits to the EAO an application for an EA certificate. The EAO has a period of 30 days in which to conduct a preliminary screening of the proponent's application to determine whether it contains sufficient information for the assessment to proceed. The EAA states specifically that 'the executive director must not accept the application for review unless he or she has determined that it contains the required information' (s. 16(3)). If the application is deficient, it is returned to the proponent for revision.

Step 6: Reviewing the Application for an Environmental Assessment Certificate

Once the application is accepted, the EAO has 180 days to complete its review and prepare and submit its assessment report to the minister or ministers. The application review normally includes 'review by government agencies, First Nations and the public; First Nation and public consultation; a formal public comment period; and opportunities for the proponent to respond to issues raised' (EAO 2003, 18). The typical formal public comment period is 30–75 days.

Step 7: Preparing the Assessment Report and Referring to the Ministers

When the EAO review of the application for an environmental assessment certificate is complete, the EAO prepares a draft assessment report and coordinates review of that report by the Working Group and other interested parties. Once the assessment report is finalized, the EAO submits it, along with the proponent's application for an EA certificate, to the appropriate minister or ministers for a decision. The EAO's assessment report may include recommendations and reasons for those recommendations.

Step 8: Deciding Whether to Issue/Not Issue an Environmental Assessment Certificate

The minister(s) have 45 days to review the assessment report and recommendations (if any) and either issue the environmental assessment certificate, refuse to issue the certificate, or order that further assessment be carried out. If a certificate is issued, it may set out conditions for the project, including requirements concerning monitoring and ongoing project management. The EA certificate will also specify a deadline for commencement of the project, between three and five years after the date on which the certificate is issued. On application by the proponent, this deadline can be extended once for up to five years. As long as the project is substantially started by the specified deadline, the certificate normally remains in effect for the life of the project, unless suspended or cancelled for non-compliance.

Post-Assessment Procedures

Other Necessary Permits and Approvals

After the EA certificate has been issued, the proponent must still obtain the other provincial permits and approvals that are required for the project. One feature of the BC assessment process that has been particularly attractive to proponents is that the proponent can apply during the assessment to have these other permits and approvals considered concurrently with the environmental assessment itself. If the proponent applies for concurrent approval and meets the conditions specified in the EAA and the Concurrent Approval Regulation (B.C. Reg. 371/2002), the time required to obtain other approvals once the environmental assessment is complete can be shortened substantially. The review of other approvals can commence during the assessment, and once the EA certificate is issued, the agencies responsible for other approvals have 60 days to either issue the approvals, refuse to issue the approvals and give reasons, or specify a later date on which the decision about approval will be made, giving reasons for the delay.

Ongoing Operations Management and Monitoring

The environmental assessment certificate may include conditions concerning on-going operations management, monitoring, and periodic reports to the EAO or another governmental authority. Other approvals and permits for the project will also typically include such requirements. According to the 'Guide to the British Columbia Environmental Assessment Process', management and monitoring by the proponent are supplemented by 'complementary government monitoring, inspection and enforcement activities to ensure the project complies with all relevant provincial laws, regulations and approval conditions' (EAO 2003, 7).

Other Features of the EAA process

Time Limits

The time limits established for particular steps in the assessment process are key to the EAA scheme and may encourage proponents to opt into the BC process even for projects that would otherwise only trigger the federal assessment process. From the proponent's perspective, having the BC time limits operating for a coordinated joint assessment could help to discourage agency delays. The Prescribed Time Limits Regulation (B.C. Reg. 372/2002) includes the following limits:

1. Thirty days for the executive director to decide whether to accept an application for an environmental assessment certificate for review
2. One hundred eighty days for the executive director to review an EA certificate and produce the assessment report
3. Forty-five days for the minister(s) to make a decision about whether to issue an EA certificate
4. Three years for the proponent to provide additional information requested by the executive director at various times during the assessment process

Although the executive director has the discretion to extend these time limits and the limits do not directly bind federal reviewers, the Prescribed Time Limits Regulation does provide an incentive for timely action.

Class Assessment

The EAA authorizes the executive director of the EAO to undertake partial or full class assessments of types of projects (s. 20). Partial class assessments look at a subset of the potential adverse effects of a type of project and, if approved, relieve proponents of future projects of that type of having to provide information on those effects, as long as the proponents comply with the requirements of the partial class assessment. Full class assessments consider all of the potential adverse effects of a type of project and, if approved, relieve proponents of future projects of the approved type of having to undergo any further assessment, as long as the proponents comply with the requirements of the class assessment.

Strategic Assessment

Under section 49 of the EAA, the MOE may direct the EAO to conduct an assessment of 'any policy, enactment, plan, practice or procedure of the government' and issue a report and recommendations to the minister. Noble's (2004) survey of Canadian practices concerning policy, plan, and program assessment indicates that as of 2003 British Columbia had not made recent use of the power to conduct strategic assessments.

Provincial Policy Guidance

The EAA includes explicit provisions dealing with the relationship between other BC government policies and the environmental assessment process. First, section 11(3) provides that an assessment must 'take into account and reflect' government policy as expressed to the executive director of the EAO by other responsible government authorities. Second, during an assessment the executive director may ask for 'clarification and direction' about any policy matter from the minister responsible for the policy area in question, and the clarification or direction must be 'reflected' in the assessment (s. 21).

Coordinated Assessment with Other Jurisdictions

In 2004, British Columbia and the Canadian federal government entered into a new Agreement on Environmental Assessment Cooperation, replacing the previous harmonization arrangement between the two jurisdictions (Canada and British Columbia 2004). The agreement allows a coordinated single assessment for projects that trigger both the BC and federal assessment processes. Generally, the lead party for cooperative EAs is the federal government for projects on federal lands that trigger the federal process and the province for projects on other lands within British Columbia that trigger the provincial process. If the project is on federal and BC lands and both assessment processes are triggered, the lead party is determined by agreement (Canada and British Columbia 2004). In practice, for a BC-led assessment, federal government representatives generally take part as members of the Working Group associated with the assessment, and the terms of reference include federal assessment requirements (Carter and Alexander 2008). The assessment results in a single report, but each jurisdiction makes a separate decision about approval of the project. The BC EAO has also entered into a memorandum of understanding with the Washington State Department of Ecology about sharing information concerning assessment requirements and project proposals for major projects near the BC-Washington border (EAO and Washington State Department of Ecology 2003).

Issues

Sustainability

Sustainability is frequently promoted as a fundamental goal of environmental assessment (see chapter 2 and Gibson et al. 2005), and the absence of this goal from the EAA and associated regulations is troubling. Instead, the legislative framework focuses on the more technical matter of identifying, and avoiding or managing, significant adverse effects. Even the 'Guide to the British Columbia Environmental Assessment Process' only refers incidentally to sustainability, first in an early reference to the benefits that EA may provide to proponents in their marketing 'in the global market place, which is placing increasing emphasis on the sustainability of commodity production of all types' (EAO 2003, 1), and then in a later reference to 'sustainable use of biological diversity' (21). The guide also refers in several places to the minister of sustainable resource management, but the Ministry of Sustainable Resource Management subsequently became part of the Ministry of Environment. In contrast, the EAO website repeatedly refers to sustainability, including in its overview description of the assessment process: 'B.C.'s environmental assessment process is important to ensure that major projects meet the goals of environmental, economic and social sustainability' (EAO 2008a).

The absence of any explicit reference to sustainability in the BC legislative framework for EA may be attributable to the provincial government's experiences in the case of *Taku River Tlingit First Nation v. British Columbia (Project Assessment Director)* (2000 BCSC 1001). In that case, the Supreme Court of British Columbia determined that an environmental assessment was inadequate, in part because the process had failed to properly assess sustainability as required by the explicit purpose statement in section 2 of the previous version of the BC Environmental Assessment Act (see Gibson 2001). The *Taku River Tlingit* decision was varied by the British Columbia Court of Appeal and then overturned in 2004 on appeal to the Supreme Court of Canada on the issue of adequate consultation of First Nations, but when the new EAA was being developed, the BC government may have wanted to close the door on this type of judicial examination in future cases.

Although the EAA is progressive in including a wide range of social, environmental, and economic effects and in assessing positive as well as negative effects, without an explicit specification of sustainability as the primary goal of an assessment, it is less likely that the courts or other reviewers will draw on principles of sustainability in interpreting the Act and evaluating the actions of assessors. The absence of this consideration also makes it more difficult to determine the basis on which trade-offs are to be made among positive and negative effects. Commitments to sustainability on government websites do not carry the same weight as a legislated statement of purpose.

First Nations

Much of the land in British Columbia is not covered by treaties with First Nations, and many of the leading judicial decisions in Canada concerning Aboriginal rights and title and requirements for consultation and accommodation originated in British Columbia. The policies and guidelines developed for the EAA include procedures for consulting and involving First Nations in environmental assessments. Proponents are encouraged to consult First Nations early and to involve them in assessments, and the EAO normally asks potentially affected First Nations to participate as members of the Working Group formed to provide technical review and advice (Carter and Alexander 2008). The document 'A Guide to Preparing Terms of Reference for an Application for an Environmental Assessment Certificate' calls for 'identification of the specific areas where the project could directly affect First Nations at any phase of project development—during construction, operations or, where relevant, decommissioning' (EAO 2007b, 21). This includes identifying traditional-use and Aboriginal rights and title issues in the vicinity of the project. First Nations are typically asked to review key documents in the assessment, as well as to participate in the review of the draft terms of reference, the application screening, the draft assessment report, and the conditions for the EA certificate (Carter and Alexander 2008).

Some First Nations, however, do not feel that British Columbia's current EA policies and practices adequately involve them or appropriately deal with their interests and concerns. For example, a poster developed by the Carrier Sekani Tribal Council and the Takla Lake First Nation for the 2007 First Nations' National Day of Action states that British Columbia's EA process is 'failing First Nations and the environment' (Carrier Sekani Tribal Council and Takla Lake First Nation 2007). A critique posted on the Carrier Sekani Tribal Council website claims there are 12 main weaknesses in the BC assessment process with respect to First Nations:

1. First Nations have no decision making authority in the process or the result.
2. The decision-making criteria under the EAO legislation does not include any mandatory First Nations criteria.
3. The 2002 amendments to the BC *Environmental Assessment Act* removed a legislated role for First Nations from the process.
4. The EAO does not measure impacts of a project from an Aboriginal perspective.
5. The Executive Director has no authority to accommodate infringements of Aboriginal and treaty rights and Aboriginal title.
6. Resources and funding inequities leave First Nations disadvantaged.
7. The Working Group format of review is not conducive to a productive discussion of infringements of Aboriginal rights and title.
8. The independence of the EAO may be compromised by political interests.
9. BC EAO has never rejected a project since its inception in 1995.
10. The EA review process in BC is focused on ensuring proper 'process', rather than 'substance' of the EA.
11. Cumulative Effects Assessment does not address Aboriginal rights and title.

12.   The time limits in the EA process are too restrictive to allow for government and First
Nations to negotiate accommodation of infringements of Aboriginal rights and title.
(Carrier Sekani Tribal Council 2007)

In 2005 the provincial government committed to a 'New Relationship' with First
Nations and Aboriginal people, and the EAO's current Service Plan indicates that it
recognizes a need for improvement in this area:

> The EAO is committed to building a new relationship with First Nations that is founded
> on reconciliation, mutual respect and trust. As part of the EAO's commitment to the New
> Relationship, the EAO is identifying different ways to ensure First Nations are fully engaged
> throughout the environmental assessment process and to help First Nation communities
> understand potential project effects. (MOE 2007, 62)

Public Participation

Public consultation is mandated in the BC assessment process by the Public Consultation Policy Regulation (B.C. Reg. 373/2002), and it entails general policy
requirements concerning notice to the public, information to be made available
during an assessment, and consultation procedures. The EAO can require additional public consultation, and consultation requirements for specific projects will
typically be specified in the procedural order and terms of reference. The EAO has
also issued a public comment policy that spells out content requirements for public comments and procedures for dealing with public submissions on proposed
terms of reference or an application for an environmental assessment certificate
(EAO 2008b). Further detail on public participation is provided in the 'Guide to the
British Columbia Environmental Assessment Process':

> In general, each assessment includes:
> - information sharing: providing information to the public on the project, the as-
>   sessment process, the consultation process, and the requirements placed on the
>   proponent;
> - notification: providing public notice about key steps in an assessment, such as when
>   an application is accepted for review;
> - participation and consultation: providing opportunities for the public to identify
>   interests and potential impacts related to the project, and to submit comments;
> - issue resolution: ensuring public issues that are relevant to the assessment are ad-
>   dressed, which may include opportunities for the public to participate in issue reso-
>   lution; and
> - reporting of public issues: ensuring any reports on public consultation activities, as
>   well as comments submitted by the public, are taken into consideration in preparing
>   the assessment report and in developing any recommendations.
> - At the end of each assessment, public issues are reported on, so they can be taken into
>   consideration by ministers in making their decision on whether or not to certify the
>   project. (EAO 2003, 24–5)

Critics have complained, however, that the public consultation provisions are pre-scribed in the form of 'general policy' rather than mandatory requirements, and therefore are discretionary (WCEL 2002). There is also concern that public con-sultation may be limited to providing information and receiving and replying to comments, rather than giving the public a more meaningful role in decision mak-ing. In addition, the EAA does not include an explicit right of appeal, so those who disagree with an assessment are restricted to the remedies available on a judicial review application.

The EAO website is an excellent resource that provides information about the BC assessment process and electronic copies of the legislation, regulations, policies, and guidebooks. In addition, the website is the location for an electronic Project Information Centre, mandated by the EAA and administered by the EAO, which includes filed documents and extensive information and records about projects that have undergone or are undergoing assessment. Again, however, there has been criticism of the broad discretion given to the executive director regarding which information and documents will be provided to the public (WCEL 2002).

Thresholds for Review and Cumulative Effects

The EAA only applies to 'major projects'. This means that smaller projects do not automatically trigger the assessment provisions, even though they may have sig-nificant environmental effects. In addition to the effects of individual smaller projects, the cumulative effects of multiple projects can be a problem. There is concern, for example, in British Columbia about proposals for multiple run-of-river hydroelectric generating projects within single watersheds, each project hav-ing a generating capacity below the 50 MW threshold that would trigger the EAA (Douglas 2007). Moreover, neither the EAA nor the 'Guide to the British Columbia Environmental Assessment Process' require those individual projects that do trig-ger the BC assessment process to assess cumulative effects. In practice, cumulative effects assessment may be required by the procedural order or terms of reference issued for a specific project, but the guide to preparing terms of reference (EAO 2007b) suggests that cumulative effects need only be assessed for projects that trig-ger the Canadian Environmental Assessment Act process.

Independence of the EAO

The EAO was established by the BC government as a separate office and is described as independent and 'open, accountable and neutral' (EAO 2003, 11). However, the provisions of the EAA that require assessments to take into account and reflect government policy, and that provide for the guidance of the executive director by ministers about policy matters, offer a challenge to the independence of the EAO. It is understandable that such policies would be taken into account in the final ministerial decision about whether a project should proceed, but the assessment process and the report from the assessor to the ministers would have more cred-ibility if they were not so explicitly steered by other government policies. The credibility and independence of the EAO are especially important when so much of the assessment process is left to the discretion of the executive director and the EAO, including the power to waive completely the requirement for assessment.

Concluding Comments

The EAA establishes for major projects in British Columbia an environmental assessment review process that is designed to be efficient and timely. Strengths of the BC process include flexibility in designing individual assessments so that they can be tailored to focus on impacts that are important for the specific project under review; consideration of a broad range of positive and negative environmental, social, and economic effects; mandated time limits; concurrent review of other provincial approvals; comprehensive guidelines issued by the EAO; and the option for proponents to apply to have a project assessed that would not otherwise fall within the statutory triggers. Proponents generally like the process, as evidenced by recent surveys conducted by the EAO in which proponents expressed a high level of overall satisfaction with the BC assessment process (BC Stats Surveys and Analysis 2008). Responses to specific questions in the 2008 survey indicated that proponents generally felt that EAO staff were knowledgeable, competent, and helpful, that the EAO effectively coordinated the input of other agencies, that the EAO website was easily accessible and had the information needed by proponents, and that information on the assessment process was readily available. Areas in which the 2008 survey indicated a need for improvement included comprehensibility of the information provided, timeliness of the review, coordination with the federal process, and overall EAO management of the review process.

In contrast to the level of approval among proponents, environmental groups and First Nations have been highly critical of the EAA (e.g., WCEL 2002; Boyd 2003; Carrier Sekani Tribal Council 2007). In a recent text on Canadian environmental law and policy, Muldoon et al. (2009, 55) go so far as to complain of the 'virtual elimination of environmental assessment law in British Columbia'. Areas of concern for critics include high thresholds for inclusion of projects in the process, lack of consideration of alternatives to the project, lack of detailed mandatory requirements for providing information and consulting the public, direct policy constraints on assessments, lack of mandatory cumulative effects assessment, and too a high level of discretion given to the executive director and the EAO to prescribe the process for individual projects. In addition, although the EAO website professes broad support for environmental, social, and economic sustainability, the EAA does not include a legislated commitment to this important goal.

References

BC Stats Surveys and Analysis. 2008. Environmental Assessment Office proponent satisfaction survey report. http://www.eao.gov.bc.ca/.

Boyd, D. 2003. *Unnatural law: Rethinking Canadian environmental law and policy*. Vancouver: University of British Columbia Press.

Canada and British Columbia. 2004. Canada–British Columbia Agreement on Environmental Assessment Cooperation. http://www.eao.gov.bc.ca/pub/can-bc_agreement/.

Carrier Sekani Tribal Council. 2007. First Nations perspectives on the BC environmental assessment process for discussion purposes. http://www.cstc.bc.ca/cstc/81/envtal+assmts.

Carrier Sekani Tribal Council and Takla Lake First Nation. 2007. BC's environmental assessment process failing First Nations and the environment. http://www.cstc.bc.ca/cstc/81/envtal+assmts.

Carter, D., and G. Alexander. 2008. Canada's and BC's environmental assessment (EA) processes. Presentation to the British Columbia Association of Professional Biologists, 8 May. http://www.apbbc.bc.ca/page_loader.php?page=53.

Douglas, T. 2007. 'Green' hydro power: Understanding impacts, approvals, and sustainability of run-of-river independent power projects in British Columbia. Watershed Watch Salmon Society. http://www.watershed-watch.org/publications/?cid=3.

EAO (British Columbia Environmental Assessment Office). 2003. Guide to the British Columbia environmental assessment process. http://www.eao.gov.bc.ca/.

———. 2005. Proponent guide to the environmental assessment review process: Working draft. http://www.eao.gov.bc.ca/.

———. 2007a. British Columbia environmental assessment process: Guidelines for preparing project descriptions. http://www.eao.gov.bc.ca/.

———. 2007b. A guide to preparing terms of reference for an application for an Environmental Assessment Certificate. http://www.eao.gov.bc.ca/.

———. 2008a. The EA process. http://www.eao.gov.bc.ca/ea_process.html.

———. 2008b. Environmental Assessment Office policy: Public comments. http://www.eao.gov.bc.ca/participation.html.

EAO and Washington State Department of Ecology. 2003. Memorandum of understanding between the Washington State Department of Ecology and the British Columbia Environmental Assessment Office. http://www.eao.gov.bc.ca/.

Gibson, R.B. 2001. Specification of sustainability-based environmental assessment decision criteria and implications for determining 'significance' in environmental assessment. Report prepared for the Canadian Environmental Assessment Agency Research and Development Programme.

Gibson, R.B., S. Hassan, S. Holtz, J. Tansey, and G. Whitelaw. 2005. *Sustainability assessment: Criteria and processes.* London: Earthscan.

Greenwood, D.J. 2004. Healthy competition: Federalism and environmental assessment in Canada—1985–1995. Master's thesis, University of Waterloo, Waterloo, ON.

MOE (British Columbia Ministry of Environment). 2007. Ministry of Environment including Environmental Assessment Office: 2007/08-2009/10 Service Plan. http://www.eao.gov.bc.ca/pub/.

———. 2008. Ministry of Environment including Environmental Assessment Office: 2007/08 Annual Service Plan Report. http://www.eao.gov.bc.ca/pub/.

Muldoon, P., A. Lucas, R.B. Gibson, and P. Pickfield. 2009. *An introduction to environmental law and policy in Canada.* Toronto: Emond Montgomery.

Noble, B.F. 2004. A state-of-practice survey of policy, plan, and program assessment in Canadian provinces. *Environmental Impact Assessment Review* 24 (3): 351–61.

WCEL (West Coast Environmental Law). 2002 (updated 2004). Deregulation backgrounder, Bill 38: The new Environmental Assessment Act. http://www.wcel.org.

Chapter 16

Alberta: Environmental Impact Assessment in a Rapid Growth Setting

Roger Creasey and Kevin Hanna

Alberta is a province where fast-paced development is encouraged by a culture that embraces growth and change. In many respects, Alberta's environmental impact assessment (EIA) process mirrors this image; it seeks to protect the province's unique and dramatic natural assets while at the same time facilitating the objective of economic growth. The Alberta economy is largely dependent on natural resources, and no resource development sector is more important than energy. Whether it is coal mining, conventional oil, natural gas, wind energy, or tar sands[1] operations, energy development in many respects defines contemporary Alberta. The significance of energy development is especially evident in the drill rigs, pumps, and pipelines that are evident across the province's landscape and in the vast tar sands developments in Alberta's northeast. Undoubtedly, energy projects are often at the core of concerns over the social and health impacts of development. But other resources have certainly also changed Alberta's appearance. Forestry still has an important role, and today the pulp and paper and timber industries support a good part of the province's economy. As well, the province's sprawling urban areas and notable tourism industry have grown, as have the resulting environmental impacts.

Given the capacity of such sectors to alter the natural environment, it might be expected that environmental impact assessment would be a major part of policy, regulatory, and development planning processes. But there is little doubt that in some quarters the Alberta EIA process suffers from the perception that it is simply a growth support tool, lending a green stamp of approval to development decisions. Such concerns may not be a sign of inadequacies in the process; rather they may reflect problems relating to the willingness or capacity of decision makers, the public, and some environmental organizations to account for the information and knowledge provided by EIA.

Within the context of an economy that is growing faster than any other in Canada and is dependent on industries that have a remarkable capacity for creating environmental impacts, the expectations and challenges of EIA in Alberta are all the more poignant. In many respects, the Alberta experience well illustrates the problems, limitations, and needs associated with EIA—which bedevil the process in many locales—suggesting the potential for better multifaceted impact assessment approaches.

The Origin of Alberta's Environmental Assessment Process

Contemporary EIA in Alberta came about in the early 1990s when the province began a review of its pollution control and environmental protection legislation (Alberta 2004a). The review resulted in a new piece of omnibus environmental legislation—the Environmental Protection and Enhancement Act (EPEA)—that was enacted in September 1993. Previously, the assessment process was based on a set of guidelines and referral processes that for the most part covered major projects. In addition to the objectives of consolidation and legislative housekeeping, the province was interested in streamlining the regulatory regime. Alberta wanted a process that not only provided an efficient and credible review of predicted impacts, but did so without 'needlessly' hindering development proposals. The EPEA is a broad piece of legislation that includes, among other things, the environmental impact assessment (EIA) process (as Part 2, Division 1).

The consolidation of environmental legislation means that the principles of environmental management are generally administered under one relatively comprehensive regulatory system. One of the cornerstones of this regulatory system is a combined project review and approval process that includes public consultation requirements and provides for appeals (Alberta 2004a). While the EPEA has been amended recently, the most important portions of the original Act remain intact. Alberta does not have a strategic environmental assessment (SEA) process for reviewing plans, policies, and programs. And the Alberta EIA process addresses only physical undertakings deemed to entail large environmental risks.

Key Objectives of EIA in Alberta

Environmental impact assessment is part of the province's overall environmental regulatory system and project evaluation process. The Alberta government's environment department, Alberta Environment, describes this system as a 'framework for sustainable industrial development' that includes 'six core business functions—project evaluation; approvals; monitoring; enforcement; setting standards, objectives and guidelines; and decommissioning and reclamation' (Alberta 2004a, 2). The province's EIA process has three basic goals, to provide information, to provide a venue for public involvement, and to support sustainable development (Alberta 2008 and 2004a).

1. To Provide Information

The Alberta government describes its EIA process as a system or planning tool that supports decision making. Alberta is quite clear about this function. The environmental assessment (EA) process is seen as an information-gathering process attached to a project. The province requires an EIA report to focus on what is described as 'the information needs of regulatory and resource management decision-makers, as well as informing the public, government agencies and industry about environmental matters' (Alberta 2004a, 2). Decisions about a proponent's project are ostensibly made·after the EIA process has provided sufficiently detailed information about the predicted impacts of a project. The EIA process provides insight to decision makers but does not in itself yield the decision.

2. To Provide a Venue for Public Involvement

As other authors in this book have emphasized, EIA should be a participatory process and an EIA system should provide opportunities for meaningful participation by those interested in, or likely to be affected by, a development proposal. The province describes participation in terms of a process that 'provides an opportunity for people who may be affected by a proposed activity to express any concerns and provide advice to proponents and government agencies' (Alberta 2004a, 3). We elaborate on participation later in this chapter.

3. To Support Sustainable Development

Sustainability has emerged in many policy processes as an objective, albeit one that can be difficult to define and implement. As in other jurisdictions, Alberta has embedded the sustainable development imperative into the requirements of its EIA system. In practice, this requirement is seen in terms of considering environmental impacts at the project planning stage. In the words of the provincial government, EIA 'provides an opportunity to examine the effects that projects may have on the relationship between a sustainable environment, a sustainable economy and a sustainable community—the three components of sustainable development.' While this is laudable, it is also more than a bit vague and fraught with difficulties in interpretation. The inherent challenge is to know first of all what constitutes a sustainable economy, environment, or community. In this regard, Alberta is hardly alone. Sustainability implies a range of large conceptual and practical challenges that make its realization difficult, even through a relatively pragmatic and well-developed tool like EIA, regardless of jurisdiction. As we noted above, Alberta does not have a strategic assessment process. The lack of a strategic process weakens the potential for realizing a form of sustainability assessment or advancing the sustainability ethic through EIA.

The EIA process

Impact assessment in Alberta is a linear process for proposals that are unlikely to pose significant environmental risks. But for large projects with inherently greater potential for impacts, particularly if the project is an environmentally sensitive area, the EIA process can be more elaborate, often requiring a relatively extensive environmental review. There are also requirements for public notification of the proposal and public access to the EIA documents; these include timelines and process steps, which are all established in provincial regulations under the EPEA.

The most comprehensive and transparent form of environmental review is the preparation of an environmental impact assessment report; such reports typically consider the activities in the area around a project; the nature of the project itself; the economic, environmental, and social impacts that may result from the project; and the more nebulous objective of resource sustainability (Alberta 2004a, 4).

The project proponent is responsible for evaluating the project-specific and cumulative environmental impacts of a project and reporting that information to regulatory decision makers (Alberta 2004a). The provincial government is responsible for ensuring that the proponent meets the requirements of the EPEA and the Water Act, the two laws that direct EIA. More specifically, the Environmental Assessment Regulation and the Environmental Assessment (Mandatory and Exempted Activities) Regulation (both are part of the EPEA) outline the reporting requirements, administration, and project applicability of the EIA process. Other provincial agencies and even local governments may have a regulatory interest in a project and can participate in reviewing a proponent's terms of reference (TOR) and EIA report (Alberta 2004a). Collaboration among interested agencies in emphasized, indeed actively encouraged, largely to ensure that the process is comprehensive and efficient. Local technical specialists are given the opportunity to participate in the determination of the topics to be covered in the EIA. Cooperative reviews among participating agencies also help move the process along. In part, the cooperation among agencies, with respect to the delivery of the EIA system, is a legacy of the multi-agency approach to impact assessment review that existed before the EPEA was enacted in the early 1990s.

Canada-Alberta Agreement for Environmental Assessment Cooperation

As in other provinces, the potential for overlapping federal/provincial EIA requirements exists. There are examples where a project, or an aspect of one, may trigger the involvement of a federal agency. Under the Canada-Alberta Agreement for Environmental Assessment Cooperation,[2] a cooperative and harmonized process is employed in those instances where the federal EIA system applies to projects that are also subject to the Alberta EPEA. In such cases, the agreement establishes a single, jointly administered assessment review process. Efficiency is certainly the overarching objective of this approach, which ultimately minimizes procedural and regulatory overlap and provides a timelier outcome for the proponent. This approach can also serve the important function of reducing the potential for conflict between the two levels of government. Ross and Creasey's discussion of the Cheviot Mine project in chapter 8 of this book offers a good example of the process in the Alberta context. While the harmonized process has been improved significantly as experience has been gained, there are still difficulties inherent in coordinating a joint EIA process.

Regional Management of the EIA Process

As with other regulatory processes in the province, under the EPEA and the Water Act, Alberta's EIA system is managed and delivered by regional offices. The province is divided into three regions—Northern, Central, and Southern. Each region has a regional approvals manager, a regional compliance manager, and a regional environmental manager (Alberta 2004b). The regional approvals managers are responsible for decisions with respect to all applications under the EPEA and the Water Act in their region, while the regional compliance managers are responsible for compliance-related decisions made in their region (Alberta 2004b).

The regional environmental managers are ultimately responsible for all EA process decisions made in their region. Together, the three regional environmental managers have been designated by the minister as directors responsible for sections 43 to 56 of the EPEA (Alberta 2004b). Alberta Environment's headquarters staff also have a role in the EA process; they provide regional decision makers with support and information, sometimes in the form of specialist advice or input on the application of government and department policy to a specific project (Alberta 2004b). Department staff participate in discussions with proponents when certain issues related to the application of the EIA process need to be clarified. While there can be advantages to a regionally based EIA system, such as a certain knowledge of and sensitivity to local conditions, there is also the potential for variable interpretations and even inconsistent application of impact assessment. This inconsistency might be seen in the interpretation of significance, in the weight given to public participation or community concerns, or even in the willingness of staff to challenge the information and opinions offered by a proponent.

Public Consultation

Public involvement in the EIA system and in decision making is required under the EPEA. Public participation occurs through specific windows or required events, and unlike the dialogue that can develop between industry and government, it is not iterative. In Alberta there are seven specific mechanisms (requirements) for consultation:

1. The project-specific terms-of-reference (ToR) document outlines expectations and requirements for public consultation and is first published in draft form and then finalized with public input.
2. Notification requirements are set out in applicable laws and regulations and should be made with the needs of the community in mind.
3. The EIA report's ToR document includes an analysis of issues in which the proponent documents the issues identified through consultation and outlines whether they have been resolved.

4. The review team considers input from the public at the TOR stage and in the review of the EIA report. This is done with a view to ensuring that appropriate factual information is available to decision makers.
5. The review team provides advice to the public and proponents about the process. This can include participation in consultation activities.
6. Public notification of the undertaking is required, and an opportunity is also provided for interested parties to comment on the need for a public hearing. A public hearing provides the public with an opportunity to participate in the hearing stage. These elements of the process are shared with the agencies responsible for the public hearing process—the Energy Resources Conservation Board (ERCB), the Natural Resources Conservation Board (NRCB), and the Alberta Utilities Commission (AUC).
7. Alberta Environment maintains a public register of information related to projects in the EA process (see box below). The register contains the information provided by the proponent and input from stakeholders.

If a project might have an effect on an existing First Nations treaty or other constitutional rights in relation to Crown lands, then a unique First Nations Consultation Plan is required (Alberta Environment 2008b, 2). When a proponent develops such a plan, it is typically expected to have seven basic ingredients (Alberta 2008b, 2):

1. Project proponent contact information
2. A list of First Nations to be consulted
3. Plain-language, project-specific information
4. Delivery methods for providing project information and direct notices to First Nations
5. Any information regarding potential adverse impacts on First Nations
6. Timelines and schedules for consultation activities
7. Procedures for reporting to Alberta Environment on the progress and results of consultation

What the Public Register of Information Contains

- Names of proponents (detailed contact information is also available)
- Disclosure documents or detailed descriptions of proposed projects
- Copies of public notices
- Copies of statements of concern and other public comments
- Screening reports
- Proposed and final terms of reference
- EIA reports
- Locations where EIA reports and supplemental information can be obtained or viewed
- Orders-in-council

Consultation is undertaken by the proponent, who is then required to outline the participation results in an EIA report, and there are opportunities for participation in the government's review of the project. Notification of a proposal is made through the print media at the local and regional level, which ideally leads to broad public awareness of the undertaking. But there can be no guarantee that such venues are always sufficient for making the public aware of a proposal. Proponents are responsible for making the scope of public disclosure as broad as they deem appropriate for the project, but in practice publication in local and regional media seems to define the minimum.

Appealing an EIA Decision

The only formal appeal process for decisions made by the regional environmental managers is a judicial review to determine whether a regional environmental manager's decision was a correct application of law (Alberta 2004b). In other words, it is not the decision itself that is reviewed, but the process used to reach the decision (Alberta 2004b). If an organization or individual does not agree with a regional environmental manager's decision that an EIA report is not required, they can ask the minister to use ministerial discretion, under section 47 of the EPEA, to require that an EIA report be prepared.

Ministerial Discretion

With respect to Part 2 of the EPEA, the minister of the environment has discretion in two areas (Alberta 2004b). First, the minister can require a proponent to prepare an EIA report for a proposed activity when the 'environmental assessment' director either has decided that an EIA report is not required or is unable to require a proponent to prepare an EIA report because the proposed activity is exempt under the Environmental Assessment (Mandatory and Exempted Activities) Regulation (section 47 of the EPEA) (Alberta 2004b). The minister may also, under section 64 of the EPEA, decide that an approval or registration for a proposed activity should not be granted, effectively preventing a proposed activity from occurring. But the minister does not have the discretion to exempt from the EA process requirements a proposed action that is considered a mandatory activity under the Mandatory and Exempted Activities Regulation (Alberta 2004b). As we noted above, once a regional environmental manager has ordered that an EIA report be prepared, the minister does not have the discretionary power to undo this decision.

Coverage

The significance of the predicted impacts is the primary determinant of whether or not a project will undergo a full environmental assessment. The complexity, scale, technology, resource allocation, and siting conditions of the project are considered with respect to the potential for significant impacts. Assessors evaluate the degree of risk of an environmental impact occurring by analysing the nature of the proposal and seeking expert opinions on where those risks may lie.

Some activities, because of the nature or scale of the development involved, have been designated as activities for which an EIA report is mandatory; such activities include projects involving pulp and paper mills, oil refineries, mines and large dams, tar sands mines, sour gas processing plants, cement plants, steel mills with coke ovens, water reservoirs (over 30 million m^3 capacity), or hazardous waste incinerators. Such projects are always subject to EIA and are (along with many other types of projects) identified in the Environmental Assessment (Mandatory and Exempted Activities) Regulation. This regulation also identifies those projects that generally do not require an EIA report. These are projects whose effects are likely to be of little significance. Small undertakings such as building a sub-surface sewage disposal system, widening an existing highway, drilling a water well, or drilling an oil well are exempt. For other projects or activities, the EPEA outlines the steps (noted below) that should be taken to determine if the EA process should be applied (Alberta 2004a).

Steps in the Alberta Environmental Assessment Process

As is outlined in chapter 1 this book, the ideal EIA process follows distinct steps or stages, although these vary according to the jurisdiction, the nature of public participation, and the applicability and coverage of the EIA system. The Alberta EIA process has four stages, each of which is composed of smaller steps (see Figure 16.1). The descriptions provided here have been adapted from Alberta 2004a (5–9).

1. Screening (does an EIA have to be done?)

The EIA process begins when the proponent, another government department, a local government, or other person informs Alberta Environment about a new project. In most cases, the initial point of contact with Alberta Environment will be the regional office. The regional staff decides if the project requires approval under legislation administered by Alberta Environment. On the basis of information about the project and the applicability of environmental legislation, the agency will either proceed to review the application or recommend that further assessment of environmental effects under Part 2, Division 1, of the EPEA may be warranted before the application can be considered further. In the latter case, the project is referred to the regional environmental manager, who considers the project information, legislative requirements, and staff recommendations. If the regional environmental manager agrees to consider the project, the proponent may be advised that the project is being considered for further assessment. This process has three potential outcomes:

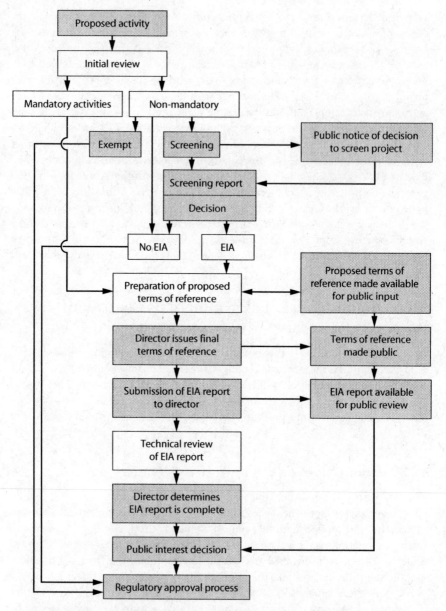

Figure 16.1 The Alberta environmental impact assessment process

1. The project is determined to be a mandatory activity, and the proponent must prepare and submit an EIA report.
2. The regional environmental manager determines that an EIA report is not required, and the proponent is advised to apply for any approvals that may be required from Alberta Environment or other provincial agencies.

3. More information is needed, and a screening report is requested to help determine if an EIA report is required. In essence, this becomes a formal screening process. Preparation of the screening report would include public disclosure of the project by the proponent and the provision of opportunities for public comment on both the merits of the project and the ultimate need for an EIA report. At this point, other government agencies and departments may be asked to provide comments. The minimum period for public comment is 30 days, after which the information provided by the proponent, the public, and government agencies is used by Alberta Environment to prepare a screening report. This report considers the complexity of the project, the nature of technology involved, the sensitivity of the location, the presence of other similar activities, public interest, and any other factors the regional environmental manager sees as significant. The objective of the screening is to identify the type of review that will follow (e.g., preparation of an EIA report, an approval application review process, or a routine regulatory approach). The screening report is made available to the public, and the agency advises the proponent and the public on whether an EIA report will be required. If, after the formal screening process, it is determined that an EIA report is not required, the proponent would apply for any approvals required (from other agencies).

2. Preparation of the EIA Report

In instances where an EIA report is required, the proponent begins the process by preparing a proposed terms of reference document. This is provided to the provincial government and made public for comment and to identify the issues that need to be addressed in the assessment process. The EPEA provides specific procedures for preparing the ToR, advertising its availability for public and government review of the proposal. The proposed ToR is published with notices of the EIA report requirement and a project description. Alberta Environment acts as the coordinating agency for the involvement of other provincial agencies in the review of the proposed ToR. The federal Canadian Environmental Assessment Agency (CEA Agency), possibly municipal government agencies with environmental responsibilities, and perhaps even other provinces are also notified that an EIA report is required. Usually, a review team is established from the several regulatory levels to coordinate comments.

The extent of inter-jurisdictional consultation depends on the nature of the project. The CEA Agency would be asked to coordinate a determination of federal interests and regulatory requirements. As noted above, Alberta has a cooperative EIA agreement with the federal government. If either a cooperative environmental assessment (where both governments require an environmental assessment by law) or potential federal involvement in Alberta's EIA process is identified, the CEA Agency will discuss participation plans and information needs with the regional environmental manager. Where a project is reviewable under the Canadian Environmental Assessment Act (CEAA), efforts would be made to ensure that the

TOR for the EIA report issued under the EPEA will also address requirements under the CEAA. The regional environmental manager considers input received from the public and from other government agencies, and issues the final terms of reference, which establish the scope of assessment for the project. The proponent uses the final TOR to prepare the EIA report. The final terms of reference are also made available to the public.

The EPEA sets out general requirements for information to be included in the EIA report. These requirements are confirmed, varied, or detailed in the final terms of reference issued by the regional environmental manager. Most EIA reports will include six basic elements:

1. **Project details:** detailed description of the project, including the nature and scale of specific activities involved
2. **Setting and baseline conditions:** the location and environmental setting for the project, as well as baseline environmental, social, and cultural information
3. **Impacts:** the potential positive and negative environmental, health, social, economic, and cultural impacts of the proposed activity
4. **Mitigation:** plans to mitigate potential adverse impacts and to respond to emergencies
5. **Consultation account:** public and First Nations consultation programs undertaken with respect to the proposed activity and actions taken by the proponent to resolve public concerns
6. **Cumulative impacts:** an assessment of 'cumulative effects', which are the combined effects of the proposed project and other activities that are occurring or may be reasonably expected to occur in the subject area. The obligation to account for cumulative impacts is a notable strength in the Alberta system.

Alberta's high growth rate in recent years has led to an increased number of environmentally significant activities and has placed particular importance on cumulative effects assessment (CEA) as a key component of an EIA report (Alberta 2004a, 7). Since the concept of CEA was incorporated into the initial 1993 version of the EPEA, several proposed projects have illustrated the difficulties inherent in such an evaluation process, not only in its application in Alberta, but also for other jurisdictions. In Alberta, the requirement that proponents document the cumulative effects in a project impact assessment has led to an examination of the intent of CEA and to improvements in how such assessments are conducted.

In areas experiencing significant industrial growth, such as in the tar sands production area of northeast Alberta, the difficulty of applying project-based assessments in areas of multiple adjacent projects becomes apparent (Creasey and Christie 2001). This cumulative impacts challenge, coupled with the lack of a strategic environmental assessment process in Alberta, can lead to duplication of effort on the part of project proponents and to frustration on the part of interest groups and the public.

In areas experiencing significant industrial growth, such as the tar sands, efforts have been made to examine regional environmental effects. For example, the Cumulative Effects Management Association (CEMA) was formed specifically to address the challenge of regional cumulative impacts from industrial development (Spaling et. al. 2000). The CEMA initiative involves industry, government, environment/conservation, and Aboriginal groups, providing a unique example of a coordinated and ongoing approach for addressing cumulative impacts, in contrast to the ad hoc mechanisms that tend to be more common. A challenge in considering cumulative impacts arises when projects that may not warrant a formal EIA process still interact in a cumulative manner with other activities in a defined area. In such instances, the Alberta government's land-use macro-planning process, known by the title Integrated Resource Planning, would play a role and would ideally provide the venue for addressing or defining an overall vision for use of public lands (Dias and Chinery 1994).

The dialogue between the proponent, government agencies, and the public should be ongoing, and the Alberta system encourages proponents to establish an ongoing 'dialogue with government agencies that will be reviewing the EIA report to ensure that the information provided will meet the needs of those agencies' (Alberta 2004a, 7). In the case of most large energy or resource development projects, the EIA report would also be part of the development approvals application submitted to the Energy Resources Conservation Board or, in the case of a non-energy proposal, the Alberta Utilities Commission (Alberta 2008a).

3. Technical Review

Environmental impact assessment reports are made available to the public. Public comments may then be made to the proponent or, if a public hearing is held at the project decision stage, to the approving board or agency. The EIA report is submitted for review to Alberta Environment, where the task is assigned to a multidisciplinary, integrated team from various government agencies. These agencies are usually discipline-based entities, thus resulting in a team of specialists for the consideration of air, water, terrestrial, and health issues (Alberta 2004a). At this stage, inter-jurisdictional review and consultation may take place. In those instances where both a provincial and federal environmental assessment is required, the EIA is conducted and the resulting report reviewed under the terms of the Canada-Alberta Agreement for Environmental Assessment Cooperation.

The objective of the review process is to identify uncertainties or risks and determine that the information provided by the proponent has

- satisfied the requirements of the terms of reference and the EPEA;
- described the nature and setting of the proposed activity;
- described the possible effects of the proposed activity on the environment in the context of 'good science';
- described mitigation measures to reduce negative effects from the proposed activity;

- described how monitoring and management of residual impacts will be done;
- shown how the proposed activity relates to existing and future activities with which it may interact; and
- explained how consultation was done with those who may be affected by the project and described the key issues discussed during consultation and how/ if they were resolved.

If the EIA report provides information that is unclear or insufficient to meet these objectives, then the review team requests additional supplemental information from the proponent (Alberta 2004a). Supplemental information is provided to those who have received the EIA report, including non-governmental interests, to help ensure a degree of transparency with respect to the information used to make a final decision. In those cases where the EIA report is part of the application to a board (e.g., the AUC or NRCB, as noted above), the request for supplemental information is coordinated through the respective board (Alberta 2004a).

4. Completeness Decision

Supplemental information is reviewed in the same manner as the EIA report to determine that the information provided by the proponent is complete. Throughout this part of the process, reference to the original ToR is important to avoid scope creep, which can occur when information other than that determined during the ToR stage is sought and the demands for this extra data become onerous. Seeking supplemental information requires a balance between the initial ToR, the need for flexibility in considering impacts, and the need to respond to new issues (that may not have been anticipated in the ToR). When the review team is satisfied that they understand the nature of the proposed activity and the proponent's description of potential effects and mitigation measures, then the team will recommend that the EIA report be deemed complete and ready to inform other aspects of the approval process. The regional environmental manager considers the recommendation of the review team and make a final, formal determination that the EIA report is complete. The proponent and the appropriate board or minister is then advised in writing that the EIA report is complete, and the report is formally referred to the board or the minister for project decision making. Once an EIA is completed, it does not mean the project is approved—it means that the proposal can proceed to the next review stage (Alberta 2008a, 4). When the EIA report is part of an application to the AUC or NRCB, it will proceed with its normal application review process, which would include assessment of the technical and economic aspects and environmental impacts. This may include a public hearing to determine if the proposed activity—given its social, economic, and environmental aspects—is in the public interest. If the EIA report is referred to the minister, it will be considered by the minister with or without advice from the Lieutenant Governor in Council (Cabinet).

Concluding Comments

Alberta's multi-step EIA process and the resulting EIA report support decision making. In instances where it is required, an EIA report provides a summary of the nature of the project, potential impacts, proposed mitigation measures, and monitoring needs. The information in an EIA report and the decisions based on the information are used to set up the 'approval terms and conditions including emission limits, monitoring requirements, research needs, siting and operating criteria, and decommissioning and reclamation requirements'; further, the EIA report and the subsequent decisions may 'be considered in regional environmental and resource management systems and multi-stakeholder forums' (Alberta 2004a, 9). As in other Canadian jurisdictions, the Alberta EIA process is about supporting decision making within the context of the best information available, one that allows for the best possible assessment and prediction of environmental impacts.

Given the rapid growth in Alberta's economy, primarily associated with the energy production sector, Alberta's EIA process may experience challenges in keeping pace. Already we have seen the growth of third-party environmental assessment, where private consulting firms and individual citizens review EIAs in a manner similar to the role once performed by public servants. The EIA process and the number of proposals currently before the government have outstripped the government's ability to provide review staff. The omnibus legislation of the 1993 EPEA provides the provincial practitioners with the tools to conduct suitable and potentially comprehensive impact analysis. The Alberta process also addresses the importance of cumulative impacts. Yet there are aspects of the EIA system that should be augmented and improved through the use of more contemporary ideas. Such ideas may come from practitioners working in impact assessment or from public involvement in other jurisdictions, particularly with respect to the need for stronger public participation and for greater attention to the social and human health impact issues—concerns that have emerged in recent years with respect to some energy developments. The lack of strategic assessment may be a major impediment to the advancement of sustainability objectives and may make it difficult for Alberta to account for the greater global impacts of the energy and other resource industries.[3]

The reliance on regional implementation may lend itself to variable and inconsistent interpretations and applications of the EIA process. There may also be a perceptual challenge with respect to Alberta EIA. If EIA is seen as a regulatory hurdle, then development proponents, and even provincial agencies, will try to avoid it, but if EIA is seen as a tool of good planning, as it is intended to be, then it should be embraced and used to the fullest extent to improve projects and support sustainable development.

As Alberta moves into the next decade in a strong economic position, it should also be in a position to implement innovative environmental protection plans without reducing the momentum or positive impacts of economic growth. It remains to be seen whether this opportunity will pass unrealized or whether new ideas related to cumulative effects assessment, environmental carrying capacity, and wise use of natural capital will influence decision making in the near future and lead to a truly sustainable economy and environment. EIA can play an important role in realizing such objectives, if it is used effectively.

Notes

1. Tar sands are natural bitumen infused sand formations. With processing they may yield oil that can be further refined to produce petroleum products. 'Tar sands' is a geological term; 'oil sands' is also commonly used.

2. This agreement was negotiated under the Sub-agreement on Environmental Assessment, which is part of the Canadian Council of Ministers of the Environment Canada-Wide Accord on Environmental Harmonization.

3. At this time, there is a discussion within the Alberta government about options for developing a strategic assessment mechanism.

References

Alberta Environment. 2004a. *Alberta's environmental assessment process.* Edmonton: Alberta Environment, Environmental Assessment Program.

———. 2004b. Alberta Environment staff, personal communication.

———. 2008a. *Alberta's environmental assessment process.* Edmonton: Alberta Environment, Environmental Assessment Program.

———. 2008b. *Frequently asked questions from industry, June 2008.* Edmonton: Alberta Environment, Environmental Assessment Program.

Creasey R., and R. Christie. 2001. Managing the effects of multiple mega-projects. Presentation to the International Association for Impact Assessment (IAIA) Annual Conference, Cartagena, Colombia.

Dias, O., and B. Chinery. 1994. Addressing cumulative effects in Alberta: The role of integrated resource planning. In A.J. Kennedy (Ed.), *Cumulative effects assessment in Canada: From concept to practice.* Edmonton: Alberta Association of Professional Biologists.

Spaling, H., J. Zwier, W. Ross, and R. Creasey. 2000. Managing regional cumulative effects of tar sands development in Alberta, Canada. *Journal of Environmental Assessment, Policy, and Management* 2 (4) (December): 501–28.

Chapter 17

Environmental Impact Assessment in Saskatchewan

Marie Ann Bowden and Bert Weichel

History

Saskatchewan introduced environmental assessment (EA) to the province in 1971 as a policy for government departments such as Mines and Highways. In 1973 the Department of Environment Act was amended to facilitate environmental assessment on a more systematic basis and enable the consideration of larger projects, including the proposed Churchill River hydroelectric project at Wintego Rapids. In that particular case, an environmental and socio-economic study led to a decision to abandon the proposal. In 1976 a formal EA policy was introduced along with the establishment of the Environmental Assessment Branch (EAB or EA Branch) within Saskatchewan Environment.[1] The province wholly embraced this new process. Between 1976 and 1980, some 152 proposals were reviewed and 52 assessments conducted. Of these, 33 developments received approval to proceed, 15 were withdrawn or deferred by the proponents, and four were found unacceptable and not permitted to proceed.

Among the major projects that fell within the scrutiny of the early process were the 1980 Nipawin Hydroelectric Project, the 1978 Cluff Lake Board of Inquiry (also known as the Bayda Inquiry), and the Key Lake Inquiry, established in 1979. The latter two assessments were initiated to review the proposed development of large uranium mines and mills in Northern Saskatchewan. The EA process garnered much credibility over the early years, aided particularly by the quality of these two major mining project inquiries.

These were formally environmental assessment inquiries, but had extremely comprehensive terms of reference. The scope of the Bayda Inquiry [Cluff Lake] went beyond the federal government's Mackenzie Valley Pipeline Inquiry approach, which limited the review to specific proposals, to consider an entire policy area. Bayda examined whether Saskatchewan should be in the uranium mining business at all, and addressed the environmental, socio-economic and ethical aspects of uranium mining. The process used was inclusive: During the Bayda Inquiry numerous public hearings were held and dissenters were funded to prepare and present their cases. As well, the inquiry disseminated specially prepared digests of testimony for the use mainly of northern native stakeholders, who would be most directly affected by mining and commercialization within their traditional lands. The question of further uranium development received extensive media and public attention, and created a great deal of controversy, especially within the New Democratic Party, where it was strenuously debated, and was a divisive issue.[2]

With a policy of EA already well established, the Saskatchewan government enshrined environmental assessment in legislation in 1980.[3] The Environmental Assessment Act was purposely divorced from environmental protection legislation owing to the unique focus of EA as a planning tool. Considered to be at the forefront of environmental legislation in Canada at the time it was enacted, the Act has remained substantially unchanged ever since. This has been defended on the basis that, as framework legislation, it has left relevant parties significant latitude to consider advances in best EA practice, including, for example, cumulative effects analysis. Nonetheless, others would argue that some of the shortcomings of the legislation might have been more adequately addressed through amendment sometime over the past 25 plus years.

During the first decade following passage of the Act, some 636 development proposals were screened, but only 80 were required to undergo full assessments, of which all but two received ministerial approval to proceed. The board of inquiry mechanism was used only once after EA was legislated, during the controversial Rafferty-Alameda dams project, a case that received national attention because of implications for the federal government's obligations in regard to EA (see chapter 14). Throughout the 1980s, criticism of EA grew; there were increasing concerns over the lack of regulations passed pursuant to the legislation since passage of the statute, while the level of public participation and confidence in the process declined. In response, the province appointed the Saskatchewan Environmental Assessment Review Commission (SEARC) in 1990. The mandate of the commission was to review the process and recommend changes to the legislation. SEARC consulted extensively with stakeholders throughout the province. Its report, *Environmental Challenges*,[4] issued early in 1991, put forward 162 specific recommendations as part of new model for EA.

In spite of responding to SEARC with a commitment to 'reform the province's environmental legislation and process',[5] the Government of Saskatchewan did not introduce new legislation and regulations to overhaul EA. Both the SEARC report and the Saskatchewan Environment and Resource Management (SERM) response languished within government. While in recent years the EA Branch has introduced some policy innovations, such as environmental protection plans (EPPs) in certain industrial sectors,[6] the only significant legislated amendment to the 1980 Act occurred in 1996 with the introduction of mandatory assessment of forest management activities.[7] No further initiatives aimed at major legislative reform are currently on the horizon.

Process

The process generally begins with the submission of a project proposal by a proponent to the EA Branch (refer to Figure 17.1). The proposals typically include a description of the undertaking and a brief statement of the anticipated impacts and possible mitigation measures. After each of these proposals (currently approximately 300–400 per year) is reviewed, feedback is provided to the proponent/developer as the case may be, usually within 30 days of submission of the proposal. If the project is not a 'development', the proponent may then seek other necessary approvals for the undertaking pursuant to applicable environmental or other applicable statutes and codes.

In accordance with the Environmental Assessment Act, the full Saskatchewan EA process applies to developments and includes any expansion, alteration, or initiation of a project, operation, or activity that is likely to trigger one or more of listed criteria within the legislation.[8] Unlike the project list model, the 'criteria' approach permits the individual proponents to measure their proposed project, activity, or operation against factors that would indicate the likelihood of negative environmental impacts. The criteria, outlined in section 2(d) of the Act, address any proposal that is likely to

(i)   have an effect on any unique, rare or endangered feature of the environment;

(ii) substantially utilize any provincial resource and in so doing pre-empt the use, or potential use, of that resource for any other purpose;

(iii) cause the emission of any pollutants or create by-products, residual or waste products which require handling and disposal in a manner that is not regulated by any other Act or regulation;

(iv) cause widespread public concern because of potential environmental changes;

(v) involve a new technology that is concerned with resource utilization and that may induce significant environmental change; or

(vi) have a significant impact on the environment or necessitate a further development which is likely to have a significant impact on the environment.

The Lieutenant Governor in Council may provide an exemption from the process in emergency circumstances only.[9] Otherwise, if any of the triggering criteria are applicable, the proposal, be it public or private, is deemed to be a development and must thus follow the EA process to its conclusion. The initial stages of a development (e.g., mining exploration or feasibility studies), though not exempt from assessment, may be permitted to occur (under scrutiny by appropriate regulatory agencies) in advance of a full EIA.[10] This may be allowed because the results obtained through such preliminary steps could form part of the information an environmental impact statement (EIS) might need to contain, but it does mean that considerable and extensive impacts may occur well before the minister gives (or does not give) his or her EIA-based approval.

In theory, since the legislation is silent on the matter, it is the proponents, and not the minister, who make the determination as to whether or not a project is

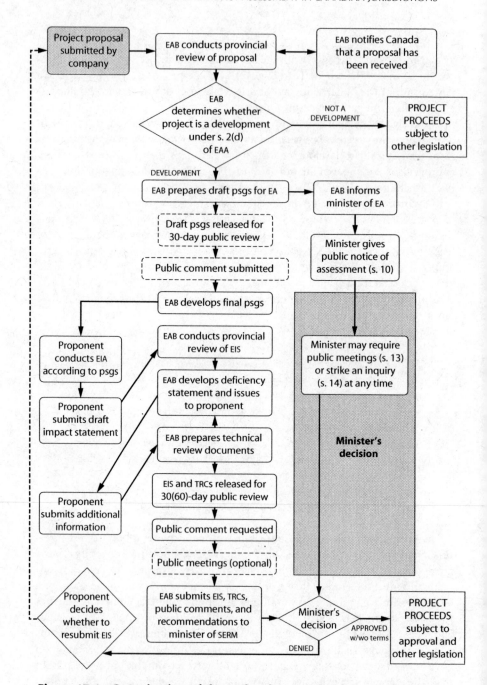

Figure 17.1   Organization of the Saskatchewan EA process as depicted (June 2004) on the Saskatchewan Environment website

a development within the meaning of section 2(d). In reality, however, it is the Environmental Assessment Branch that plays a pivotal role in the decision, and does so for a number of practical and statute-related reasons. First, in accordance with section 8(1) of the Act, 'no person shall proceed with a development until he has received ministerial approval'—this, notwithstanding any other required licence, approval, permit, or other form of consent. In keeping with the fundamentals of environmental assessment, the EA approval is perceived as being a part of the planning process that facilitates examination of the environmental impacts of the undertaking prior to irrevocable decision making. Those who proceed with a development in contravention of section 8 risk project shutdown[11] as well as fines up to $5,000/day or, in the case of a continuing offence, $1,000/day on summary conviction.[12] Considering the possible development costs associated with delay, the low monetary penalties alone would not provide sufficient deterrence in the case of larger projects facing critical construction deadlines. Project delay is a far more effective as a deterrent. Further deterrence also exists in the legislation in the form of liability for damages associated with proceeding without development approval or exemption as set out in s.23. This remedy is available to any person suffering loss, damage, or injury as a result of the unapproved development.[13] Section 23(2) reverses the burden of proof in such cases by placing the onus upon the person proceeding with the development to establish that the loss, damage, or injury was not caused by his or her project. Moreover, negligence is not a prerequisite to liability. Given the potential consequences of not assessing a development, it is understandable that most proponents not only evaluate their project relative to the criteria, but also rely upon the opinion of the EA Branch before proceeding.

Second, it is also common among proponents to submit a mitigation plan or formal environmental protection plan (EPP) along with their proposal. In so doing, they may be able to avoid having the project designated as a 'development' within the meaning of the Act by simply eliminating the threat of an impact or impacts likely to trigger the designation. Thus, proponents whose project might otherwise go through a full EA can avoid the legislation at the initial screening stages by simply ensuring their project is not a development within the meaning of the Act. Although some might argue that mitigation plans would be premature, such plans would clearly indicate that the proponents had turned their minds to the potential impacts of a proposed undertaking. The early submission of an environmental protection plan, again a desirable goal, may ensure timely approval and the screening out of projects that might otherwise be subject to more lengthy evaluation. The EPP mechanism has been formalized for some sectors, including oil and gas projects and intensive livestock operations (ILOs).[14] According to Saskatchewan Environment,

> The EPP should be complete and accurate so that SE[RM] can form the correct opinion as to whether or not the project is a 'development'. Without adequate information, SE[RM] must conclude that a project is a 'development' and hence subject to full EIA if it appears 'likely' that the project will trigger one or more of the criteria in the Act. Proponents are encouraged to discuss EPP scope with SE[RM] before starting to prepare it.[15]

Should the determination be made that a proposed project is a development and approval of the minister is required, the proponent is expected to conduct an environmental assessment and prepare an environmental impact statement (EIS).[16] The scope and terms of that statement are determined by the EA Branch, often in consultation with the proponent, and are spelled out in the form of project-specific guidelines. Although comments may be invited, there is no legislated requirement that there be public input into those guidelines. Pursuant to section 10 of the Act, the public receives formal notification of the assessment 'when the Minister becomes aware that an assessment is about to be conducted . . . in any manner that may be prescribed in the regulations.' Unfortunately, since passage of the statute almost 30 years ago, there have been no regulations promulgated pursuant to the legislation. Instead, it is the policy of the EA Branch to announce any assessments in local and province-wide newspapers and to list them on the department's website. The project-specific guidelines are often made available at the time of notification along with contact information within government and on behalf of the proponent. In addition, the EA Branch prepares quarterly reports that outline the history and status of projects submitted during the period.[17]

When a project is designated as a development, the EA Branch gathers a technical committee specific to the project to review the proposal, assist in development of the terms of reference, and review the adequacy of the EIS document. The committee membership is drawn from a standing panel called the Saskatchewan Environmental Assessment Review Panel (SEARP), whose members, assigned to the panel by their ministry or agency, boast the expertise necessary to assess the individual development.[18] There are no legislated requirements that obligate the proponent to follow the project-specific guidelines or consult with the EA Branch during EIS preparation. However, it is clearly in the best interests of the proponent to maintain a rapport with the committee and the EA Branch throughout the process, particularly in scoping the assessment (i.e., guideline preparation), so that delays can be avoided.

There are no regulations regarding the generic content of an EIS, a situation that led to litigation in *Stop Construction of the Rafferty Alameda Project Inc. (SCRAP) v. Swan*.[19] In the *SCRAP* case, the court held that, in the 'unfortunate' absence of legislative definition or regulation to explain the terms, the court was the only forum that could and would address questions relating to both the content and adequacy of an environmental impact statement.[20] Generally, an EIS describes and explains the rationale for the project, identifies alternative methods of undertaking the described development, describes the existing environment, and identifies the known or predicted impacts associated with the development (both negative and positive), including cumulative impacts, mitigable and non-mitigable negative impacts, and monitoring and follow-up plans. Although consultation is not specifically part of the legislation, it is anticipated that the developer will engage in a process of meaningful public involvement during the preparation of the EIS.

Preparation of the EIS document can take anywhere from six months to three years, depending on the complexity of the development. Once the document is complete, it is submitted to the EA Branch, and consequently to the technical committee, for review of the statement and all supporting documentation. Technical review comments are provided to the developer, and appropriate steps are taken to respond to the concerns raised regarding the EIS.[21] Once satisfied with the completeness of the statement, the minister must provide notice to the public and make the EIS, the technical review, and all other documentation available for inspection.[22]

It is noteworthy that the public right includes scrutiny not only of the statement but of the review as well. According to the Saskatchewan Court of Appeal, since the word 'review' is unqualified in the statute, 'it must be taken in its broadest sense to mean that anything that underlies it, any information, or documents relating to a development in the possession the Minister, must be made available for public inspection.'[23] With regard to confidential information, if in the minister's opinion it is in the public interest or the interest of any person to withhold or limit the production of any information or document, she or he may do so in accordance with the regulations. The exception to this discretion is information relating to pollutants, public health, or human safety; information of this sort must be released.[24] No regulations have been developed to address the issue.

Once the EIS and its review are released to the public for comment, the public normally has 30 days to provide written submissions to the minister. The minister may extend that time frame by an additional 30 days when he or she 'considers it appropriate'.[25] Section 13 of the Environmental Assessment Act provides that 'any time prior to making his decision whether to approve a development the minister may: (a) cause an information meeting to be conducted relating to the development and (b) direct the proponent to make experts available to attend the meeting.' Current policy is to hold such meetings during the public review period. The meetings are conducted by the proponent, with EA Branch personnel in attendance as observers. The presentation by the developer may be verbal or may use alternative media, such as poster boards or a self-conducted open house format. Question and answer periods are often included.

In the alternative—or addition—to the information meeting option, section 14 of the legislation provides that the minister may, at any time prior to decision making, appoint a board of inquiry to examine any or all aspects of the development. The inquiry must be conducted in accordance with the province's Public Inquiries Act;[26] in essence, it must be a full public hearing into the EIS and project review. The minister, however, is in no way fettered in his or her decision making by the recommendations of such an inquiry. As mentioned earlier, there has been only one inquiry in Saskatchewan under the Environmental Assessment Act, that being for the Rafferty Alameda Dam development.

Once the proponent is deemed to have met all the requirements of the Act and the public participation requirements have been satisfied, the minister must, within a reasonable time, determine whether or not the development is to proceed in accordance with his or her approval.[27] Terms or conditions may be attached and become requirements of the ministerial approval.[28] Notice of the decision must be forwarded, along with written reasons, to the proponent, to anyone who makes a written submission during the public review period, and to any other person the minister considers advisable.[29] Approval in hand, the developer may then seek other approvals, licences, and additional regulatory requirements.

Although there has been some evolution in the way EA is practised, the Saskatchewan Environmental Assessment Act itself has remained virtually unaltered since passage in 1980. In 1996, in the only legislative change, section 9.1 was inserted in the Act to address the contentious issue of forest management plans. The section defines a series of actions that fall within the definition of 'forest management activities'. These include harvesting of forest products, site preparation and improvement, reforestation or renewal, camp or road construction, forest improvement, application of chemical or biological agents, forest protection, and any other activity prescribed in the regulations. Unlike other activities, the section specifically provides that forest management activities undertaken in pursuance of a 20-year forest management plan (as defined in the Forest Resources Management Act)[30] are deemed to be a development for the purposes of the Environmental Assessment Act. In the absence of approval of a forest management plan (not the activity), no one shall proceed with any forestry activity within the plan area. Section 9.1 goes on, however, to provide two exceptions to this requirement:

1) holders of 20-year forest management plans approved prior to the coming into force of the section must only seek ministerial approval under the statute for the next revised 20-year forest management plan due after the coming into force of the section and may continue with their activities until the expiry of their earlier plan or until their development under the EA Act is denied approval; and

2) once approved, barring any change in the development that does not conform with terms or conditions of the ministerial approval, no further ministerial approval is required for revision of the 20-year plans.

In terms of evolving practice, one of the major changes in the EA process relates to the relationship among governments and to the formalization of coordination between the provincial and federal levels.[31] A significant early example was the Joint Federal-Provincial Panel on Uranium Mining Development in Northern Saskatchewan, which assessed several major development proposals during the period 1991 to 1997. In 1999 Saskatchewan signed the Canada-Saskatchewan Agreement on Environmental Assessment Cooperation, thereby translating the general principle of an earlier bilateral Sub-agreement on Environmental Assessment. The agreement was renewed in 2005.[32] The specific administrative framework outlined in the agreement articulates the role of each government in projects that require an environmental assessment by both the Government of Canada and the Government of Saskatchewan. As with all such agreements that fall within the Canada-Wide Accord on Environmental Harmonization, the underlying goal is to promote efficiency and effectiveness in areas of overlapping constitutional jurisdiction. Projects that fall within the scope of the agreement will undergo a single assessment, administered cooperatively by the two governments. Although one government may take the lead in the assessment process, no delegation of powers will take place and the parties retain their individual responsibilities and decision-making powers pursuant to their own EA legislation.

Critique

The Saskatchewan Environmental Assessment Act applies to private developments as well as to those of the Crown.[33] Decision making rests with the environment minister, who is accountable in writing to the proponent, to those who have made written submissions during the review stage, and to the public through traditional democratic means. The mechanics of the process seem to work efficiently; the efficacy of the process in ensuring environmental protection is more problematic.

There is no mechanism for policy-level or strategic assessment within the present EA legislation or practice in Saskatchewan.[34] Although forest management plans now fall within the purview of the legislation, this inclusion is very resource-specific and relates to an activity that in the opinion of many, already fell within the purview of the legislation. Although a policy (strategic) assessment mechanism was outlined by the Saskatchewan Environmental Assessment Review Commission in 1991, no formal or informal attempts have been made to assess systematically the environmental sustainability of government policies, plans, or programs.

Saskatchewan's environmental assessment legislation has now been in place for 28 years. In that time, and in spite of numerous references within the legislation, no regulations have been passed to place flesh on the bare-bones statute. The contents of an environmental impact statement, for example, still remain the subject of policy making and of negotiation between the government and proponents. In lieu of regulation, policies within the EA Branch have been relied upon to address such issues as notification and access to documents. Court actions, primarily initiated by individual stakeholders, have provided guidance in other areas, such as access to information and the scope of an environmental review. At this time, there appears to be no appetite for a more a systematic response to the shortcomings of the statute. In fairness to the Environmental Assessment Branch, however, fiscal and human resources have been significantly eroded over the years by an overall lack of government commitment to the environment department.

The success of any environmental assessment process is dependent upon active and meaningful public participation; such participation should begin early in the process and continue throughout. Unfortunately, the Saskatchewan EA process falls short in this area for a number of reasons.

First, in accordance with the formal requirements of section 10 of the Act, the public often becomes aware of a proposal only after the decision has been made regarding its status as a development. There is little or no opportunity for public input into the initial screening decision, and there is no requirement for public contribution regarding the scope and content (i.e., project-specific guidelines) of an EIS should the 'development' designation follow. As a result, concerns have been expressed over whether the process is vigorously applied and applied consistently to public and private sector proponents.

With regard to the scope of application, the case of *Irving v. Kelvington Super Swine*[35] focused on the definition of scope as determined by the development criteria within section 2(d) of the Act. The Saskatchewan Court of Appeal provided a very narrow interpretation of the section, maintaining the following with regard to widespread public concern criteria:

> A great deal of information is contained in the affidavits and the supporting material concerning the number of public meetings which were held and the number of petitions which were circulated and signed by people in the area of the project. It is fair to conclude that while there has been some public concern expressed about the possible environmental effect of these proposals in the Kelvington area by some groups, that concern is not widespread. From the material we examined it is doubtful the number of people expressing concern may not even represent a majority of the residents in the area of the project. While there is local interest in the proposal and local concern about possible environmental effects, those concerns are not wide-spread.

As a result, post-*Irving,* no intensive livestock operations (ILOs) in the province have been subject to the full EA process. In fairness, perhaps the ILO projects that would otherwise trigger the 'development' designation have been withdrawn, redesigned, and/or moved to a better site, thus essentially meeting the planning objectives of the legislation. Nonetheless, the fact remains that the primary responsibility for the establishment and operation of ILOs (including the environmental aspects of the industry) rests with Saskatchewan Agriculture.[36]

The narrow focus on the scope of projects and activities subject to EA has resulted in most proposals being 'screened out' at the initial stages. This is done without any formal public input: often the public is unaware that a proposal has been submitted or that the Act might even apply. The only option available to members of the public who might wish to challenge the determination is through judicial review; a costly and time-consuming proposition.

The second reason that the Saskatchewan EA process falls short in the area of public participation has to do with the fact that while it is expected that the proponent will seek public input during the course of EIS preparation, in the absence of regulation, this step is guaranteed only if it is included in the project-specific guidelines or is demanded by the technical committee. The nature of the activities in public participation should include not only technical input but also the seeking of values and preferences. In some respects, the definition of 'environment' in the statute would seem to ensure that public input would be sought so that the impacts on all environments could be properly assessed (the definition includes the biophysical environment and the 'social, economic and cultural conditions that influence the life of man or a community'[37] as they relate to the biophysical environment). However, mitigating against such participation are a number of factors, including timeliness issues of the proponent and language barriers. The matter arises most notably with regard to First Nations people, who find themselves unable to fully participate in the process. This is particularly the case in relation to government-initiated developments in Saskatchewan's north.[38]

Third, the 30-day public review period at the development review stage has proved a hindrance to effective public participation. In spite of the opportunity for a 30-day extension at the discretion of the minister, the size and complexity of a typical EIS and associated review documents place an unreasonable burden on stakeholders, who must access, evaluate, and meaningfully assess what might amount to volumes of materials. Directly associated with this problem is the lack of stakeholder funding available to facilitate participation.

The fourth problem, also associated with the public review stage, is the failure of the legislation and policy to provide adequate opportunity for meaningful dialogue between the stakeholders. The written comments of reviewers disappear into a bureaucratic void, and the information meetings are often a 'show and tell' exercise as opposed to a 'give and take'. Government officials maintain passive roles as they view these sessions from the sidelines. This is not to suggest that public hearings need be the norm, but the option has been underutilized to date.

It should also be noted that public participation in the process must not be confused with the superadded duty to consult with regard to the interests of First Nations people. The latter duty arises from a number of sources, including treaty and TLE agreements and section 35 of the Constitution Act.[39] The parameters of the duty to consult have been and continue to be the subject of litigation[40] and negotiation[41] both nationally and within the province.

Two smaller concerns are worthy of brief mention. First, there are few timelines included within the legislation or project-specific guidelines. This places all the parties at a planning disadvantage and can lead to particular frustration for proponents who are anxious to proceed with their development in a timely fashion. Second, the penalties within the legislation are dated and consequently provide an inadequate deterrent, especially in light of growing public concern about environmental protection.

In conclusion, a major shortcoming in the present system is weakness in the area of monitoring and follow-up. The approval of the minister does not come in the form of a licence or permit, and it is difficult for members of the public and government officials themselves to monitor developments and ensure that the terms and conditions included within the EA approval are actually satisfied. Similarly, owing to budget constraints, there has been little or no systematic follow-up to assess the efficacy of the process in achieving broader EA objectives. Recently, EA approvals have incorporated the concept of a commitments table for the EISs so that the documents would become auditable items in an ISO system; the goal of the initiative is to capture the significant EA commitments in a single location as part of the ministerial approval document. Theoretically, this will make follow-up and monitoring easier to track.

Notes

1. As of 2008, the provincial environment department has again been renamed and is currently the Ministry of Environment. In this paper, Saskatchewan Environment (SE) is also called Saskatchewan Environment and Resource Management (SERM) and Sas-

katchewan Environment and Public Safety (SEPS), depending on the year and the government organization at the time.

2. Eleanor Glor, The government tango: Communications and co-ordination, in *Is innovation a question of will or circumstance? An exploration of the innovation process through the lens of the Blakeney government in Saskatchewan, 1971–82.* http://www.innovation.cc/Book/chapter8.htm (accessed 19 September 2008).

3. Environmental Assessment Act, S.S.1979–80 (effective 25 August 1980) as amended by S.S.1983 c.77: 1988-89 c.42 and c.55: 1996 c. 19.1: and 2002 c.C-11.1. http://www.qp.gov.sk.ca/documents/English/Statutes/Statutes/E10-1.pdf (accessed 19 September 2008).

4. SEARC, *Environmental challenges* (Regina: Saskatchewan Environment and Public Safety, 1991).

5. Saskatchewan Environment and Resource Management, *Environmental assessment reform: Summary document* (Regina: Saskatchewan Environment and Resource Management, 1994).

6. EA Branch, *Guidelines for preparation of an environmental protection plan (EPP) for oil and gas projects: Procedures under the Environmental Assessment Act (Saskatchewan)* (Regina: Saskatchewan Environment and Resource Management, 2000).

7. It is noteworthy that the 1994 SERM reform document suggested a proposed automatic project review list. This list included select forestry activities, waste management projects, and mine and energy developments among its contents. See note 5 above.

8. See note 3, s. 2.

9. Ibid., s. 4.

10. Ibid., s. 8(3)

11. Ibid., s. 18.

12. Ibid., s. 21.

13. Ibid., s. 23(1).

14. In the case of ILOs, Saskatchewan Agriculture and Food (SAF) is the lead agency. Pursuant to the Agricultural Operations Act, S.S. 1995, c. A-12.1, SAF requires the submission of a waste management and waste storage plan with every ILO application. Any applications involving an operation greater than 1,000 animal units for hogs and 3,000 for beef operations are then forwarded to the EA Branch for review. The manure management plan is very similar to an EPP in terms of the underlying environmental objectives. http://www.publications.gov.sk.ca/details.cfm?p362. (accessed 28 September 2008).

15. See note 6 above.

16. See note 3 above, s. 9(1).

17. Saskatchewan Environment, Environmental Assessment Quarterly Status Reports http://www.environment.gov.sk.ca/Default.aspx?DN=91075781-7ce2-4bc4-97ea-f09ce968e8db (accessed 26 September 2008).

18. Saskatchewan Environment, A guide to the environmental assessment process: A process overview (Regina: EA Branch). http://www.environment.gov.sk.ca/Default.aspx?DN=1aeacd42-0d54-49d9-8d44-2c36fbbf712a (accessed 26 September 2008).

19. [1988] S.J. No. 292 (Q.B.).

20. Ibid., IV.

21. See note 20 above.

22. See note 3 above, s. 11.

23. *Saskatchewan Action Foundation for the Environment Inc. (SAFE) v. Saskatchewan (Minister of the Environment and Public Safety)* [1992] S.J. No.3 (C.A.), at p. 11.

24. See note 3 above, s. 7. The narrow scope of the minister's power to withhold information was confirmed (see note 23 above) in the *SAFE* case.

25. See note 3 above, s. 12(b).

26. R.S.S. 1978. c.P-38. http://www.publications.gov.sk.ca/details.cfm?p=787 (accessed 29 September 2008).

27. See note 3 above, s. 15(1).

28. Ibid., s. 17.

29. Ibid., s. 15(2).

30. S.S. 1996, c. F-19.1, as amended by the Statutes of Saskatchewan, 1997, c. W-13.11; 1998, c. W-13.12; 2000, c. 46 and 50; 2002, c. 31; 2004, c. T-18.1; 2005, c. M-36.1; and 2007, c. P-13.2 and c. 29. http://www.publications.gov.sk.ca/details.cfm?p=525 (accessed 28 September 2008).

31. See chapters 1, 9, and 14 in this book.

32. Government of Saskatchewan, news release, 30 May 2005, 'Canada-Saskatchewan sign renewed agreement on environmental co-operation' http://www.gov.sk.ca/news?newsId=412d51dc-265b-4067-aa08-21578136ced7 (accessed 30 September 2008).

33. See note 3 above, s. 3.

34. See chapter 6.

35. [1997] S.J.No.739 (C.A.).

36. The Saskatchewan Agriculture website describes the regulatory responsibility of Saskatchewan Environment as follows: 'Saskatchewan Environment may have requirements for domestic waste, dead stock disposal, shoreline development or other activities which may have impacts on the environment' (Government of Saskatchewan, Saskatchewan Agriculture, The regulation of intensive livestock operations in Saskatchewan. http://www.agriculture.gov.sk.ca/Regulation_ILOs_SK [accessed 1 October 2008]).

37. See note 3 above, s. 2.

38. For a recent publication promoting First Nations involvement in Saskatchewan EA, see International Institute for Sustainable Development, *Environmental assessment and Saskatchewan's First Nations: A Resource Handbook,* June 2008, http://www.iisd.org/pdf/2008/environmental_assessment_sask.pdf (accessed 1 October 2008).

39. Being Schedule B to the *Canada Act 1982* (U.K.), 1982, c. 11, http://laws.justice.gc.ca/en/const/annex_e.html (accessed 26 September 2008).

40. *Haida Nation v. British Columbia (Minister of Forests)*, 2004 SCC 73 and the cases that refined the requirements set out therein.

Chapter 18

Environmental Assessment: Manitoba Approaches

Kenton Lobe

> The objective of this Branch is to ensure that development maintains sustainable environmental quality.
>
> —*Manitoba Environmental Appeals Branch mission statement*

Introduction

Environmental assessment (EA) is a central feature of project decision making in Manitoba. Applying to both public and private undertakings, the legislation for assessments is marginally progressive, particularly when looked at in conjunction with the creation of a provincial environmental oversight commission and functioning mechanisms to fund public participation in the process. There are, however, significant shortcomings in both the legislation and its practical implementation.

This chapter begins with a look back at the origins of environmental assessment in Manitoba, highlighting the current legislation governing the EA process, the Manitoba Environment Act. Following this background overview is a discussion of the activities and projects that are required before an EA can be undertaken. The chapter then outlines the actual process of assessment, identifying the three mandatory and two discretionary steps involved. The final section touches on a critique and analysis of several issues, including staged assessment, the impact of the discretionary powers of both the director of the Environmental Assessment and Licensing Branch and the minister of conservation, and some highlights of the public participation process. The chapter concludes with a brief summary of the current status of EA in Manitoba and a list of sources of further information.

Origins of Environmental Assessment in Manitoba

Manitoba's environmental assessment process can be traced through several key pieces of legislation. The current legislation governing the EA process in Manitoba, the Manitoba Environment Act (S.M. 1987–88, c. 26—Cap. E125, hereafter referred to as the Act), replaced the Clean Environment Act of 1968 and the Environment Assessment and Review Process, which had been adopted as provincial Cabinet policy in 1975. The establishment of the new Act was due in part to frequent amendments to the Clean Environment Act, to changing environmental values and public expectations, as well as to the growing realization that the environment was coming under increasing pressure as a result of Manitobans' demand for economic development in their province.

According to the *Guide to the Manitoba Environment Act*, the Manitoba government identified six basic principles that needed to be reflected in the new Act:

- *Licensing process.* The Act tied EAs to a licensing process, issuing both public and private developments with a licence addressing the significant environmental impacts identified with a particular development. Developments would be reviewed on different levels or classes, depending on the nature and location of the development.
- *Public consultation and participation.* The Act strengthened and formalized the consultation and participation roles of the public in the environmental decision-making process, primarily through the establishment of the Clean Environment Commission.
- *Environmental scope.* The Act expanded the definition of environment and required that all actions that significantly affect the environment be subject to public scrutiny. Environment is defined in section 1(2) of the Act as 'air, land, and water, or plant and animal life, including humans'.
- *Pollution control.* The Act provided for site-specific limits and standards for actual or potential pollution to be addressed by the province through the licensing process.
- *Non-polluting environmental damage.* The Act mandated the province to address all types of environmental damage, including non-polluting, and outlined a process of EA to determine these impacts.
- *Enforcement.* The Act provided enforcement responses for non-compliance with licensing conditions to ensure compliance with laws to protect the environment.

The Manitoba Environment Act was eventually passed by the provincial legislature in 1987, becoming law on 31 March 1988. The purpose of the Act is stated in the opening paragraph:

> The intent of this Act is to develop and maintain an environmental management system in Manitoba which will ensure that the environment is maintained in such a manner as to sustain a high quality of life, including social and economic development, recreation and leisure for this and future generations. (*Manitoba Environment Act* 1(1))

As outlined in the six principles above and identified in the opening paragraph of the Act, better tools and processes, including environmental impact assessment (EIA), were recognized as valuable components of an improved environmental management process that would be tasked with identifying and mitigating the impacts of development. However, in contrast to other provincial EA processes, the purpose of the Manitoba Environment Act is to provide a legislative framework for a number of aspects of environmental management—not only the EA process. Environmental assessments of projects that are likely to have significant effects on the environment therefore represent only one component of the Act, but one that is directly linked to the actual licensing process. Several additional components of the Act provide a mechanism for state-of-the-environment reports (*Manitoba Environment Act* 4(1)) and recognize the emerging role of mediation in environmental decision making (though the state-of-the-environment reports ended in 1997 and mediation remains largely at the edge of the agenda).

The Manitoba Environment Act further sets out the functions and the general terms of reference for the Department of Conservation (formerly the Department of Environment), the responsible authority under the Act to do environmental assessment in the province. These functions are carried out by the Environmental Assessment and Licensing Branch of the Environmental Stewardship Division and include administration and enforcement of the Act, development and implementation of environmental quality standards and objectives, involvement of the public in environmental decision making, acquisition of knowledge and data, and development of environmental strategies and policies (*Guide to Manitoba Environment Act*). Many of the functions entail a good deal of discretionary decision-making authority, which rests with either the elected minister of conservation or the appointed director of the Environmental Assessment and Licensing Branch— the licensing arm of the provincial Department of Conservation.

As identified above, in order to provide a mechanism for public consultation and involvement, section 6 of the Act establishes the Clean Environment Commission (CEC), which consists of up to 10 commissioners appointed by the lieutenant-governor. The CEC is intended to be an arm's-length provincial body that is responsible for providing advice and recommendations to the minister, conducting formal investigations into environmental issues in the province, holding public hearings in the review and assessment process for proposed developments, and acting as a mediator between two or more parties to an environmental dispute. Its primary role in decision making consists of reporting back to the minister and providing recommendations based on its work. The CEC's role in the public participation process is discussed below.

There are other mechanisms for public consultation as well. One such example is a clause on public review in section 41(2) of the Act that requires that the minister provide opportunity for public participation and seek advice and recommendations on the development of new regulations or on amendments to the Act.

Coverage of Manitoba's Environmental Assessment Process

One of the characteristics that differentiates the EA process in Manitoba from that in other provinces is the inclusion of both public and private projects. Those activities that are subject to EA are defined as developments in the Manitoba Environment Act. The definition of *development* is found in section 1(2) of the Act and serves as the key definition around which scoping of projects for assessment occurs:

> Any project, industry, operation or any alteration or expansion of any project, industry, operation or activity which is likely to cause
>
> a)   the emission or discharge of any pollutant into the environment
> b)   an effect on any unique rare or endangered feature of the environment
> c)   the creation of by-products, residual or waste products
> d)   a substantial utilization or alteration of any natural resource in such a way as to pre-empt or interfere with the use or potential use of that resource for any other purpose
> e)   a substantial utilization or alteration of any natural resource in such a way as to have an adverse impact on another resource
> f)   the utilization of a technology that is concerned with resource utilization and that may induce environmental damage
> g)   a significant effect on the environment or will likely lead to a further development which is likely to have a significant effect on the environment
> h)   a significant effect on the social, economic, environmental health and cultural conditions that influence the lives of people or a community in so far as they are caused by environmental effects. (*Manitoba Environment Act* 1(2))

Developments are further categorized into three separate classes that serve in part to determine the process tracks used in the EA (discussed in the following section). In practice, these three classes also serve to define the amount of detail required from the regulator in the environmental impact statement (EIS), with Class 1 developments requiring significantly less than Class 2 or 3. Defined broadly in the Act, Class 1, Class 2, and Class 3 developments are differentiated by their perceived impact on the environment, with Class 1 having the least effect and Class 3 the most. While the Act provides some general definition in section 1(2), the Classes of Development Regulation (E125—M.R. 164/88) includes an exhaustive list of the kinds of projects included in each class of development. Thus, a development is exempt from the EA process, or its inclusion is at the discretion of the minister, unless it is specifically defined in the regulation.

Table 18.1 provides a summary definition and a list of categories of development for each of the three classes. Each category is further defined in the regulation to include a listing of the specific types of development for which a proposal must be submitted to the province. Where categories are present in more than one of the classes, one example from the regulation is included in parentheses.

The Process

The actual EA process in Manitoba is divided into five steps, three of which are mandatory and two that remain at the discretion of the director of the Environmental Assessment and Licensing Branch or the minister of conservation:

table 18.1 Categories of developments

Class of Development	Definition— Section 1(2)	Examples of Categories of Development (as defined in regulation)
Class 1	Effects that are primarily the discharge of pollutants	Agricultural; energy production (e.g., steam plants); fisheries; forestry (e.g., sawmills); manufacturing; transportation; waste disposal (e.g., water treatment plants)
Class 2	Effects that are primarily unrelated to pollution or are in addition to pollution	Energy production (e.g., stations generating less than 100 MW); forestry (e.g., pulp and paper mills); habitat modification; mining; recreation; transportation and transmission (e.g., transmission lines between 115 kV and 230 kV); waste treatment and storage and scrap processing (e.g., sewage treatment plants); water development and control (e.g., land drainage projects not greater than 500 km²)
Class 3	Effects that are of great magnitude or that generate a number of environmental issues	Energy production (e.g., stations generating greater than 100 MW); mining; transportation and transmission (e.g., electrical transmission lines greater than 230 kV); water development (e.g., land drainage projects greater than 500 km2)

1. Proponent files a proposal.
2. Proposal is screened.
3. Proponent is required to provide further information.
4. Opportunity is provided for public hearings.
5. A licensing decision is made.

The process is described in some detail in the Act, the corresponding regulation, and in the September 2002 *Information Bulletin* (97–01E) put out by Manitoba Conservation. Some of the highlights of each step are presented below and in Figure 18.1.

Step 1: Proponent Files a Proposal (Mandatory)

All proponents planning a development that is likely to have a significant effect on the environment are required to file a written proposal with the Department of Conservation. As outlined above, the specific types of development that require a proposal are listed in the Classes of Development Regulation. It is worth noting that proposals are often undertaken by private sector consultants hired by the proponent.

The actual requirements for a proposal are outlined in the Licensing Procedures Regulation (C.C.S.M. c. E125) and in the Environment Act Proposal Form. The proposal form requires the proponent to submit details regarding the location of the proposed development and the designated land use and a description of the potential impacts of the development on the environment. The description of potential impacts must include, but is not limited to,

- the type, quantity, and concentration of pollutants to be released into the air, water, or on the land
- impact on wildlife
- impact on fisheries
- impact on surface and groundwater
- forestry-related impacts
- impact on heritage resources
- socio-economic implications resulting from environmental impacts

The proposed environmental management practices and mitigation measures employed to address the impacts must also be identified by the proponent, as must be any government grants of funding applied for or received for the development. Finally, a schedule indicating the dates for construction, commencement of operation, and any staging of development must be included. All completed proposals are sent to the director of Environmental Assessment and Licensing, Manitoba Conservation, where they are advertised to the public, sent to the Public Registry, and assigned a government contact person.

Step 2: Proposal Is Screened by Public and Technical Advisory Committee (Mandatory)

The public is provided an opportunity, usually for a period of 30 days, to comment on all proposals received by the province. Proposals are further screened by an intergovernmental Technical Advisory Committee (TAC) to determine the form of assessment required. At this point in the process, the Manitoba Environment Act outlines specific requirements for the different classes of development. These differences are highlighted in sections 10, 11, and 12 of the Act.

Once the province receives a proposal, the public review process is initiated by either the director or the minister (depending on the class of development being proposed). A summary of the proposal is filed on the Public Registry, which is accessible to the public at several locations as well as electronically on the province's website. The province may also use advertisements in the local newspaper or radio to inform the public of the proposal and invite written comments. Both of these activities are required by regulation to occur within 30 days of the receipt of a proposal for Class 1 and 2 developments and within 45 days for Class 3 developments.

The Technical Advisory Committee generally consists of representatives from different departments in the provincial and federal government who assist in the

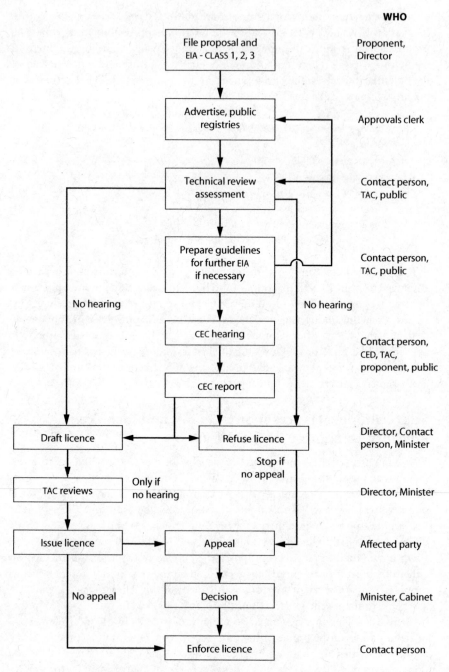

Figure 18.1   Manitoba Environment Act proposal process

screening of proposals (for a full discussion of harmonization between the federal and provincial EA processes, see chapter 9). The review is most often conducted informally by faxing proposal information to all TAC members and soliciting feedback on the need for further information or a formal EA process. This process may result in a recommendation to the minister to change the class of development. Again, timelines for this process are defined in the regulation and range from 30 days for Class 1 and 2 developments to 45 days for Class 3 developments.

At the conclusion of the screening process, which can last anywhere from 60 to 120 days depending on the class of the proposed development, a decision is made on the form of assessment required for the proposal, and the proponent is notified. The two discretionary steps, public hearings or the requirement for further information, may be initiated at the discretion of the director at this point in the process. These two steps will be discussed below, before we move to the fifth and final step—a licensing decision.

Step 3: Proponent May Be Required to Provide Further Information (Discretionary)

As a result of the screening process, either the director of the Environmental Assessment and Licensing Branch or the minister (depending on the class of development) may require further information from the proponent. This information can be obtained in one of two ways: through direct questions to the proponent or through the issuance of guidelines and instructions for the preparation of a full EIS on the development. Depending on the class of development and the discretion of the minister, the guidelines are prepared with assistance from the Technical Advisory Committee and at times with input from the public.

The director may also require the proponent to carry out public consultations as part of the assessment, request that the environmental impact statement be reviewed by the public, or request that the minister direct the chairperson of the CEC to conduct a public hearing.

Step 4: Province May Require Public Hearings on the Proposal (Discretionary)

The public hearing process is not mandatory under the Act. It is a discretionary power of the minister that can be used when a proposal is of general interest to the public, when a large number of Manitobans will be affected, or when, through the screening process, significant public concerns are identified.

In the case of a Class 1 proposal, the public has 30 days to provide the director with written comments and objections regarding submitted proposals. If the director receives objections to a proposal, he or she may, within the time frame specified in the regulation, recommend that the minister direct the Clean Environment Commission to hold a public hearing. According to existing legislation in the Act, this decision remains at the discretion of the director and, finally, the minister. In the event that objections are received and a public hearing is not called for, the director is required to provide written reasons to the objectors and inform them that the decision may be appealed to the minister.

The minister may also require the CEC to hold public hearings on the proposed development. According to the Act, the role of the CEC, upon receiving such direction, is to notify the public of the hearing process, open the participant assistance process, conduct the actual hearings, and, finally, provide recommendations to the minister based on the evidence received during the hearing. The scope of the hearings generally remains at the discretion of the minister (or director, depending on the class of development), and the recommendations of the CEC remain just that, recommendations that the minister is not required to adopt. In the CEC hearings of 2003 regarding Winnipeg's waste water system, the CEC did expand the scope of the hearings, but recommendations remained simply recommendations. However, if the recommendations that emerge from the CEC hearing process are not included in the eventual environmental licence issued by the province, the minister is required to provide written reasons to the proponent and the public by way of the Public Registry.

The actual hearings are usually conducted in the community immediately impacted by a project, but sometimes they are carried out in multiple locations to ensure that all interested members of the public have an opportunity to attend and/or participate. The general format includes presentations by the proponent, submissions by members of the public, and a question period during which both the public and the presiding commissioners have an opportunity to pose questions to the proponent. Both proponents and participants are required to file relevant documentation with the CEC.

Public submissions to the Clean Environment Commission are in the form of written letters or, alternatively, in-person presentations. In the case of presentations, speakers are given 15 minutes to provide their input unless they have made additional arrangements for a more formal presentation. Members of the public (participants—i.e., lawyers, experts, businesses, ENGOs [environmental non-governmental organizations], etc.) who have received participant funding from the province to participate in the hearings (details of which are discussed in the following section) are given additional time during the hearing and are encouraged to obtain outside professional assistance, such as might be provided by scientists, lawyers, engineers, or environmental experts, to aid them in making their submission. Finally, upon completion of the hearings process, the CEC has 90 days to submit a recommendations document to the minister.

Step 5: Licensing Decision (Mandatory)

Following the conclusion of the assessment and review process, the Department of Conservation makes a decision either to issue a licence with limits, terms, and conditions or to refuse a licence owing to the adverse environmental impacts likely to occur as a result of the proposed development. If a licence is granted, the Environmental Assessment and Licensing Branch of the department remains the responsible authority for enforcing its terms, limits, and conditions through the work of environment officers located in the regional offices throughout the province. The proponent may appeal the refusal of a licence.

Critique and Analysis

Now that the background to the Manitoba Environment Act and the relationship of the EA process to the Act have been outlined, it is useful to examine some of the perceived strengths and weaknesses in both the Act and the process. This section begins with one of the crucial definitions omitted from the Act and then goes on to examine the discretionary power of the province, the implications of staged assessment, and issues surrounding public participation in the process.

Effects

In the early portions of the Manitoba Environment Act, the word *significant* appears at many of the crucial decision points in descriptions of the EA process and legislation, yet does not appear in the definitions section of the Act or anywhere in the regulations. As noted above, the definition of *development* in section 1(2) of the Act refers to significant effects on the environment and on the social, economic, environmental health, and cultural conditions. With no definition of this key qualifier, the determination of *significance* remains at the discretion of the minister.

Discretionary Power

The absence of a definition for *significance* links with the larger issue of discretionary power wielded by the director of Environmental Assessment and Licensing and the minister of conservation. This power is apparent in the frequent use of 'minister may' clauses throughout the Act. Outlined above in steps 3 and 4 of the EA process, the decision to hold public hearings regarding a proposal remains at the discretion of the minister as outlined in the Act in sections 10(7), 11(10), and 12(6). Further, the decision as to whether or not further information is required also remains a discretionary power. Even when a formal EA is required, the form of assessment with regard to guidelines, public involvement, and review remains in the hands of the minister.

Members of Winnipeg's environmental community who have participated extensively in the EA process in Manitoba comment that this discretionary power has allowed the province to fast-track proposals for development. One of the ways in which this happens is through collaboration between the province and the proponent during the early stages of the proposal process. An example of the discretionary power of the director occurred in early 2003, when CEC hearings were called to examine the City of Winnipeg's $750-million plan to upgrade the existing waste water treatment facilities. The proponent (City of Winnipeg) had conducted almost 10 years of research and study to determine the correct course of action for the planned upgrades, but the director had not required a formal environmental impact statement. During the course of the first round of CEC hearings in January 2003, members of the public successfully petitioned the CEC to suspend the hearings until the proponent had in fact completed and submitted a full EIS for the proposed developments and allowed the public a chance to comment on the proposal.

Staged Developments

In the proposal form completed by proponents, there is provision for the staging of developments and for the actual assessment to be staged as well. Further, section 13 of the Manitoba Environment Act outlines the power of the minister to issue environmental licences in stages, with specifications, limits, terms, and conditions. This allows for a phased-in approval process that coincides with project planning and development. While this is seen as an attractive option for proponents planning larger projects, it serves to fragment a development and limit the scope of the assessment process. Two recent examples illustrate the difficulty with staging in the assessment and licensing process.

In 1999, Maple Leaf Pork, a subsidiary of the Toronto-based Maple Leaf Foods, began the process of gaining approval for the construction of a $120-million hog-slaughtering and -processing facility. The proposed development, to be located in Brandon, Manitoba, would, at capacity, slaughter and process 54,000 hogs per week. The construction of a $13.5-million waste water treatment plant, to be owned by the City of Brandon, was also slated for construction to deal with the 5,725 m³ of waste that would be discharged from the plant on a weekly basis. The development consisted of two major components, each of which was ultimately licensed and assessed separately. As identified above, the first component consisted of significant upgrades to the waste water treatment system in the area. This entailed the clearing of land (pre-construction), the construction of the facility, and, finally, the operation of the facility. Each stage was assessed independently of the other, thus de-linking the potential impacts of what was clearly one larger development. The second component of the development, the actual hog-slaughtering and -processing plant, went through a separate process that was staged as pre-construction, construction, and operation. The environmental effects of each of the components were considered independently even though they clearly originated from a single project that, as a unified whole, would create significant environmental effects.

A second, more common example of staged EA can be found in the construction of major hydroelectric developments in the province. In Manitoba, the actual construction of dams, with all of the associated environmental impacts, are regularly separated from the construction of transmission lines. Local environmental groups involved in the EA process continue to lobby the government to eliminate the use of staging in EA. The current provincial New Democratic Party (NDP) government has acknowledged the problem of staged assessment but has yet to address the issue through the legislative process.

Public Participation

One of the strengths of Manitoba's EA process is that it allows the possibility of an increased role for public participation in environmental decision making. While the discretionary power of the minister still largely determines the extent to which the public can participate in the process, the Public Registry system, the provision for participant assistance in public hearings, and the CEC itself warrant further discussion.

One of the benefits of the Manitoba Environment Act is that it instigated the creation of a series of publicly accessible registries located around the province. These registries were intended to provide the public with easy access to information about development proposals, the assessment process, assessment reports, environmental licences, and other pertinent things to facilitate more meaningful public involvement in environmental management decisions. However, there are persistent problems concerning the physical accessibility of documents, including limited hours of operation, lack of harmonization among registry locations, poor conditions of documents, incomplete indices, and expensive copying charges (Kidd 2001). The Internet has facilitated the development of the registry further, with all of the information currently accessible through the Department of Conservation's website. In addition, the Manitoba Eco-Network, a local environmental non-governmental organization, provides e-mail updates of all new postings to the registry through a listserv that they maintain.

The Act makes provision for participant funding during CEC hearings, something only done by Manitoba and Canada with regard to EA. This process is further outlined in the Participant Assistant Regulation (C.C.S.M.c. E125). While the funding remains at the discretion of the minister, when utilized, it provides an opportunity for more effective participation by the public. Notice of the issuance of participant funding is given through advertisements in the local media; these notices alert the public that applications are being accepted and indicate that they should be sent to the director of the Environmental Assessment and Licensing Branch. When implemented, the program provides funding for those individuals and organizations that qualify on the basis of a submitted application. The five-page application form (available online) includes relevant applicant details, organizational information, description of financial need, other funding information, and a tentative budget. Once completed, the form is vetted by a Participant Assistance Committee (such committees are established through the CEC) and also by the actual proponent of the development in question. The applicant is required to meet with these two parties and may be questioned on their submitted proposal for assistance (a process point that has been criticized by many applicants as somewhat confrontational). Again, final decision-making authority rests with the minister.

While underused from 1997 to 2001, the CEC hearing process is currently quite active in the province, though hearings still occur for less than 1 per cent of licences granted by the province and generally only for Class 2 and 3 developments. In April 2003 the minister of conservation released the terms of reference for the

CEC hearing regarding Manitoba Hydro's construction of the Class 3 Wuskwatim generating station and transmission lines projects—a major dam in northern Manitoba. The CEC was given the mandate to consider the justification of, need for, and alternatives to the proposed development, as well as the potential environmental, socio-economic, and cultural effects of the construction and operation. Close to $1 million was set aside for participant funding—a broad mandate for participation, in part because this was a blended CEC/Public Utilities Board hearing. Fifteen groups submitted applications for participant funding, and 11 of them received funding for their requests. The top two recipients, the Consumers Association of Canada and the Pimicikamak Cree Nation, received $190,000 and $160,000 respectively (see chapter 4). One of the ENGOs that received funding commented that 'everyone is on a learning curve on this' and indicated that while almost every applicant was given less than the amount requested, the process seemed to be providing the public a chance to take a 'sober second look' at proposed developments.

Another critical piece of any meaningful public participation initiative requires that notice of proposed developments and of opportunities to participate be given early and include all those potentially impacted by the development (Sinclair and Diduck 2001). Sinclair and Doelle (2003) suggest that EA legislation should facilitate meaningful and effective public participation, thereby ensuring that all of the affected and interested publics are notified of proposed developments early in the decision-making process. They further state that notice provisions should avoid situations such as the one that emerged during the construction of the Maple Leaf hog-slaughtering and -processing plant in Brandon in 1999. In this case, notice of a development whose effects included discharging large volumes of effluent into the Assiniboine River (with Winnipeg downstream) was limited to a small notice placed in the local Brandon newspaper. Notice does not necessarily come early and does not always reach all of Manitoba, as the province determines how widely the process is advertised.

Current Status of EA in Manitoba

In the fall of 1997, a multi-stakeholder process was initiated by the sitting Progressive Conservative government of Manitoba to develop recommendations on how best to implement the guidelines on sustainable development in land-use planning and environmental decision making. The recommendations (1999) that emerged from the Consultation on Sustainable Development Initiative (COSDI) were accepted by the new NDP government in the summer of 2000. The broad scope of the recommendations that were presented required changes to several pieces of legislation, including the Environment Act, a process that is yet to be completed. This section highlights several of the proposed amendments that arose from the COSDI recommendations.

Under the current Act, the EA process attempts to identify and mitigate the environmental effects of a development. The scope of the review is accordingly limited to these environmental effects and to the social and economic impacts associated with them. The COSDI report, however, recommends that this process be expanded to an effects assessment that would include all the components of sustainable development, including social, health, cultural, economic, and cumulative impacts. Several amendments to the Act have been identified as necessary to fulfil this central recommendation.

The expansion of the EA process to an 'effects assessment' would require a more substantial review process than is currently required by the Act. Highlights of an expanded process would include sections in the final document dealing with assessment methodology as well as with the results of public consultation. Another major shift would be the inclusion of an analysis of the need for, and of alternatives to, the project, including an analysis of the 'do nothing' option. Finally, a review of cumulative effects and project sustainability rounds out the list of some of the critical new elements of an effects assessment.

Manitoba Conservation's September 2001 document 'Building a Sustainable Future: Proposed Changes to Manitoba's Environment Act' provides a complete list of amendments and potential issues arising from them; here are some examples:

- Provision of increased capabilities and scope for the CEC to investigate development proposals on its own
- Establishment of clear Terms of Reference with respect to the scope of issues to be addressed by the CEC
- Clarification of the participant assistance provisions of the Act with reference to conditions for use, funding caps, and its relationship to larger planning initiatives
- Expansion of the capability and capacity of the Public Registry
- Amendment of the staged licensing process to make it contingent on a determination that environmental impacts of the project as a whole are not significant
- Coordinate multiple review processes
- Enshrine TAC in legislation and provide public access to it
- Proposals be consistent with larger planning initiatives

Sinclair (2002) outlines the process surrounding COSDI, highlighting some of the critical strengths and weaknesses of the process that resulted in the proposed amendments. Despite the government's own efforts and the concerted effort of a multi-stakeholder committee struck to consider what changes should be made in the Manitoba Environment Act, no changes in this regard have been made to the Act, and given the time that has passed, none are likely. One participant in the consultation recently indicated that 'industry likes the Act the way it is'.

Further Information on EA in Manitoba

Use of the Internet has contributed significantly to the available information on EAs in Manitoba. While there remains room for improvement, particularly on the provincial website, a wealth of websites—from the province's to those of the ENGOs—provide useful and up-to-date information on Manitoba's EA process and serve as an ongoing resource on the current realities in Manitoba.

Manitoba Conservation (Environmental Assessment and Licensing Branch)

http://www.gov.mb.ca/conservation/eal/index.html
- Manitoba Environment Act and associated regulations
- Public Information Bulletins
- Participant Assistance Application Form
- Public Registry
- Environment Act Proposal Form

Clean Environment Commission

http://www.cecmanitoba.ca
- summaries, transcripts, and proceedings of CEC public hearings
- updates for ongoing public hearings

Manitoba Eco-Network

http://www.mbeconetwork.org/
- updates to the Public Registry
- umbrella organization for Manitoba ENGOs

Canadian Parks and Wilderness Society (CPAWS)

http://www.cpawsmb.org/
- current information on EA process in Manitoba, particularly regarding provincial parks

References

Government of Manitoba. *Environment Act,* S.M., 1987–88, c. 26 – Cap. E-125, s. 1.
———. 1988. *Classes of Development Regulation* (E125—M.R. 164/88).
———. 1991. *Participant Assistant Regulation* (C.C.S.M.c. E125).
———. 1999. Report of the consultation of sustainable development initiative. Sustainable Development Coordination Unit, Winnipeg.
Kidd, Scott. 2001. My adventures at the Public Registry. *Manitoba Eco-Journal* 11 (1): 5.
Manitoba Conservation. 2001. *Building a sustainable future: Proposed changes to Manitoba's Environment Act—a discussion paper.* Winnipeg: Government of Manitoba.
———. 2002. *Information Bulletin* (97–01E).
Manitoba Environment. *Guide to the Manitoba Environment Act.* Winnipeg: Government of Manitoba.
Sinclair, A.J. 2002. Public consultation for sustainable development policy initiatives: Manitoba approaches, *Policy Studies Journal* 30 (4): 423–43.
Sinclair, A.J., and A.P. Diduck. 2001. Public involvement in EA in Canada: A transformative learning perspective. *Environmental Impact Assessment Review* 21 (2): 113–36.
Sinclair, A.J., and M. Doelle. 2003. Using law as a tool to ensure meaningful public participation in environmental assessment. *Journal of Environmental Law and Practice* 12 (1): 27–54.

Chapter 19

The Ontario Environmental Assessment Act

Sonya Graci

The Ontario Environmental Assessment Act (EAA) came into force in 1975. It was Canada's first environmental impact assessment (EIA) law and has been the subject of controversy and criticism ever since (Estrin and Swaigen 1993). The Ontario EAA has been used as a model for other environmental impact legislation worldwide, and over its 34-year history it has supported the planning of several thousand projects.

History of the Ontario Environmental Assessment Act

The EAA had limited application when proclaimed in October 1976. At that time, the Act was initially applied only to Ontario government department and agency undertakings. Proclamation was accompanied by the publication of 200 pages of orders exempting individual undertakings, such as the Darlington Nuclear Generating Station, and whole classes of significant undertakings, such as all projects by conservation authorities and municipalities (Estrin and Swaigen 1993).

Within the public sector, the first expansion of the application of the EAA was to conservation authorities, in 1977. Municipal undertakings were also made subject to the EAA, following years of lobbying to delay implementation. The EAA was rarely applied to the private sector, although over the years the tendency for certain private sector projects to be designated under the EAA has increased (Estrin and Swaigen 1993). Despite Ontario's being the first province to develop EAA legislation, environmentalists have criticized the provincial government for its failure to apply the EAA to the private sector, its exemption of significant public undertakings, the lack of requirements for information sharing and public involvement, and its failure to provide participant funding at the planning stages of assessment (Estrin and Swaigen 1993). Proponents have also criticized the process for being too bureaucratic and costly. In the past, numerous studies had to be conducted without a clear set of rules to guide them.

The EAA was modified in 1996 to become the Environmental Assessment and Consultation Improvement Act (still abbreviated as EAA). The modified Act introduced a number of legislative and program changes to improve its effectiveness. Key features of the revised EAA include

- requirements for terms of reference (TOR);
- regulated timelines for reaching a timely decision;
- mandatory public consultation;
- the ability to refer matters to mediation;
- the ability to focus the Environmental Review Tribunal hearings on outstanding issues; and
- the entrenchment of the class EA process and Part II orders.

These changes were significant and have strengthened the process, addressing many of the criticisms that have risen since the EAA's inception. With such improvements, the Ontario EAA has been able to maintain its reputation as perhaps entailing the most comprehensive environmental assessment laws in Canada.

The Ontario Environmental Assessment Act

The Ontario EAA governs a decision-making process that concerned parties may use to promote good environmental planning by assessing the potential effects of certain activities on the environment (Ontario Ministry of the Environment 2003). The EAA provides a systematic method of evaluating proposed undertakings that have the potential to affect the environment and determining whether or not these undertakings should be approved (EAAB 2001b).

The purpose of the EAA is to provide for the protection, conservation, and wise management of Ontario's environment. The definition of *environment* is broad and encompasses the natural, social, cultural, economic, and technical aspects of the environment. The EAA defines the environment as

1. air, land (includes enclosed land, land covered by water and subsoil) or water,
2. plant and animal life, including human life,
3. the social, economic and cultural conditions that influence the life of humans or a community,
4. any building, structure, machine or other device or thing made by humans,
5. any solid, liquid, gas, odour, heat, sound, vibration or radiation resulting directly or indirectly from human activities, or
6. any part or combination of the foregoing and the interrelationships between any two or more of them, in or of Ontario. (*Ontario Environmental Assessment and Consultation Improvement Act* 2002)

This comprehensive definition ensures that a breadth of environmental impacts will be considered prior to development.

The purpose of the environment assessment (EA) process is to ensure that decisions are made following a rational and objective planning process. An EA must be rational, consistent, traceable, reproducible, and fair (Ontario Ministry of the Environment 2003). This requires that the EA show continuity in the development of the assessment process, provide sufficient information for any interested person to trace the decision-making process, demonstrate rational decision making at each stage of the process, and demonstrate a process that does not have a predetermined result (Estrin and Swaigen 1993). The proponent meets these requirements by providing clear and complete documentation throughout the EA process. As the EA process is self-assessed in nature, due process must be followed.

Application of the EAA

The Ontario EAA applies to public and designated private sector undertakings. Public sector undertakings by Ontario government ministries and agencies, Ontario municipalities, and other public bodies defined by Ontario Regulation 334 under the EAA (e.g., the Conservation Authorities, Ontario Power Generation, and Hydro One) are subject to the EAA. Public sector undertakings are often infrastructure developments such as public roads and highways, transit facilities, waste management facilities, sewage and water works, electrical generation and transmission facilities, and flood protection works (EAAB 2001b). The minister of the environment (the minister) may, with Lieutenant Governor in Council (Cabinet) approval, declare that the EAA does not apply to a specific proponent or undertaking for any or all of the requirements of the EAA. However, the minister must be satisfied that this would be in the public interest and may impose conditions to ensure that the environment will be protected.

Private sector undertakings are only subject to the EAA if specifically designated by regulation. Private sector undertakings are projects of non-government organizations (NGOs), private companies, and individuals, and are only subject to the EAA if the Lieutenant Governor in Council makes a regulation designating individual private sector projects or general classes of projects as being subject to the EAA. Typically, environmentally significant private projects (e.g., landfills, waste transfer processing stations, and incineration projects) are subject to the EAA. Electricity projects are also subject to Ontario Regulation 116/01, which identifies the scope of projects subject to the EAA (Ontario Ministry of the Environment 2003; McLennon 2003).

Ontario Environmental Assessment Act Program Areas

Ontario's EAA has four program areas. These four areas target all significant undertakings planned in the province of Ontario:

1. Individual environmental assessments
2. Declarations
3. Designations
4. Class environmental assessments

The Ontario Ministry of the Environment's Environmental Assessment and Approvals Branch (EAAB) completes the managing and reviewing of projects. The role of EAAB is to advise the minister on EAA program areas; to coordinate government reviews of EAA program-area undertakings; to facilitate issue resolution; to provide advice and guidance to proponents, the public, and other stakeholders on EAA requirements; and to develop policy to support the EA program (EAAB 2001b).

Individual Environmental Assessments

Individual environmental assessments are undertaken to assess the potential for significant environmental effects (McLennon 2003). The types of projects that are subject to an individual EA are generally waste disposal facilities (incinerators, landfills, large transfer stations); transitways and highways; and other projects that are not covered by a blanket environmental assessment (waterfront trails, water supply, and cogeneration facilities) (EAAB 2001b).

Projects that are required to undertake an individual EA must follow a process prescribed through legislation. This enables a rational, reproducible, fair, and traceable planning and decision-making process. If a project must undergo an individual EA, the proponent must apply to the minister of the environment for approval to proceed with the project. An application for approval must include the terms of reference and an EA document (*Ontario Environmental Assessment and Public Consultation Improvement Act* 2002; EAAB 2000a). Figure 19.1 outlines the process for submitting an individual EA.

Terms of Reference

One change to the EAA introduced in the 1996 legislation is the requirement for the preparation, submission, and approval of terms of reference prior to in-depth studies being conducted on an undertaking. An approved ToR represents an agreement between the proponent and the minister about how the EA document will be structured and completed to ensure proper assessment of potential effects on the environment. Once approved by the minister, the ToR sets out a framework that will guide and focus the preparation of the EA. In the EA planning and approval process, the approval of the ToR is the first statutory decision by the minister. An approved ToR, however, does not guarantee approval of the proposed undertaking (Ontario Ministry of the Environment 2003).

The purpose of the ToR is to provide a framework with which to proceed with an impact assessment. The ToR defines the content of the EA document, enables the proponent to focus the EA on potential environmentally significant issues, describes how the proponent will consult with the public and review agencies when preparing the EA, and establishes a framework for evaluating the EA. The ToR also provides the government, public, and other stakeholders the opportunity to review and comment on the proposed undertaking prior to any major studies being

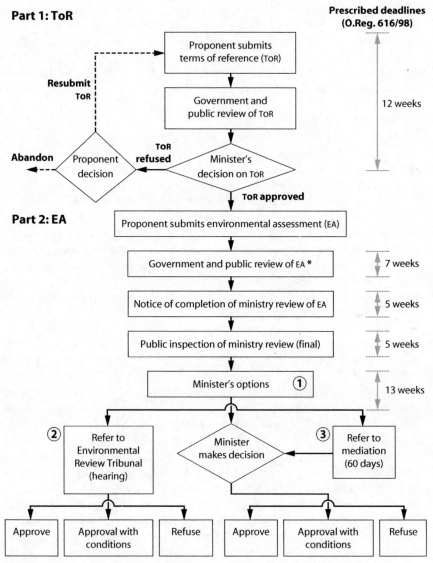

Part 1: ToR

Proponent submits terms of reference (ToR)

Resubmit ToR

Government and public review of ToR

Abandon   Proponent decision   **ToR refused**   Minister's decision on ToR

ToR approved

Part 2: EA

Proponent submits environmental assessment (EA)

Government and public review of EA *

Notice of completion of ministry review of EA

Public inspection of ministry review (final)

Minister's options   ①

Minister makes decision

② Refer to Environmental Review Tribunal (hearing)

③ Refer to mediation (60 days)

Approve | Approval with conditions | Refuse | Approve | Approval with conditions | Refuse

Prescribed deadlines (O.Reg. 616/98)

12 weeks

7 weeks

5 weeks

5 weeks

13 weeks

① The minister has three options: (1) refer all or part of application to the tribunal; (2) make a decision; or (3) refer to mediation.

② If referred to the tribunal, the minister has 28 days in which he or she may review the tribunal decision. The tribunal has the same decision options as the minister (approve, approve with conditions, or refuse).

③ If referred to mediation, the minister shall consider the mediator's report when making a decision.

* The director may issue a Deficiency Statement. If the deficiencies are not remedied, the minister may reject the EA.

Notes: Self-directed mediation may occur at any time.

The minister may refer an EA application to mediation (referred mediation) any time during the EA process (60-day maximum).

Figure 19.1   The Ontario environmental assessment process

Source: Environmental Assessment and Approvals Branch, Ontario Ministry of the Environment, Toronto, 2003.

undertaken (EAAB 2000a). This stage in the process provides a form of reassurance to all stakeholders but mostly to the government and proponent, as it ensures that the undertaking is on an acceptable path.

There are three methods for preparing a ToR and subsequently an EA. Section 6(2)a of the EAA states that the proposed ToR must indicate that the EA will be prepared in accordance with the requirements set out in subsection 6.1(2) of the EAA. This subsection stipulates that the EA shall be prepared in accordance with the approved ToR and must consist of the following:

1. description of the purpose of the undertaking;
2. description of and a statement of the rationale for the undertaking, the alternative methods of carrying out the undertaking and alternatives to the undertaking;
3. description of the environment that will be affected or that might reasonably be expected to be affected, directly or indirectly, the effects that will be caused or that might reasonably be expected to be caused to the environment, and any actions necessary to prevent, change, mitigate or remedy the effects on the environment by the undertaking, the alternative methods of carrying out the undertaking and the alternatives to the undertaking;
4. evaluation of the advantages and disadvantages to the environment of the undertaking and alternatives to the undertaking; and
5. description of consultation conducted pertaining to the undertaking and the results of the consultation. (*Ontario Environmental Assessment and Consultation Improvement Act* 2002)

Since the inception of the EAA, EAs have typically followed this format, which has thus been deemed traditional. This method is used when plans are still at the conceptual stage or when no previous related work has been completed on the project (EAAB 2001b).

A ToR can also be submitted under section 6(2)b, which states that the proponent must indicate that the EA will be prepared in accordance with the requirements prescribed for the type of undertaking the proponent wishes to proceed with. This particular type of undertaking occurs on a regular basis but is not covered by an existing class EA process. The appropriate requirements for the preparation of the EA are to be prescribed on an individual project basis by regulation. To date, this method has not been used (*Ontario Environmental Assessment and Consultation Improvement Act* 2002; EAAB 2001b).

The third method consists of submitting a focused ToR. Section 6(2)c of the EAA states that a proposed ToR must set out in detail the requirements for the preparation of the EA. This enables the proponent to define the boundaries of what will be considered in the EA, a useful exercise for proponents at an advanced stage in their planning process. The proponent can define the starting point of the EA and focus on alternatives that address the project's specific needs and circumstances, rather than having to undertake an exhaustive analysis. This method has normally been used by private sector proponents that have obtained a particular project site and are in the business of implementing a particular undertaking. For example, a private sector proponent that required an EA to build a landfill site on a particular plot of land used this method, as it was not deemed economically feasible for the proponent to consider alternatives or alternative sites (EAAB 2001b; Desautels 2003). On 17 June 2002, however, the Superior Court of Justice, Divisional Court decision on the Richmond landfill undertaking deemed that the minister could not approve any focused ToR submitted under section 6(2)c of the EAA. The reasoning was that section 6(2)c is open to interpretation and the minister does not have the power to reduce the requirements of the EAA.

The ToR is not a draft EA; rather it is a plan of action for addressing the requirements of the EAA. Prior to the ToR being submitted formally to the ministry, proponents are encouraged to consult the ministry on the ToR contents. The draft ToR should be circulated to the EAAB, the government review team (GRT), and other interested parties, such as the public, NGOs, and First Nations, to ensure that potential issues have been identified and addressed. The GRT is composed of representatives of federal, provincial, and municipal government agencies and ministries (including technical reviewers from the Ontario Ministry of the Environment) that may be interested in the proposed undertaking. Consultation prior to formal submission is recommended, as it is difficult to amend a ToR after submission (EAAB 2000a).

Once the ToR has been submitted formally for review, the minister must make a decision on the application within 12 weeks. To announce the formal submission, a 'Notice of Submission' must be given. This notice is normally published in the local newspapers, but other methods of notice can be used. The minister cannot make a decision about the proposed ToR until proper public notice has been given and any resulting submissions considered. As identified in Figure 19.1, the formal public consultation stage of the ToR is 30 days. During this public and government consultation period, the ToR is reviewed and commented upon. The submissions are considered in the minister's final decision. Once the formal consultation period has ended, the Environmental Assessment and Approvals Branch reviews the submissions and the ToR and makes a recommendation to the minister. At this time, the minister may decide to approve the ToR, approve the ToR with amendments, refuse the ToR, or order mediation. The minister's decision on the ToR is final (EAAB 2000a).

Once the ToR is approved, the proponent begins to prepare the EA. An approved ToR sets out the minimum requirements for what the proponent will do in the preparation of the EA, but a proponent may undertake more than the ToR requires. No further amendments can be made once the minister has made a decision about the ToR (McLennon 2003; EAAB 2000a). If in the course of preparing the EA the proponent discovers a need to change the ToR, a new ToR must be submitted.

There are many benefits to the ToR. First of all, the ToR was introduced to expedite the EA process; prior to this amendment, a proponent would submit an EA without any direction. The ToR provides certainty for the proponents, public, government review team, and other interested parties with regard to the preparation of the EA. Further, it requires and facilitates consultation with interested parties before irreversible decisions are made. It is also cost-effective, as it enables the GRT to identify any unreasonable proposals before government and proponents invest their time and money.

The EA Document

The EA document is prepared in accordance with the approved ToR. An approved ToR does not have an expiry date but may become dated. Government standards may change, and it is the responsibility of the proponent to verify that the ToR is up to date. The director of the EAAB can deem an EA deficient if it does not meet the requirements of the approved ToR. To date, only one EA was considered deficient in Ontario (Desautels 2003). The EA for the Bennett Incineration undertaking was deemed deficient, as it did not provide the requisite amount of technical information and data for the GRT to make a decision on the environmental effects of the project (Amodeo 2003).

When a formal EA document is submitted to the ministry, the proponent is required to give public notice of the submission. The review of the EA document takes 30 weeks. (See Figure 19.1 for details about the steps in the EA submission and review process.) Once the above notice has been given, the public, the government review team, and other interested parties have seven weeks to review the EA and submit their concerns regarding the project to the ministry.

Once the public and the GRT have completed their review of the EA, the EAAB has five weeks to review the EA document and all public and GRT submissions. The EAAB then publishes a document reporting any shortcomings identified by the GRT and the public. This document also assesses how well the proponent has addressed the requirements of the EAA and the concerns of the public and the government, and whether the preparation of the EA document has been carried out in accordance with the approved ToR and the EAA (Ontario Ministry of the Environment 2003). The ministry publishes a 'Notice of Completion' in major newspapers and on government websites to notify the public of the release of the ministry's review document and of the final public comment period. This second five-week comment period gives the public the opportunity to make written requests to the minister (perhaps to suggest what issues are outstanding and how these might be resolved through specific conditions of approval, or to request a hearing and/or mediation) and one last chance to comment on the EA and bring their concerns to the minister's attention. At the end of this period, the EAAB reviews the public and GRT submissions and addresses outstanding concerns by negotiating conditions of approval with the affected stakeholders. Such negotiations facilitate resolution of the contentious issues between affected stakeholders and the proponent. Enabling the public to play a role in the decision-making process and instituting parameters that will reduce the undertaking's impacts on the environment are the types of conditions that are implemented. Following the negotiation of conditions, the EAAB provides a recommendation to the minister.

The minister then has 13 weeks to decide whether to approve an undertaking. The minister will consider the submission made by the proponent, the recommendations of the GRT, the ministry review, comments received from the interested parties, and the EA's consistency with the approved TOR. In making the decision, the minister has five options: to approve the undertaking, to approve the undertaking with conditions, to refuse to approve the undertaking, to defer the decision and refer some or all of the undertaking to mediation, or to refer some or all of the undertaking to a hearing (*Ontario Environmental Assessment and Consultation Improvement Act* 2002). With a decision involving mediation or a hearing, the decision on whether or not to approve the undertaking must still be made. Once a minister makes a decision on an undertaking, it must be approved by the Lieutenant Governor in Council. In most cases, the minister will approve an undertaking with conditions. To date, no undertakings have been approved without conditions, and only one undertaking has been rejected by the minister; in 2001 the South Simcoe landfill environmental assessment was the first EA refused by the minister for not being in compliance with the EAA (Connelly 2003). In addition, no projects have been referred by the minister to mediation, and since the EAA was revised in 1997, only one project, the Adams Mine landfill site, was referred to a hearing (McLennon 2003; Connelly; 2003).

Timelines

One of the most fundamental changes to the EAA since its inception has been the introduction of Ontario Regulation 616/98. This regulation was introduced to prescribe timelines in the individual EA process. Prior to this regulation, there was little certainty about when a decision on an EA would be made. There have been several instances where the EA process took up to three years before a decision was made. It was due to the lack of certainty and to an increase in delay that this regulation was promulgated. The regulation prescribes timelines for each step in the TOR and EA process. Figure 19.1 outlines the regulated timelines for each step. As these timelines are instituted for the government staff and minister, the timely review of a submission is ensured. There are no penalties for not meeting the timelines, as a decision is not invalid if it is late, but the consequences for the government are proponent pressure, public and auditor criticism, and lack of credibility. It is very rare for the ministry not to meet the regulated timelines (Ontario Ministry of the Environment 2003).

Public Consultation

Consultation with the public, government agencies, First Nations, and other affected parties is now a required part of the EA process and a key element of the EAA. In 1997 amendments to the EAA formally recognized the benefits of early consultation by legally requiring proponents to consult with the public prior to submitting an environmental assessment.

Consultation in the EA process comprises 'the activities carried out by a proponent to provide a two-way communication process to involved interested stakeholders in the planning, implementation and monitoring of an undertaking' (EAAB 2000b, 6). In the EA process, consultation is intended to help proponents achieve public acceptance of the final proposal. While consultation alone cannot ensure that a proposal is in the public interest, it has been demonstrated over the years that proponents who are responsive to concerns are more likely to develop an acceptable proposal (EAAB 2000b).

Effective consultation must be tailored to the unique needs of the project and community involved. The ministry has developed a draft 'Guideline on Consultation in the EA Process, 2000' that identifies the key elements of effective consultation, such as transparency and information sharing. This guide also discusses consultation methods, tools of effective consultation, and issues to be considered when evaluating the EA proposal. Effective consultation could include town hall meetings, open houses, newsletters, alternative dispute resolution, and mediation (EAAB 2000b).

Declaration Orders

The EAA allows the minister, with Lieutenant Governor in Council approval, to declare, through an order, that the EAA or part of the EAA or a regulation does not pertain to a certain proponent, undertaking, or activity. The declaration order changes a proponent's project status from being subject to the EAA to being exempt from some or all of the requirements of the EAA. Declaration orders are sometimes considered in cases of emergency, where the proposal is in the public interest and potential environmental impacts are minimal or where environmental impacts are already addressed adequately through other legislation.

For an undertaking to be declared exempt from the EAA, a proponent must make a written submission to the minister requesting that the undertaking be declared not subject to the provisions of the EAA. The declaration request is posted for a minimum of 30 days for public comment on the Environmental Bill of Rights Registry prior to the minister's and consequently the Lieutenant Governor in Council's decision. The request is evaluated on the basis of the ability of other legislation to deal with issues or concerns, the nature of ministry or agency concerns, public comments, and the potential for significant environmental effects. If a declaration order is granted, it is normally with conditions of approval. These conditions are developed to ensure that the environment will be protected and to address issues such as the period for which the declaration order will be in effect, the specific studies that must be conducted, and the consultation that should be undertaken (Ontario Ministry of the Environment 2003). An example of undertakings that have been declared exempt from the EAA is the renewal of the Ontario Ministry of Natural Resources Forest Management Class EA. Undertakings in this area were declared exempt with conditions of approval in June 2003, as the Ontario government assessed that their environmental impacts were being addressed adequately through current Ministry of Natural Resources legislation (Rohaly 2003).

Designation Regulations

A designation regulation is issued by the Lieutenant Governor in Council. This regulation changes a proponent's project status from being 'not subject' to the EAA to being 'subject' to the requirements of the EAA. Designation requests involve private sector activities, such as waste-related projects that have the potential for significant environmental effects. Any interested person can request that a project be designated. When a designation request is received, the public and the GRT are given an opportunity to comment on the proposal. As with a declaration request, most proposals are placed for consultation on the Environmental Bill of Rights Registry for a minimum of 30 days. When assessing designation requests, the ministry considers the same issues as with a declaration request and makes a determination to grant or deny the request. The final decision is made by the Lieutenant Governor in Council and posted on the Environmental Bill of Rights Registry. As many proponents realize that their projects will be designated, they can voluntarily enter into an agreement with the ministry to undertake the EA process without a designation regulation. It is an informal policy in the ministry to enter into voluntary agreements with private proponents to designate certain waste undertakings, such as private hazardous waste facilities, landfill sites, and incinerators. Table 19.1 identifies the types of undertakings that are voluntarily designated. These voluntary designations ensure that a timely process is undertaken and unnecessary delays do not hinder the EA process (Ontario Ministry of the Environment 2003; McLennon 2003).

Class Environmental Assessments

Not all undertakings subject to the EAA must undergo the individual EA review and approval process as previously described. Some groups or classes of projects are undertaken routinely, have predictable and mitigable environmental effects, and therefore do not warrant an individual EA. These are known as 'class environmental assessment' (class EA) projects. The class EA process was not entrenched in legislation prior to1997, when the changes to the EAA were entrenched in legislation.

Table 19.1   Types of projects that are voluntarily designated to be subject to the EAA

Waste Type	Landfill	Incineration	Processing	Transfer
Municipal waste	>40,000 m³	>100 tpd	>200 tpd	>300 tpd
Hazardous/hauled liquid industrial waste	all	all	>200 tpd	>300 tpd

Note: Tpd means metric tonnes per day.

Source: EAAB 2001a, 6.

Ontario's class EA document sets out a standardized planning process that covers routine activities. It is submitted and reviewed under the previously described review and approval process for individual EAs. Approval, if granted, applies to the entire class of undertakings. Thus, a proponent who receives approval for a class of undertakings does not need to obtain separate approvals under the EAA for each specific project, provided the class EA planning process is adhered to (Ontario Ministry of the Environment 2003). Currently, Ontario has approved a total of 11 class EAs:

1. Municipal Engineer's Association Municipal Class Environmental Assessment
2. Ministry of Natural Resources Class Environmental Assessment for Timber Management on Crown Lands
3. Ministry of Natural Resources Class Environmental Assessment for Small Scale Projects
4. Ministry of Natural Resources Class Environmental Assessment for Parks and Conservation Reserves
5. Ministry of Transportation Class Environmental Assessment for Provincial Transportation Facilities
6. GO Transit Class Environmental Assessment Document
7. Ontario Realty Corporation Class Environmental Assessment for Realty Activities
8. Ontario Power Generation Class Environmental Assessment for Modifications to Hydroelectric Facilities
9. Ontario Power Generation Class Environmental Assessment for Shoreline and Riverbank Modifications
10. Hydro One Class Environmental Assessment for Transmission Facilities
11. Conservation Authorities of Ontario Class Environmental Assessment for Remedial Flood and Erosion Control Projects

These 11 documents are called 'parent' class EAs. Projects within the parent class EA are pre-approved providing they follow the approved planning process. To ensure that the environmental effects are considered for each project, proponents are required to follow the planning and design procedures, including public consultation, that are set out in the approved class EA.

ToRs are required for the preparation of the parent class EA but are not necessary for individual projects approved under a class EA. Occasionally, projects that warrant being individually dealt with through the EAA's review and approval process may be carried out in accordance with a class EA. Under the class EA, a Part II order (formally known as a 'bump-up provision') is the mechanism under which any affected or interested party may request that the minister order that an individual EA be prepared for a particular project within the approved class of undertakings.

Part II Orders

In accordance with section 16.1 of the EAA, any individual, group, or agency that has significant environmental concerns with a class EA project, or with the manner in which a proponent is meeting the requirements of the relevant class EA process, may request that the minister order a project to undergo an individual EA. Within the class EA, the level of assessment varies for each category of project. Thousands of projects are undertaken annually under various class EAs, but projects are only reviewed by the minister when a Part II order is requested. The appropriate district office of Ministry of the Environment is involved with every class EA project through consultation conducted by the proponent (Ontario Ministry of the Environment 2003).

The review of a Part II order is under strict timelines. The timeline for reviewing each parent class EA Part II order varies, but generally the time allotted is between 45 and 66 days. During this time, the EAAB will review the request, accompanying documentation, and any work already undertaken by the proponent, and make a recommendation to the minister. Prior to making a decision on the request, the minister may require the proponent to undertake an individual EA, determine that an individual EA is not required with or without conditions, or order mediation.

Future Direction for Class EAs

Each parent class EA is unique in terms of procedure, but they all have the same planning premise. In the future, all parent class EAs will be revised to be more consistent (e.g., regarding notices and timelines). Moreover, class EAs will be more focused on project effects and mitigation measures than on processes. Another future direction will be to examine the Part II order procedure to find ways to implement regulated timelines for the review of requests. In addition, harmonization with other processes (federal and provincial) will be examined so that the various approval processes may be streamlined. Auditing procedures for class EA projects will be mandatory in the future, and the types of projects to be included under class EAs will be re-examined. The new generation of parent class EAs will have additional accountability and will provide greater flexibility to proponents (EAAB 2001b).

Mediation

Mediation is an option in the decision-making process that may be utilized to address contentious issues. It can take place at any point during the EA process and should follow informal attempts that have not been successful in resolving a dispute. While not all disputes are amenable to the process, when properly approached, mediation can strengthen a proponent's public consultation process, increase trust and accountability among participants, and facilitate timelier EA preparation, review, and approval. Mediation is conducted by a neutral third party.

Proponents and participants may jointly choose self-directed mediation, which does not involve the minister. This type of mediation follows a mutually agreed upon process that includes jointly selecting a mediator. Although participants are encouraged to conduct self-directed mediation, the EAA provides for minister-directed mediation, bound by a 60-day timeline. Minister-directed mediation may be at the request of interested participants or of the proponent, or as recommended by the ministry. The minister may appoint one or more persons to act as a mediator to resolve identified contentious issues. The EAA also permits the minister to initiate mediation with respect to the TOR and/or with respect to the minister's decision on approval of the application (Ontario Ministry of the Environment 2003). To date, no projects have been referred by the minister to mediation.

The Environmental Review Tribunal

The Environmental Review Tribunal (formally the Environmental Assessment Board and the Environmental Appeal Board) is a quasi-judicial tribunal that holds decision-making power on EA projects referred to a hearing. The minister may refer all or part of an EA application to the tribunal for a hearing and decision. With the revisions to the EAA in 1997, the minister can now refer parts of a project to a hearing and set timelines for the hearing process. Previous to these amendments, most hearings were long and drawn out and could take several years. The hearing on the Ministry of Natural Resources Class EA for Timber Management took six years, lasting from 1988 to 1994. This hearing consisted of 411 days for the presentation of evidence and close to two years for the tribunal to come to a decision (Rohaly 2003). These lengthy hearings were very costly for the proponent, the interested parties, and the government. The revisions to the EAA ensure that only the contentious issues of a project are referred to a hearing. This serves to expedite the process and help the stakeholders focus on the issues. The first scoped hearing—on the Adams Mine landfill site in Kirkland Lake, Ontario—was referred to the tribunal in December 1997. The decision on this hearing was to be made in May 1998. So that the hearing might be focused, only issues surrounding the hydraulic containment design of the proposed landfill site were examined. The minister issued a timeline of 15 hearing days, and the total time allotted for this hearing was approximately six weeks. The tribunal reached a decision on the hydraulic containment design in June 1998. With the tribunal decision, the minister and the Lieutenant Governor in Council decided to approve the undertaking with a series of conditions (McLennon 2003).

Critics of the EA process believe that scoped hearings with timelines are inadequate to fully assess a contentious EA project. Because of the complexity of many of the projects, interested parties often want the Environmental Review Tribunal to assess and make a decision on the entire project. While this may be optimal in some cases, the purpose of the EA process is to ensure that most of the environmental impacts and issues are assessed and addressed before the tribunal becomes involved. The tribunal's role is to provide a recommendation on the most contentious of issues—thus the rationale for the scoped hearing.

Any person may request that the minister refer all or part of an EA application to the Environmental Review Tribunal. The minister makes the decision as to whether this request is warranted. The tribunal will conduct the hearing and reach a decision, but the minister, with the approval of the Lieutenant Governor in Council, may vary the decision, substitute his or her decision for the tribunal's decision, or request a new hearing (Ontario Ministry of the Environment 2003).

Streamlined and Consolidated Hearings

Streamlined hearings were introduced via three regulations. These regulations were promulgated in the 1980s to reduce duplication and provide a more efficient hearing process. Previously, it had been possible for one project to be subjected to more than one hearing, depending on the approvals it required. The Consolidated Hearings Act was designed to streamline the approval process by consolidating two or more hearings. This Act created a new tribunal, the joint board, consisting of members of both the Environmental Review Tribunal and the Ontario Municipal Board. The intention of the Act is to eliminate a multiplicity of hearings before different tribunals concerning matters relating to the same application (Estrin and Swaigen 1993).

The EAA's hearing requirements have been streamlined with those of the Environmental Protection Act (EPA) and Ontario Water Resources Act (OWRA). This regulation ensures that if an undertaking has met the requirements of the EAA and the proponent has made a commitment to comply with the conditions set out in the EAA approval, the proponent does not have to proceed with an EPA or OWRA hearing. For example, a waste disposal site or waste management system undergoing an individual EA is not subject to an EPA hearing. EAA requirements, however, must be met prior to the issuance of a certificate of approval under the Environmental Protection Act. Similarly, a sewage works proceeding under an individual or class EA is required to meet the OWRA list but is not subject to an OWRA hearing. These methods of streamlining were introduced to reduce duplication, time, and cost for all the parties involved (EAAB 2001b).

Minister's Decision

A final decision by the minister or the Environmental Review Tribunal cannot be appealed. The only recourse is to launch a judicial review where only points of law pertaining to the process can be argued. A judicial review enables the courts to examine whether the EAA was properly followed, but it does not examine contentious issues related to a project (Rohaly 2003).

The minister can reconsider a decision if there is new information or a change in circumstances concerning an application. However, if a proponent wishes to change an undertaking after receiving an approval, the project is considered to be a new undertaking and the EA process must begin again.

Harmonization

Over the past several years, Ontario has been working with the federal government to harmonize their EA processes. Currently, undertakings by the federal government are not subject to the Ontario EAA. The Canadian Environmental Assessment Act (CEAA) applies to projects for which the federal government holds decision-making authority—whether as a proponent, land administrator, source of funding, or regulator. It is due to this that some undertakings require both provincial and federal EA approval. Although it has not been legislated, informal harmonization has occurred. For example, the Ottawa-Carleton Light Rail Transit EA required both federal and provincial EA approval. With this project, the province took the lead in the assessment and ensured that additional CEAA requirements, such as the assessment of cumulative effects, were incorporated into the EA process. The project went through the provincial process but received both approvals (EAAB 2001b).

Weaknesses of the Environmental Assessment Act

The EAA, though strong in many respects, has some fundamental weaknesses that must be addressed for its integrity to be maintained. The EA process is not so much flawed as challenged—in many respects. The first fundamental challenge of the EA process is its inability to monitor, audit, and enforce the conditions of approval on a project. The EAA does not include an auditing mechanism to ensure that the EA is followed, that monitoring and mitigation occurs, and that conditions of approval are met. The provincial auditor has criticized this weakness in the Act, and this has led to internal discussions and plans within the EAAB to rectify the situation. As yet, no auditing has occurred, but there are plans for it in the future. One problem is that it is difficult to audit all the undertakings that have been approved with conditions, as many of them are class EA projects that have been submitted for Part II order review. Approximately 200 projects are submitted to the EAAB annually, of which a large majority proceed through the process and are approved with conditions. Since auditing has not been done in the past, there has accumulated 32 years worth of projects that need to be audited. This is an immense task for the province to undertake, even with its good intentions.

A further weakness in the Act can be seen in the difficulty in enforcing a breach of condition legally, as the ability to prosecute would depend on the wording of the condition. Under the Statute of Limitations of the Provincial Offences Act (POA), any person who contravenes any provision of the EAA is guilty of an offence and is subject to a fine of $10,000 for a first offence and $25,000 for a second offence. Under section 76 of the POA, a prosecution of an offence under the EAA must commence within six months of the offence being committed or being alleged to have been committed. The ability to prosecute would depend on the wording of the condition, but the time and effort required to prepare for the prosecution of the offenders would be significant, and given the six-month Statute of Limitations, prosecution would most likely be impossible to pursue. In addition, the case law associated with these types of prosecution is non-existent (EAAB 2001c). There may be other courses of action—such as placing a hold on other provincial approvals, red flagging proponents for future audits of all their projects, and posting the names of the proponents who have not complied with conditions on the EAAB website for public viewing—but these options may not effectively ensure that conditions are implemented. Future EAs should include an auditing policy, but it seems that past projects are on the honour system.

The EAA also does not account for cumulative effects. Despite the fact that the EAA is seen as having been progressive over the years as one of the world's leading pieces of EIA legislation, it has never been strongly associated with the idea of assessing cumulative impacts. The CEAA process evaluates cumulative effects, and this inclusion is key to any environmental impact assessment. There are no plans to include cumulative effects in any EAA legislation, which leads to the question, how comprehensive can an assessment be without assessing the total impact of the undertaking?

Furthermore, the EAAB is facing many challenges having to do with the introduction of new technologies. The EAA does not provide for the development of new technologies (e.g., energy derived from petroleum coke), and the EAAB is struggling with how these undertakings fit under the EAA. It is possible to develop new regulations that cover new technologies, but the risk comes with the transition period, since projects will proceed without there being full understanding of the environmental effects of the undertaking (Amodeo 2003).

Lack of participant and intervenor funding is another criticism of the Ontario EAA. Whereas intervenor funding at the hearing stage was part of the EAA in the past, participant funding has never been supported. There have been ongoing debates over the past three decades concerning the need for funding. The province feels that if mandatory public consultation is entrenched in the legislation and occurs early in the process, issues are resolved prior to the need for intervenor funding. In terms of participant funding, the province does not generally support this concept, as it wants to remain impartial and believes that funding is not necessary for public consultation and issue resolution to occur. If the EA process is conducted in a consultative manner, it will not need to reach the hearing stage and hence intervenor funding will not be needed. The main concern with intervenor funding has to do with who receives the funding and how the province manages to remain impartial with regard to many of the groups that wish to participate in a hearing. This is not to say that intervenor funding is not a good idea; however, further analysis of the benefits of such funding is required.

Another weakness of the EAA is the lack of policy decisions surrounding some of the issues that arise. Over the years, the EAA has undergone many small policy changes—through, for example, the development of prescriptive planning rules, the waste management master planning process, the determination of what a willing host is and how this plays a role in the EAA, and the banning of incineration facilities. With each government, policy implications change, and it may be preferable to have a standard prescriptive process in place to deal with contentious issues than to deal with these issues on a case-by-case basis. The ad hoc method of policy development and decision making depends on the whim of the current government and has not been beneficial to the EA process.

In order to address these weaknesses, the Ontario government in 2004 set up an advisory panel of expert EA practitioners to develop proposals on possible approaches to the EA process for waste management facilities, transit and transportation projects, and clean energy. It is the task of the advisory panel to identify key impediments to obtaining timely approvals for projects subject to the EA process, as well as identify potential improvements by category of activity (e.g., in the areas of guidance, the review process, the EA approvals process, other approvals necessary following EA approval). This panel, which is made up of representatives from the municipal, waste management, and clean energy sectors, the environmental community, academics, the consulting industry, and the legal community, will try to address some of the weaknesses identified within the EAA. The 1997 amendments to the EAA were beneficial in some instances but may have led the EA pendulum to swing too far to one side. It is hoped that the advisory panel will recommend actions that will bring the pendulum back to a balance that will ensure that increased environmental protection is at the forefront of the EA process (Hennessy 2004; Ontario Ministry of the Environment 2004).

Strengths of the Environmental Assessment Act

Despite these challenges, Ontario's Environmental Assessment Act has several strong points. The Act provides a systematic method of evaluating the potential environmental effects of an undertaking. In its 34 years of existence, it has been key to ensuring that the environment of the province of Ontario is protected. Its major strengths are its provision for mandatory consultation with the public, its broad definition of the environment, its systematic method of evaluating the net environmental effects of an undertaking, its foresight in including mediation as a means of facilitating agreements between the public and proponents, and its stipulation that clear, complete documentation be made available to the public. A further strength is the EAA's power to ensure that public and other affected stakeholder concerns are identified and addressed early on in the process. This guarantees a comprehensive and complete planning and decision-making process.

This chapter has outlined many of the strengths of the EAA and has provided an overview of the EA process in Ontario. The Ontario EAA may not be perfect, but through it the stakeholders in the province have gained much experience and many other jurisdictions have capitalized on these experiences. It is important to note that the EA process, like many legislative processes, can be affected by the political climate of the time, but despite political interference at times, the Ontario EAA has maintained its ideals and principles and ensures that its purpose—the protection, conservation, and wise management of Ontario's environment—is maintained.

References

Amodeo, Piero. 2003. Supervisor, Project Coordination Section, Environmental Assessment and Approvals Branch, Ontario Ministry of the Environment, Toronto. Personal communication, 1 August.

Connelly, Gemma. 2003. Senior Project Officer, Project Coordination Section, Environmental Assessment and Approvals Branch, Ontario Ministry of the Environment, Toronto. Personal communication, 1 August.

Desautels, Solange. 2003. Senior Project Officer, Project Coordination Section, Environmental Assessment and Approvals Branch, Ontario Ministry of the Environment, Toronto. Personal communication, 26 June.

EAAB (Environmental Assessment and Approvals Branch). 2000a. *A guide to preparing terms of reference for environmental assessments.* Toronto: Ontario Ministry of the Environment, 15 December.

———. 2000b. *Guideline on consultation in the environmental assessment process.* Toronto: Ontario Ministry of the Environment, 15 December.

———. 2000c. *Guideline on mediation in the environmental assessment process.* Toronto: Ontario Ministry of the Environment, 15 December.

———. 2001a. *Procedures for the development of certificates of approval.* Toronto: Ontario Ministry of the Environment.

———. 2001b. Introduction to environmental assessment. Ontario Ministry of the Environment presentation, March.

———. 2001c. The auditing of bump-up denials under the Environmental Assessment Act—A discussion paper. Ontario Ministry of the Environment, May.

Estrin, D., and J. Swaigen (Eds). 1993. *Environment on trial—A guide to Ontario environmental law and policy.* Toronto: Emond Montgomery Publications Ltd.

Harrison, Michael. 2003. Project Officer, Project Coordination Section, Environmental Assessment and Approvals Branch, Ontario Ministry of the Environment, Toronto. Personal communication, 1 August.

Hennessy, Mary. 2004. Technical Support Manager, Eastern Region, Ontario Ministry of the Environment, Toronto. Personal communication, 29 May.

McLennon, Catherine. 2003. Project Officer, Project Coordination Section, Environmental Assessment and Approvals Branch, Ontario Ministry of the Environment, Toronto. Personal communication, 24 July.

Ontario Environmental Assessment and Consultation Improvement Act, 1996. Revised 1 April 2002.

Ontario Ministry of the Environment. 2003. Ontario's Environmental Assessment Act activities website. http://www.ene.gov.on.ca/envision/env_reg/ea/english/general.

———. 2004. *Media backgrounder: Waste management strategy.* Toronto, 5 April. http://www.ene.gov.on.ca/envision/news/2004/040502mb1.htm.

Rohaly, B. 2003. Senior Project Officer, Director's Office, Environmental Assessment and Approvals Branch, Ontario Ministry of the Environment, Toronto. Personal communication, 24 July.

Chapter 20

Environmental Assessment in Quebec

Darren R. Bardati

Introduction

Environment assessment (EA) has been established in Quebec for 30 years. It is somewhat distinct from EA in any other province because of ongoing jurisdictional debates with the federal government, the role of the permanent Public Hearings Bureau (or BAPE) that oversees matters of public participation, the existence of enormous hydroelectric development projects in the northern region, as well as the provincial government's evolving relationship with the northern Aboriginal inhabitants.

This chapter examines environmental assessment in Quebec, its legal basis and evolution, the role of public participation, the range of activities covered, and the formal procedures to which these activities must be subjected. One section also discusses the special status of the province's vast northern region. Throughout, emphasis is placed on highlighting the strengths and weaknesses of EA in Quebec.

Legal Basis

In Quebec, the legislative and regulatory framework for environmental assessment is comprised of the following five statutes:

1. Environment Quality Act (*Loi sur la qualité de l'environnement*, L.R.Q. 1978, c. Q-2)
2. Regulation Respecting Environmental Impact Assessment and Review (*Règlement sur l'évaluation et l'examen des impacts sur l'environnement*, R.R.Q., 1981, c. Q–2, r. 9)
3. Rules of Procedure Relating to the Conduct of Public Hearings (*Règles de procédure relatives au déroulement des audiences publiques*, R.R.Q., 1981, c. Q–2, r. 19)

4. Regulation Respecting Environmental Assessment and Review Applicable to a Part of Northeastern Quebec (*Règlement sur l'évaluation et l'examen des impacts sur l'environnement dans une partie du Nord-Est québécois*, R.R.Q., 1981, c. Q–2, r. 10)

5. Regulation Respecting Environmental and Social Impact Assessment and Review Applicable to the Territory of James Bay and Northern Québec (*Règlement sur l'évaluation et l'examen des impacts sur l'environnement et le milieu social dans le territoire de la Baie-James et du Nord québécois*, R.R.Q., 1981, c. Q–2, r. 11)

The Environment Quality Act (EQA) is the legal basis for all matters pertaining to environmental protection inside the territorial boundaries of the province of Quebec within the context of its jurisdiction defined in the British North America Act of 1867. Although originally adopted in 1972, the EQA was greatly modified in December 1978 following the events surrounding the James Bay hydroelectric development project in northern Quebec. Chapter 1 of the EQA deals with all of Quebec in general, while chapter 2 details the provisions applicable to the James Bay and Northern Quebec region, which itself, given its size and cultural factors, is divided along the 55th parallel. Consequently, the regulations and procedures for impact assessment and public hearings are considerably different in the north than in the south, as will be discussed later in the chapter.

Overall, the language of the EQA is very progressive. The starting point for environmental protection is captured in article 19.1 (environmental rights): 'Every person has a right to a healthy environment and to its protection, and to the protection of the living species inhabiting it, to the extent provided for by this Act and the regulations, orders, approvals and authorizations issued under any section of this Act.' While the first part of the statement speaks broadly of our right to protect environmental quality for humans and non-humans, the second part narrows the extent of the right to those things specified in the Act.

Regarding impact assessment, accountability is achieved through article 22 (certificate): 'No one may . . . carry on activity . . . if it seems likely that this will result in . . . a change in the quality of the environment, unless he first obtains from the Minister a certificate of authorization.' To obtain the certificate of authorization, applicants must undergo the formal 'environmental impact assessment and review procedure' (EIARP), which is formulated in the Regulation Respecting Environmental Impact Assessment and Review. The methods and obligations relating to public information and consultation set out in the EQA and regulations will be discussed first.

A Role for Public Participation

The Environment Quality Act acknowledges that public participation, particularly in the form of public hearings and formal public consultations, lies at the very heart of impact assessment and decision making in Quebec. Impact assessment is conceived as an open and accountable process that is concerned with protecting the environment and people's quality of life and in which the public has an important role to play.

The 1978 revision of the EQA created the Public Hearings Bureau (Bureau d'audiences publiques sur l'environnement or BAPE). The BAPE is a permanent, quasi-judiciary body within the Ministry of the Environment that is officially mandated to 'give the public the means by which it can inform itself, to collect opinions from the public and to incorporate the public's concerns relating to a project in question' (BAPE 2002, 1). The BAPE has powers to conduct public hearings and environmental mediations.

Like the EQA, the BAPE makes a strong case for the effectiveness of environmental impact assessment (EIA):

> Because of its preventative nature, environmental assessment is a true exercise of planning sustainable development of land and resources. It allows decision-makers to consider, ana-lyze and interpret all the factors affecting ecosystems, resources and the quality of life of the individual and communities. Moreover, by encouraging dissemination of information and consultation with the public . . . projects are better conceived and their impacts are limited. (BAPE 2003)

The BAPE is composed of at least five full-time members, or 'commissioners' (at the time of writing, June 2008, there were seven), including a president and a vice-president. All commissioners are appointed by the government for a renew-able period of five years. There is an additional bank of some 50 part-time com-missioners, representing all types of disciplines, to assist in the various functions of the bureau; they are appointed by government on an as-needed basis. All BAPE commissioners, whether full time or part time, must adhere to a code of ethics and duty. This code ensures that the commissioner is not in any conflict of inter-est and, so that even the appearance of a conflict of interest may be avoided, has no particular interest in the dossier in his or her charge. The commissioner must also demonstrate political neutrality, showing no favouritism towards any party, interest, or position. In essence, the commissioner must, to the best of his or her ability, remain impartial to the public opinions the BAPE receives and objective in the analysis and reporting of those opinions (BAPE n.d., 1–2).

Since its creation, the BAPE has seen a steady increase in the mandates it receives from the minister of the environment for public consultations and hearings. The average annual number of mandates for public consultations has grown from about 14 in the 1980s, to 18 in the 1990s, to over 20 since 2002. The average annual number of public hearings has seen a similar increase, from three in the 1980s to seven in the 1990s to 14 more recently. This increase is probably the result of a combination of interrelated factors: a larger number of projects being subject to the procedures, a favourable economic context for resource development, and an increased sensitivity on the part of the general population towards the quality of the environment and the repercussions of development projects.

There is little criticism about the procedural functions of the BAPE. In fact, the bureau is recognized within and outside Quebec as one of the strongest governmental bodies in support of public participation. Its strengths were mentioned in a 5 May 2003 Canadian House of Commons debate on Bill C-9, an amendment to the Canadian Environmental Assessment Act (CEAA). The honourable member from Hochelaga-Maisonneuve (in Montreal) states:

> I would remind the honorable members—and those who are familiar with Quebec know this—that when environmental assessment legislation is mentioned, one thing and one thing alone comes to mind and that is the BAPE. People know the BAPE; they know its strength. . . . Why is this act [the Quebec Environment Quality Act] better? Why does this act deserve to be more complied with? First, because it is more transparent. From the beginning to the end, it associates the Bureau des audiences publiques sur l'environnement with our fellow citizens, who can be heard and who can file submissions. A tabled report is made public. A whole influence process is possible with the BAPE. (House of Commons 2003, 5783)

Of course, the strength of the quote's glowing support of the BAPE must be tempered by the knowledge that the words were spoken by a Bloc québécois parliamentarian arguing against some of the components in the proposed Bill C-9. Nonetheless, his points are valid. The BAPE is well known for its impartial treatment of public opinion, its accurate portrayal of public concerns in its reports, its extremely transparent process, and the accessible documentation it produces. All BAPE reports are available to the public on the Internet (http://www.bape.gouv.qc.ca/).

However, what is more debatable is the claim that a 'whole influence process is possible with the BAPE'. While it is true that the BAPE allows public influence to shape the content of its report to the minister to a great degree, the actual degree to which the public, through the BAPE, can influence the outcome of the process is less certain. The BAPE has always been advisory only, having no binding decision-making power (as is the case with EIA procedures in all Canadian jurisdictions). By extension, one might argue, the public has no real say in the matter, despite the stainless appearance of the government's commitment to public participation. For better or worse, the minister routinely permits authorizations for projects that the BAPE has made recommendations against. Rarely is the opposite true. This brings up the topic of the minister's discretionary powers, which have increased in recent years owing to amendments to the EQA.

Amendments to the Environment Quality Act

A dozen or so years after the 1978 EQA was written, there had yet to be any substantial changes made to it regarding impact assessment. However, in 1992, following two commissioned reviews of the EIA procedures (Lacoste Report 1988; Gouvernement du Québec 1992), the Quebec government proposed Bill 61 to once again amend the EQA. The bill called for significant changes regarding EAs and addressed major objectives. The first was to ensure that 'consideration be given to any environmental questions involved in policies and programs of the Government' or its agents (Gouvernement du Québec 1992). The second was to ensure that there would be public consultations or formal public hearings. The third was to provide intervenor funding in order to encourage 'individuals, groups or municipalities to take part in public hearings.' In addition, the bill included provisions that would enable government to treat several projects with a common objective as a single project (sec. 31.8); assign the right to prescribe measures for monitoring project effects (sec. 31.9.9); and require that proponents produce 'attestations of conformity' at various predetermined stages of the project development process (Meredith 2004). The bill also proposed that wider discretionary powers be given to the minister of the environment and the Government of Quebec, including the ability to remove certain projects from public participation or even from the entire EIA procedure.

The proposed changes were hotly debated by different actors representing social, economic, and environmental interests (Leduc and Raymond 2000). In the decade after Bill 61 was first proposed, ambitions were tempered, and in 1996 and 1999 revisions were made. However, some key issues have not been fully resolved. Policies and programs have been dropped from subsequent drafts of the bill, as have expressed concerns over intervenor funding. Despite the appearance of progressive changes, the current legislation contains a certain purposeful vagueness that leaves the determination of which projects are to be subject to the procedures somewhat unclear. This shows just how subjective EIA really is, leaving much to the discretion of the minister.

The 1995 version of the public document, entitled 'Environmental Assessment in Québec' and produced by the Environmental Assessment Directorate (Direction des évaluations environnementales), states: '[T]he procedure has become an integral part of Québec's environmental democracy, and is considered by many people to be one of the best in the world, thanks to its *non-discretionary* and non-judiciary form, and its inclusion of public participation' (page 4; emphasis added). Interestingly, the 2001 version of the same document includes the same sentence but omits the words 'non-discretionary'. Although it is questionable whether the 1995 statement was accurate in the first place, the word change perhaps reflects the degree to which the minister's discretion has become more visible.

The wording of the law also reinforces this point. The very notion of *environment* in the new law appears to have been reduced to mean only the 'bio-physical aspects in the ambient milieu' (Leduc and Raymond 2000).[1] This tendency to reduce the scope of the definition of environment could possibly lead to conflicting interpretations of the law's applicability to certain projects. For example, one might ask if projects that have large social implications but limited biophysical impacts are exempted for the EIARP. Again, the minister's interpretation may potentially overrule any other interpretation. While the full implication of these changes is still to be felt, the increased discretionary power of the minister is currently one of the most contentious issues in the application of the law.

Regarding the issue of activities subject to environmental assessment, the recent revisions to the EQA have added projects to the list, including, for example, the construction of natural gas pipelines over 2 km long. Clearly, this is a positive change that shows that the intent of the law is to limit damage to the environment. But even this change has not been without controversy. One would suspect that this larger, specified list would put more constraints on the private sector than did the earlier version of the law. However, the EQA's wording restricts applicable projects to 'the cases provided for by regulations of the Government' (article 31.1). In practice, some might argue, this change actually limits the projects subject to EIARP to very specific regulatory provisions, to precise areas of activity, or to specific sites, thereby effectively weakening the EQA goal of environmental protection while giving the appearance of strengthening it.

One last amendment to the EQA worth noting deals with Quebec's relations with other parts of Canada. It is well known that, in discussions with other provinces and the federal government, including the Canada-Wide Accord on Environmental Harmonization (CCME [Canadian Council of Ministers of the Environment] 1998), Quebec has always aimed at clarifying jurisdictional divisions. Discussions over proposed changes to environmental assessment are no different, as was alluded to in the earlier discussion about Quebec's position on Bill C-9. The Canada-wide standards were not endorsed by Quebec, though Quebec acts within its area of jurisdiction in a manner consistent with the Canadian standards. Furthermore, with respect to environmental assessment, Quebec did include a clause in the EQA's 1999 revision that pointed the way towards cooperation with other jurisdictions in EA matters, as follows:

> Section 31.8.1. Where a project referred to in section 31.1 is to be carried out in part outside Québec and, as a consequence, the project is also subject to an environmental assessment procedure prescribed under an Act of a legislative authority other than the Parliament of Québec, the Minister may make, as provided by law, an agreement with any competent authority to coordinate the environmental assessment procedures, which may include the establishment of a unified procedure.

In essence, the spirit of the CCME harmonization accord may well be put in practice with Quebec's involvement after all, even though for political reasons Quebec had refused to sign on.

Projects Covered

According to the EQA, activities that are subject to environmental impact assessment in Quebec must fall into the category of 'major development project' likely to disturb the environment to a significant degree and give rise to widespread public concern. The projects themselves and the thresholds at which they become subject to the procedure are identified in the EQA and the various associated regulations.

According to the Regulation Respecting Environmental Impact Assessment and Review (30 December 1980), projects include '[r]oad construction, construction of electric and transmission lines and stations, dredging or filling, construction of dams, dykes, ports or wharfs, rerouting or diversions of rivers, establishment of airports, oil pipeline construction, aerial pesticide spraying and toxic waste disposal facilities.'

According to the Act Respecting the Establishment and Enlargement of Certain Waste Elimination Sites (8 June 1993), projects include '[a]ll projects to establish or enlarge sanitary landfill sites or dry material dumps'.

According to the Regulation to Amend the Regulation Respecting Environmental Impact Assessment and Review (24 January 1996), projects include '[c]ertain clearly-defined mining, industrial and natural gas pipeline construction projects'.

The Environment Quality Act, Schedule A, includes a list of 17 types of activities that are 'automatically subject' to the EIARP. Beyond those already listed, these include

> [g]ravel pits and quarries over 3 hectares; storage and water supply reservoirs for electricity; forestry roads of at least 25 km in length; all land use projects which affect more than 65 km²; all access roads to a new development; all wood and pulp and paper mills; all new parks and ecological reserves; all outfitting camps designed to accommodate 30 persons or more.

Finally, the Environment Quality Act, Schedule B, includes a list of 14 types of activities that are 'automatically exempt' from the EIARP:

(a) all hotels or motels of 20 beds or less and all service stations along highways;

(b) all other structures intended for dwellings, wholesale and retail trade, or intended for offices or garages, or intended for handicrafts or car parks;

(c) all fossil-fuel fired power generating plants having a calorific capacity below 3,000 kW;

(d) all school or educational establishments, rest areas, observation points, banks, fire stations or immovables intended for administrative, recreational, cultural, religious, sport and health purposes or for telecommunications;

(e) all control or transformer stations of a voltage of 75 kV or less, or electric power transmission lines of a voltage of 75 kV or less;

(f) all water and sewer mains, and all oil or gas mains of less than 30 cm in diameter with a maximum length of 8 km;

(g) all testing, preliminary investigation, research, experiments outside the plant, aerial or ground reconnaissance work and survey or technical survey works prior to any project;

(h) all forestry development when included in plans provided for in the Forest Act (chapter F–4.1);

(i) all municipal streets and sidewalks;

(j) all maintenance and operation of public and private roads;

(k) all repairs and maintenance on existing municipal works;

(l) all temporary hunting, fishing and trapping camps and all outfitting facilities or camps for less than 30 persons;

(m) all small wood cuttings for personal or community use;

(n) all borrow pits for highway maintenance purposes.

While the list of activities subject to the EIARP is impressive, there are some activities not listed (and thereby subject to the minister's discretion) that can radically alter environmental and social conditions, leaving the list subject to criticism.

The Controversial Case of High-Density Pork Production

High-density pork production operations are the most notable examples of projects that draw criticism because they are not covered under existing EIA legislation. While the interprovincial and international linkages and associated controversies surrounding Quebec pork production are beyond the scope of this chapter, it suffices to note that public outcry over the rapidly expanding high-density pork production industry has increasingly been making itself felt over the past few years, as citizens and municipalities have expressed concern that aesthetic, health, and environmental factors are being overlooked. In response to this fallout, the minister of the environment mandated the BAPE to investigate public opinion on high-density pork production. The inquiry began in October 2002. Over the course of the following year, the inquiry conducted 138 public hearings in 18 communities, involving 9,100 people. The report, tabled in September 2003, describes the considerable ecological and social impacts of the pork production industry and strongly recommends that it become subject to EIA procedures as soon as possible (BAPE 2003, 223). The guidelines in Quebec now regulating the environmental impacts of the hog farming industry are among the strongest such guidelines in Canada. However, at the time of writing (2008), no changes have been made to provincial EIA regulations since the tabling of the report in 2003. Of the 51 environmental assessments in which the BAPE was involved since that time, not one has dealt with high-density pork production.

Environmental Impact Assessment Procedures

The administrative procedure for environmental impact assessment in southern Quebec includes a number of stages, divided into six phases (Figure 20.1). Each is described briefly below.

Phase 1: Instructions

The first phase includes two main stages: (1) filing the 'project notice' and (2) preparing the instructions. When the proponent files a project notice with the minister of the environment, the official administration procedure begins. The eight-page notice must contain basic information about the proponent, outline the project objectives and plans, include a basic description of the impacts, and describe how it plans to inform the public.

Needless to say, the depth of detail in the eight-page notice is minimal, yet the notice must contain enough information to assist in the determination of whether the project is subject to the procedures. Once completed, the proponent forwards

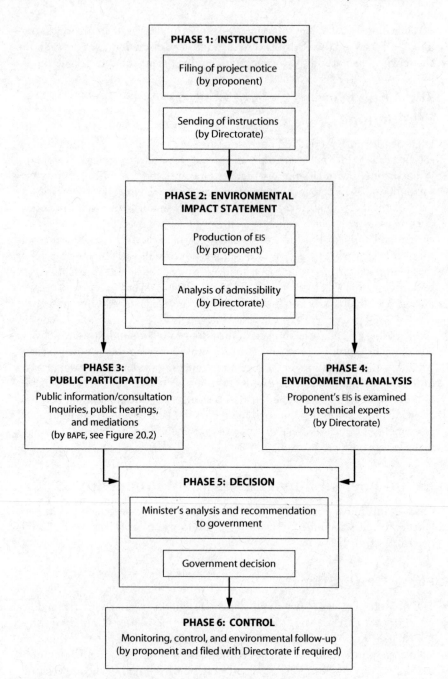

Figure 20.1   Procedure for southern projects

Source: Adapted from Environnement Québec 2001.

the project notice to the Environmental Assessment Directorate, where it is examined. If EIA is applicable, the project notice is posted on the directorate's Internet website (http://www.menv.gouv.qc.ca/evaluations/inter_en.htm) and the directorate prepares preliminary instructions for the environmental impact statement (EIS), identifying the main points to be addressed. These instructions are prepared in consultation with other government departments and public bodies concerned with the project, according to their respective jurisdictions. The final instructions, which reflect the comments made during the intergovernmental consultation process, are then sent officially to the proponent by the minister of the environment.

Phase 2: Environmental Impact Statement

The second phase also includes two stages: (1) production of the environmental impact statement or EIS (*étude d'impact sur l'environnement*) and (2) analysis of the statement admissibility (Figure 20.1). The proponent is charged with producing an EIS in accordance with the instructions received. Dialogue between the proponent and the Environmental Assessment Directorate is usually maintained to ensure that all the elements required by the minister's instructions are dealt with to the satisfaction of the parties.

The EIS is a planning tool aimed at identifying environmental concerns related to each phase of the project. It is intended to help the proponent ensure that the project is environmentally safe while remaining technically and economically feasible. There are a number of steps to be followed in the preparation of an EIS:

1. Project overview (executive summary), including background of the proponent and justification for the project
2. Description of the environmental and social setting
3. Description of project and project options
4. Impact assessment of retained option(s)
5. Risk management considerations
6. Proposal of monitoring and follow-up program

According to the regulations, the EIS is supposed to be clear and concise (although it may often contain thousands of pages in several volumes). The methods and criteria used to prepare the impact statement—including their reliability, range of accuracy, and interpretation limits—should be identified and explained. Maps at the appropriate scale, air photos, and environmental inventories are encouraged.

When the EIS is completed, the proponent sends 30 copies, together with all the other documents that make up the authorization file, to the minister. The Environmental Assessment Directorate then produces a notice of admissibility, informing the minister of the quality of the EIS and providing sufficient information for the minister to decide whether the statement should be made public and whether phase 3, public participation, should be initiated.

Issues relating to the confidentiality of information and data come into play here. The proponent may request that the minister withhold from public consultation information contained in the EIS. In accordance with section 31.8 of the Environment Quality Act, 'the Minister may withdraw from a public consultation any information or data concerning industrial processes.' However, the proponent must demonstrate that this information or data concerns an industrial process and that its disclosure would harm the proponent. Before making public his or her decision on whether or not to withdraw the information, the minister must first inform the proponent about the decision.

Phase 3: Public Participation

The public participation phase enables all those affected by, or believed to be affected by, the project to have access to the available information about the project and to express their opinions about the project. This phase is under the responsibility of the BAPE, discussed above, and it proceeds in three stages: (1) public information and consultation, (2) public hearings or mediation, and (3) the filing of the report. Phase 3 is the only phase that is subject to time limits established by regulation (Figure 20.2).

The public participation phase begins with a 45-day period of public information and consultation, during which any individual, group, or municipality can consult the file of the application for authorization and ask the minister to hold public hearings on the project.

When the BAPE is instructed to hold a hearing, its president forms a committee of five to nine people to analyse the project, drawing from the bank of 50 or so BAPE commissioners appointed by government discussed earlier. The BAPE committee is given four months to conduct the hearings and produce a report for the minister. The hearings proceed in two stages: the presentation of information and the hearing of briefs.

In the information stage, the BAPE encourages applicants to explain why they requested a public hearing. This stage also allows the proponent to explain the project in detail and offers the public the opportunity to ask questions about the project. The minister of environment or his or her representative is usually present to clarify the administrative decisions taken with respect to the assessment procedure (admissibility, applicable regulations, etc.). Specialists from government departments or in support of proponents are also on hand to discuss technical issues.

In the second stage, the hearing of briefs, the BAPE committee hears briefs (*mémoires*) and oral presentations on the project, the impact statement, and any other documentation in the file brought forward by individuals, groups, or bodies wishing to be heard. To be admissible at the public hearing, these briefs must be prepared in accordance with prescribed forms and be submitted by the determined deadlines.

In certain circumstances, the minister may ask the BAPE to hold mediation sessions instead of public hearings—such as when the number of applicants is small,

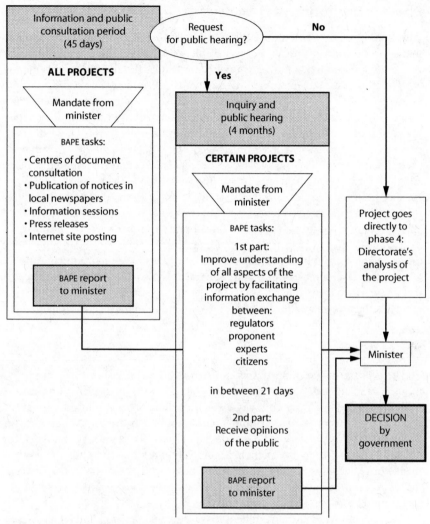

Figure 20.2   Public participation (phase 3 of southern procedure)

Source: Adapted from BAPE 2002.

the scope and number of questions raised in the requests are limited, or there is a possibility that the parties (proponents and hearing applicants) could come to an agreement.

After the hearings or mediation sessions, the BAPE committee produces a report containing a summary of the information and opinions expressed and an analysis of its observations. The report is sent to the minister, who must make it public within 60 days of receiving it. All BAPE reports are available, in French only, on the BAPE Internet website (http://www.bape.gouv.qc.ca/).

Phase 4: Environmental Analysis

During the public participation phase, a parallel process takes place—the environmental analysis of the project. This is why phases 3 and 4 are side by side in Figure 20.1. The Environmental Assessment Directorate does the analysis in consultation with other concerned government departments and bodies. The stated aim is to 'produce an objective view of the environmental acceptability of the project' (Environnement Québec 2001, 9). The analysis enables the minister to consider whether the option proposed by the proponent is the option of least impact, whether the project's impacts are acceptable from an environmental point of view, whether the project complies with applicable legislation, regulations, and government policies, and, in light of these justifications and impacts of the project, whether the project should be carried out. On the basis of the analysis, the directorate produces the environmental analysis report. Most of the recent (since 2001) environmental analysis reports are available on the directorate's Internet website (http://www.menv.gouv.qc.ca/programmes/eval_env/index.htm), while the older reports are only available for public consultation at the Ministry of the Environment office in Quebec City. The directorate's environmental analysis report and the BAPE report (see phase 3) are the two main points of reference for the minister when he or she evaluates the project and formulates a recommendation to the government.

Phase 5: Decision

The fifth phase of the impact assessment procedure consists of three stages: (1) ministerial analysis, (2) government decision, and (3) ministerial authorization. In the first stage, the minister analyses the two documents produced in phases 3 and 4 in order to prepare a recommendation to the government concerning the application for authorization. Procedures and criteria for this stage are not defined or made public. When the analysis is complete, the minister sends the government a brief and a draft order-in-council containing the Ministry of the Environment's recommendation on the acceptability of the project and any conditions that should be imposed.

In the second stage, the government makes the final decision and either issues a certificate of authorization for the project, with or without modifications, or refuses authorization. The government's decision is then made known to the project proponent and the participants who made presentations at the public participation stage.

If the project is authorized (stage 3), the Ministry of the Environment endeavours to ensure that the project plans and specifications are in compliance with the content of the government's decision. The ministry prepares a certificate of authorization to be issued by the minister.

Phase 6: Control

Phase 6 deals with observations made during the construction and operation phases. The aim of phase 6 is to ensure that the project remains in compliance with the government decision. There are three stages to the control phase: (1) monitoring, (2) control, and (3) follow-up.

Monitoring, stage 1, is performed by the proponent. It consists in ensuring that the construction and operational activities, including proposed measures to mitigate identified environmental impacts, comply with the certificates of authorization and the relevant legislation and regulations as required by the government decision. A monitoring report is filed with the minister of the environment. In addition to reporting on the above issues, the report must also describe any changes made to attenuate new disturbances resulting from the project, should any be identified.

Control, stage 2, is carried out under the authority of the regional branch of the Ministry of the Environment. It serves to verify the implementation and effectiveness of the proponent's monitoring program, as well as ensure that governmental and ministerial authorizations are complied with.

Environmental follow-up, stage 3, is performed by the proponent, who must examine, for a given period, the nature, intensity, and development of particular natural processes or phenomena presumed to have been disturbed by a project and for which the impact statement and current knowledge are insufficient to allow a reasoned decision on the probable impact. This stage is not required unless there is evidence that a follow-up is necessary. In such cases, a follow-up report is filed with the Ministry of the Environment.

None of the documentation resulting from activities performed in phase 6 is readily available to the public. Likewise, there is no formal recognition of the role of public participation in the control phase. This situation has left the procedure open to some criticism. If there are no guarantees for public participation in the control phase of the project and there is the potential that an already stretched governmental office may not fulfil its overseeing role adequately, there is an open door to abuse. Future improvements to the procedure will have to address this problem.

The Special Status of James Bay and Northern Quebec

The situation in northern Quebec is more complex than in the south because of the James Bay and Northern Quebec Agreement (JBNQA) signed in 1975 between the Cree and Inuit people, the federal and provincial governments, and Hydro-Québec and its subsidiaries. As the first modern comprehensive land claim agreement in Canada, the JBNQA provided land, cash, and the power to administer cultural matters (education, health, and social services) to Aboriginal peoples. In exchange, the Cree and Inuit surrendered their claim to the land and agreed to allow construction of Hydro-Québec's massive, multi-phase hydroelectric development project.

Much has been written about the history of the JBNQA and Hydro-Québec's mega-projects and their implications for social and environmental conditions (e.g., Berkes 1981, 1988; Connor-Lajambe 1990; McCutcheon 1991; Scott 2001).

Here, this chapter offers a brief account. Under the JBNQA, two committees and one commission were created to evaluate and review development projects within the jurisdiction of Quebec:

- The Evaluating Committee (Comité d'évaluation, or COMEV) is a tripartite Quebec/Canada/Cree agency responsible for assessing and drawing up guidelines for the impact study of projects located south of the 55th parallel.
- The Review Committee (Comité d'examen, or COMEX) is a bipartite Quebec/Cree agency responsible for reviewing projects located south of the 55th parallel.
- The Kativik Environmental Quality Commission (KEQC), composed of Quebec and Inuit representatives, is responsible for assessing and reviewing projects located north of the 55th parallel.

For projects within federal jurisdiction, there are provisions for bipartite (Canada/Cree or Canada/Inuit) committees.

According to the JBNQA, an administrator is charged with making the final decision on the assessment and review of development projects, basing that decision upon the recommendations of the committees and commissions. This person is either the minister of sustainable development, environment and parks if the project is provincial in nature, the chairman of the Federal Environmental Assessment Review Office if the project is federal in nature, or the administrator of the appropriate Cree band council.

Within the controversial history of the James Bay hydroelectric development project, which began in the early 1970s, the exact nature of both impact assessment procedures and Aboriginal participation has long been debated. Of course, at the time of the JBNQA negotiations, there was no environmental impact legislation in Quebec. In order to ensure protection of the lands and resources, the Cree and Inuit insisted on negotiating an environmental and social impact assessment regime that would apply to all future development in the region (Feit and Beaulieu 2001). The Aboriginal peoples also fought for increased participation in the decision-making bodies that oversaw natural resources and environmental management. As a consequence, clauses relating to EA review procedures and Aboriginal participation were included in the JBNQA. Despite this, the Aboriginal peoples have often expressed dissatisfaction with the environmental results of the development projects and have consistently sought to negotiate for better social and environmental conditions (Diamond 1990).

The controversy was heightened in the early 1990s, when plans for the second phase of the project, James Bay II or the Great Whale Project, were undergoing environmental assessment and review procedures. At that time, an attempt by the then-Liberal Quebec government to split the EIA review into two parts—one dealing with the impacts of project infrastructure (e.g., roads, airports), the other dealing with the hydroelectric complex itself—was successfully contested by the

Aboriginal peoples, who won a court injunction preventing this division of the review process. Interestingly, in the court case, both proponents and opponents made reference to the JBNQA to support their position. The debate only served to attest to the vagueness and ambiguity of the text of the JBNQA.

The voice of Aboriginal opposition was strong throughout the 1990s. In 1995 the Cree were instrumental in 'shelving' the Great Whale Project as a consequence of their very visible public demonstrations in New York City. Soon after the protests, the American purchaser, New York Power Authority, cancelled its $17-billion contract with Hydro-Québec, prompting the newly elected Parti québécois government to put the former Liberal government's capstone project on the back burner.

In the late 1990s, the discourse seems to have shifted away from the Great Whale's potentially enormous implications for the biophysical environment, turning more towards political issues such as control over resources and the benefits of development. In attempting to forge new alliances with northern communities, which would enable greater cooperation in future development, the government signed new agreements—dubbed 'La Paix des Braves' (translation: 'The Peace of the Braves')—with the Aboriginal peoples in early 2002. Worth $3.6 billion and $470 million to the Cree and Inuit respectively, these agreements provide immediate cash benefits to Aboriginal peoples, enable partnerships in economic development, and strengthen local involvement in environmental planning (Meredith 2004). With respect to the last, the agreements stipulate that a 'local environment administrator' be given the powers and functions of the deputy minister of the environment. Final decision-making authority, of course, remains with the minister. The agreements appeared to signal real progress towards local control and social equity, and the new alliance was heralded as a victory for the environment. However, some doubted this optimism. As Meredith states, '[I]t may merely make exploitation of the northern rivers more feasible' (2004, 491).

Indeed, in November 2006, Quebec's minister of sustainable development, environment and parks announced that his department had issued a certificate of authorization to Hydro-Québec for construction of the Eastmain 1 powerhouse and diversion of the Rupert River. The certificate's issue followed the unanimous recommendation in its favour by the bipartite Quebec-Cree agency Review Committee responsible for assessing the environmental and social impacts of projects located on James Bay territory. The Eastmain-Rupert diversion project promises an investment in the order of $4 billion and is expected to create the equivalent of 27,000 jobs with significant economic spinoffs for the Cree.

As mentioned earlier, the environmental assessment procedures established for northern projects vary according to whether the project is located south or north of the 55th parallel. Moreover, the membership of the assessment and review committees varies according to whether the project is provincial or federal in nature. Therefore, four distinct procedures are possible, but they all follow the same five-phase process (Figure 20.3).

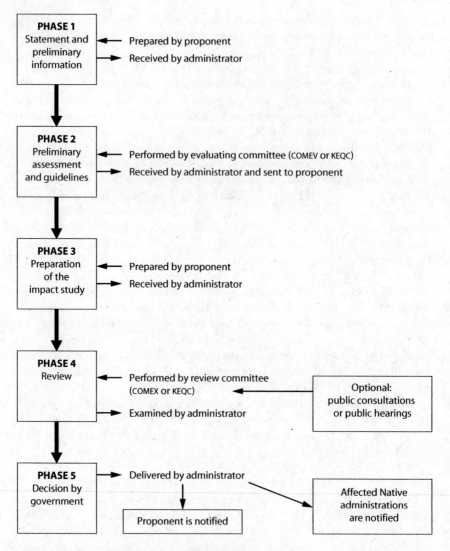

Figure 20.3   Procedure for northern projects

Source: Adapted from Environnement Québec 1994.

Phase 1: Proponent's Statement and Preliminary Information

Phase 1 is comparable to the project notice component of the southern procedure. The procedure starts when the proponent sends a 'notice of intent' to the administrator (who acts as the deputy minister of the environment in the north, discussed earlier). This short document contains preliminary information, including project objectives, the nature and scope of the project, as well as the various sites under consideration and the various development alternatives.

Phase 2: Preliminary Assessment and Guidelines

The preliminary information is then sent to the committee responsible for defining the nature and extent of the impact study. This would be the Evaluating Committee for projects south of the 55th parallel and the Kativik Environmental Quality Commission for projects north of the 55th parallel. When a project is neither automatically subject to nor exempt from the procedure, according to the law and regulations, then the committee recommends to the administrator whether or not the development project should undergo the procedure.

If a project is considered to be subject to the procedure, the committee formulates guidelines outlining the extent of the impact study that the proponent must prepare. These guidelines are forwarded to the administrator, who then forwards them to the proponent, with or without changes.

Phase 3: Preparation of the Impact Study

In this phase, the proponent formulates an impact study in accordance with the administrator's guidelines. The Regulation Respecting the Environmental and Social Impact Assessment and Review Procedure Applicable to the Territory of James Bay and Northern Quebec (r. 11) defines what elements must be included in the impact study. These elements are not substantially different from those required of a project proponent in southern Quebec in the preparation of its EIS.

Phase 4: Review

The proponent submits the impact study to the administrator, who then submits it to the appropriate committee. If the project is south of the 55th parallel, the study goes to the Review Committee (COMEX). If the project is north of the 55th parallel, it goes to the KEQC. At this point, the Aboriginal band councils, municipalities, and anyone from the general public may make presentations to the committee. The committee may also hold formal public hearings if it is deemed necessary. The advisory committee then recommends that the development project be either rejected or authorized.

Phase 5: Decision

The administrator reaches the decision to grant or refuse authorization for the project on the basis of the recommendations of the committee. If the administrator disagrees with the recommendation, he or she must consult the committee before making a final decision and informing the proponent. At this time, the Aboriginal administrations affected by the decision are informed of the final decision.

Comparison between Northern and Southern EIA Procedures

When one compares the northern procedure with the southern one, a few major differences are noticeable. First, the Environment Quality Act provides time limits for most steps in the northern procedure, ranging from 30 to 90 days, while only the public participation aspect, under BAPE, is subject to time limits in the south. Although the reasons for these differences are historically rooted in the complex nature of the administrative arrangements in northern Quebec, it would seem on the surface that northern projects are processed more easily and rapidly, especially when public hearings are not included, than southern projects. Justifications for these differences should be made more explicit so that they are understood by the general public across the province.

Second, in the northern procedure, the public is included in decision making primarily through representation on the advisory committee. Because two of the six active members of each advisory committee are Aboriginal, the northern Aboriginal public might actually have more influence on the decision than does the Aboriginal public in the south. Recall that the BAPE, whose commissioners are appointed directly by the government, is simply a mirror of public opinion with no decision-making authority. Certainly, public hearings do take place on occasion in the north, but there appears to be less emphasis on consultation with the general public. There may be several geographical reasons for that, including the low population density in the north and the vast distances that must be travelled if public hearings are to be held in every community. There may also be socio-economic and cultural reasons that have not been explored here. Nonetheless, large controversial projects, like the Great Whale, were subject to public hearings. Despite this, the northern procedure has generally been less accessible to the general public of the north and certainly very remote from the people of the south. Very few of the documents relating to northern projects are available online.

The third major difference is that there is no formal control phase (monitoring and follow-up) in the northern procedure. Neither the Environment Quality Act nor the applicable regulation respecting environmental assessment of northern projects contains any information about monitoring and follow-up. This does not mean, of course, that monitoring and environmental follow-up do not take place in Quebec's northern projects; rather, it signifies that the provincial legal and regulatory backing to this phase is not made explicit.

This deficiency in environmental follow-up is not new or unique to northern Quebec. Similar failings have been exposed by different people and organizations elsewhere in Quebec and in Canada (Lacoste Report 1988; Gouvernement du Québec 1992; Sadler 1996; CEA Agency 2001). Improvements in the post-decision phase of environmental assessment are something to watch for in the future.

Conclusion

This chapter has provided an overview of environmental assessment in the province of Quebec. It has highlighted the strengths and weaknesses in the evolution of the legal basis, the role of public participation, the kinds of projects covered, and the procedures for both southern and northern projects. Based on this overview, three concluding observations have been made.

First, the Environment Quality Act contains strong language in support of Quebec's commitment to principles concerning environmental protection and the preservation of quality of life in a healthy environment. The right to a healthy environment is combined with the right of the public to be informed and the responsibility of the public to participate in the environmental assessment process. The separate functions of the BAPE within the assessment process allow a degree of openness and transparency to be achieved. Recent amendments to the EQA and EIA processes have shown, however, that not everyone agrees on how to implement these principles. The debate and fallout over these changes have shown the volatile and controversial nature of EA in Quebec. Moreover, the Quebec government's often tenuous relations with the federal government, other provinces, and its own Aboriginal citizens demonstrate that the province will have to continually re-examine its priorities if it wants to continue to act according to the environmental principles outlined in the Act. The dynamic connection between natural ecology and the pursuit of environmental equity, with the consequent implications for environmental assessment, does not allow for political isolationism.

The second observation deals with the public's role in environmental assessment. Quebec has formally recognized the important role the public has played in environmental decision making, and the BAPE has acted as the flagship of Quebec's efforts. However, it has been pointed out that the BAPE's strengths are overshadowed by the non-binding nature of its recommendations. Too often, members of the public are disheartened to find out that, after devoting many unremunerated hours preparing detailed submissions and participating in formal public consultations and hearings, their hard work has, in the end, been completely ignored by government decree. When the disheartening spreads across a people, they begin to feel helpless and disempowered. That feeling eventually turns into either anger or apathy. The result is a society divided into those who seek to influence decision making through more radical means and those who just do not care anymore. In democratic societies like Quebec's, neither reaction is healthy.

It has been suggested that the situation might be improved if the BAPE's toolbox for public participation were expanded to include other mechanisms, like round table negotiations and alternative dispute resolution methods. The latter tools tend to be less adversarial in nature and more conducive to two-way communication. Another suggestion is to improve avenues for public participation in the implementation (or post-decision) stage of the process; at present, such participation appears to be very weak, especially for northern projects.

The final observation concerns the overall role of environmental assessment in government decision making. Although the language of the law, as we have seen, recognizes EA as a valuable tool to help identify and mitigate potential negative repercussions on the biophysical and human landscape, the inclusion of this tool in government decision making is non-binding. Consequently, even scientifically sound EAs may not necessarily lead to better outcomes. This situation has been a bone of contention for those most closely involved in EA. According to a statement by the Quebec Association for Impact Assessment,

> It has often been recognized that, while the professionals in the various fields of impact assessment could easily convince each other of the necessity of impact assessment and agree on the ethical rules which must prevail to insure its credibility and usefulness, those views of insiders are not necessarily shared by decision-makers, administrators, politicians and managers in other sectors. (Association québécoise pour l'évaluation d'impacts 2003, 1)

Although environmental assessment in Quebec is on a trajectory of continued improvement, it is not without flaws. While it is fully accepted as a tool for identifying possible impacts of proposed development projects, its usefulness for limiting environmental damage is still in question. The degree to which environmental assessment has any real input into the decision making is—perhaps more so recently—very much open to debate. Despite the 30 years of environmental assessment practice in the province, there appears to be ample room for improvement.

Note

1.   In French, the EQA (sec. 1.1.4) defines *environment* as 'environnement: l'eau, l'atmosphère et le sol ou toute combinaison de l'un ou l'autre ou, d'une manière générale, le milieu ambiant avec lequel les espèces vivantes entretiennent des relations dynamiques.'

References

Association québécoise pour l'évaluation d'impacts. 2003. Position paper. Retrieved 20 June 2008 from http://www.aqei.qc.ca/aqei/orang.htm#anchor295195.

BAPE (Bureau d'audiences publiques sur l'environnement). N.d. *Code d'éthique et de déontologie des membres du Bureau d'audiences publiques sur l'environnement.* Québec: Gouvernement du Québec.

———. 2002. *Rapport annuel 2001–2002.* Québec: Gouvernement du Québec.

———. 2003. *Consultation publique sur le développement durable de la production porcine au Québec—Rapport principale.* Québec: Gouvernement du Québec.

Berkes, F. 1981. Some environmental and social impacts of the James Bay hydroelectric project, Canada. *Journal of Environmental Management* 12:157–72.

———. 1988. The intrinsic difficulty of predicting impacts: Lessons from the James Bay hydro project. *Environmental Impact Assessment Review* 8 (3): 201–20.

CCME (Canadian Council of Ministers of the Environment). 1998. *Canada-Wide Accord on Environmental Harmonization.* Retrieved 20 June 2008 from http://www.ccme.ca/ initiatives/ environment.html.

CEA Agency (Canadian Environmental Assessment Agency). 2001. *Financing of research activities on environmental follow-up.* Ottawa: CEA Agency.

Connor-Lajambe, H. 1990. Societal impacts of utility overinvestment: The James Bay hydroelectric project. *Utilities Policy* 1 (1): 78–87.

Diamond, B. 1990. Villages of the dammed: The James Bay Agreement leaves a trail of broken promises. *Arctic Circle,* November/December, 24–34.

Environment Quality Act of Quebec. Retrieved 20 June 2008 from http://www.canlii.org/en/ qc/.

Environnement Québec. 1994. *L'évaluation environnementale des projet nordiques.* Direction des évaluations environnementales. Québec: Gouvernement du Québec.

———. 2001. *L'évaluation environnementale au Québec. Procédure applicable au Québec méridionale.* Direction des évaluations environnementales. Québec: Gouvernement du Québec.

Feit, H.A., and R. Beaulieu. 2001. Voices from a disappearing forest: Government, corporate and Cree participatory forestry management practices. In C.H. Scott (Ed.), *Aboriginal autonomy and development in northern Quebec and Labrador,* 119–48. Vancouver: University of British Columbia Press.

Gouvernement du Québec. 1992. *Commission de l'aménagement et des équipements sur la procédure québécoise.* Québec: Gouvernement du Québec.

House of Commons. 2003. *House of Commons Debates.* 5 May 2003, 37th Parliament, 2nd session. Vol. 139, no. 095. Official Report. Ottawa: Parliament of Canada. Retrieved 22 April 2004 from http://www.parl.gc.ca/PDF/37/2/parlbus/chambus/house/debates/ Han095-E.PDF.

Lacoste Report. 1988. *L'évaluation environnementale: Une pratique à généraliser, une procédure d'examen à parfaire.* Rapport du Comité de révision de la procédure d'évaluation et d'examen des impacts environnementaux (Rapport Lacoste). Québec: Gouvernement du Québec.

Leduc, G.A., and M. Raymond. 2000. *L'évaluation des impacts environnementaux: Un outil d'aide à la décision.* Montréal: Edition Multi-Mondes.

McCutcheon, S. 1991. *Electric rivers: The story of the James Bay project.* Montreal: Black Rose Books.

Meredith, T.C. 2004. Assessing environmental impact assessment in Canada. In B. Mitchell (Ed.), *Resource and environmental management in Canada,* 467–96. Toronto: Oxford University Press.

Sadler, B. 1996. *Environmental assessment in a changing world: Evaluating practice to improve performance.* Final report of the International Study of the Effectiveness of Environmental Assessment. Ottawa: CEA Agency.

Scott, C.H. (Ed.). 2001. *Aboriginal autonomy and development in northern Quebec and Labrador.* Vancouver: University of British Columbia Press.

Chapter 21

Atlantic Canada: A Story of EIA Adaptation

Hendricus A. Van Wilgenburg

In this chapter, I explore the administration of environmental impact assessment (EIA)[1] in Atlantic Canada. The history of EIA in the Atlantic provinces is a story of adaptation to the challenges and opportunities of economic development in the region. During the last half-century or more, the Government of Canada and its provincial counterparts in Atlantic Canada have invested considerable time, money, and energy into stimulating development to improve the overall well-being of citizens living in the Atlantic provinces. People generally concede that development can improve the economic circumstances of individuals, a community, or state, and consequently the prospective health and well-being of a population.[2] Nonetheless, while the benefits of development are not lost on members of the public, many are apprehensive about its more objectionable or destructive impacts.

In the past, Atlantic Canadians were often obliged to accept the environmental costs of each new venture in the spirit of economic development (Pushchak 1985). The familiar Sydney Tar Ponds in Cape Breton—the by-product of a steel plant and coke ovens that operated without pollution controls from 1901 through to 1988—is an exemplary case in point. Economic development in Cape Breton has come at considerable cost to both the environment and the health and well-being of individuals living in close proximity to the tar ponds (Guernsey et al. 2000; Rainham 2002). Mounting public anxiety over the impact of economic activity on the health of the environment and humans has generated demands for greater environmental accountability (Sadler 1979; L.G. Smith 1984). Subsequently, the provincial agencies charged with preserving and protecting the environment have been forced to take a more comprehensive approach to EIA—an approach capable of accommodating non-economic values and concerns that are broader and more dynamic (CCREM 1978).

In this chapter, I first briefly examine the historical development of EIA in Atlantic Canada—Nova Scotia, Newfoundland and Labrador (NL),[3] New Brunswick, and Prince Edward Island—and then chart the legislative changes, institutional reforms, administrative reorganization, and legislative provisions for EIA within those four provinces. Second, I describe the current institutional framework for EIA within each jurisdiction. I then move on to examine differences and similarities within each jurisdiction regarding current government policy, stages, processes, and procedures (analytical/comparative). Before bringing this discovery to a close, I provide a summary of the current status (any pending or recent changes) of EIA legislation/regulations and a discussion of recent or ongoing comprehensive EIAs in Atlantic Canada. It is to the historical origins of EIA in Atlantic Canada that I now turn.

The History of Environmental Impact Assessment

Prior to 1980, specific EIA legislation did not exist in Newfoundland and Labrador, New Brunswick, Nova Scotia, and Prince Edward Island (Emond 1985). In EIA's infancy, it was first required in response to a policy, to non-specific legislation, or to an administrative procedure. However, most efforts to carry out an EIA amounted to little more than environmental protection rather than a process specifically concerned with identifying and addressing the potential effects of an undertaking. Typically, EIAs in Atlantic Canada were carried out in response to federal legislative triggers (e.g., the Navigable Waters Protection Act).[4] Under the Environmental Assessment and Review Process (1973), or EARP, any proposed undertaking/project receiving federal commitments (e.g., land transfers) and/or funding was required to undergo an environmental assessment.

Under the EARP, the first projects submitted for review, and hence subjected to an EIA, were the Point Lepreau nuclear generating station (proposed by the New Brunswick Electric Power Commission) and the Wreck Cove hydroelectric project (proposed by the Nova Scotia Power Corporation) (see Figure 21.1). These projects were generally perceived as making an important economic and social contribution to the respective provinces.[5] Despite good intentions, in each case the EIA process fell victim to considerable political interference.

For the Point Lepreau project, the NB Department of Energy, Mines and Resources received federal funding and the NB Electric Power Commission (the proponent) received a federal land transfer (Emond 1978). For political reasons, the Point Lepreau nuclear generating station was initially planned for the Bay of Chaleur. That site was rejected because of environmental concerns, and the alternative site of Point Lepreau chosen. The review panel vetting the project ordered the preparation of guidelines to identify the scope of environmental issues that should be addressed in the EIA study and to outline the necessary procedural and reporting requirements. The federal and provincial departments of the Environment prepared the guidelines collaboratively. Preparations at the project site and for power lines to the site were well underway as the first set of guidelines was being revised. However, political pressure forced the EIA process off the rails. Emond writes: 'In fact, the issuance of the 1974 guidelines was the last time EARP was complied with. As a result, the environment impact assessment process was an empty and meaningless one inasmuch as the public was not meaningfully involved' (1978, 250). When the revised guidelines were finally made public, the review panel allocated little time for public review and input. The project appeared to be a foregone conclusion, especially in light of some of the political announcements at that time.[6]

The Wreck Cove hydroelectric project, proposed for the Highlands of Cape Breton in Nova Scotia, also suffered from political meddling. This project was well underway before the review process was complete. When public involvement did take place, it was often fragmented. In one instance, public involvement looked more like a partisan political rally than a meeting to hear public remarks and observations. In fact, before the panel's final hearing, the minister of the environment for Nova Scotia announced that the project would go forward regardless of the outcome of the hearing (Emond 1978). Political interests trumped environmental concerns.

Flawed public participatory processes were not the only problems; it was also evident that there were difficulties with the methodology used to evaluate the impacts of proposed projects. The methods used in the public sector often mimicked those employed in the private sector. The relevant agencies and administrators assessed the merits or demerits of a project by using concepts such as cost-benefit analysis, conceivably ignoring the qualitative features of the environment (Lang and Armour 1977). It was shifting public attitudes that spurred change in public decision making. Succeeding governments either established administrative procedures or relied on existing legislation to address the unwanted impacts of development (Pushchak 1985). Established approaches for evaluating the environment were either adapted or modified to respond to public demands for greater environmental accountability. Each succeeding decade has seen environmental quality placed higher on the public agenda. Public involvement and public input are now widely accepted as essential to the creditability and quality of the EIA process. Before exploring EIA procedures as currently legislated, I will situate EIA in its historical context in Atlantic Canada.

The Introduction of EIA Legislation, Policy, and Regulations

Environmental impact assessment is now formally legislated in all Atlantic jurisdictions. In addition, all jurisdictions refer to EIA as a decision-making tool used to promote sustainable development by allowing stakeholders to evaluate the potential environmental effects of major developments before the project in question is undertaken. However, the current scope and format of EIA legislation, policy, and regulations in Atlantic Canada vary somewhat from province to province. For example, differences exist in the process stages (e.g., the public review or decision-making time frames) and in the requirements for EIA documentation. However, one feature that has remained constant is the weight of ministerial authority in the decision-making process. The power to make crucial determinations at important phases in the EIA process (e.g., project approvals or EIA requirements) still rests with the minister. Significant changes in some legislation include specific requirements for public review and review by government and various governmental agencies.

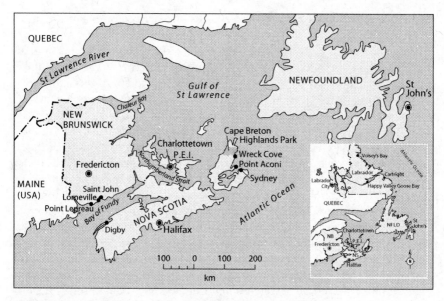

Figure 21.1   Atlantic Canada projects subjected to EIAs

Newfoundland and Labrador (NL)

The first jurisdiction in Atlantic Canada to formally legislate EIA was Newfoundland and Labrador. In 1978 the Government of NL approved in principle the preparation of draft legislation specific to EIA. Two years later, the Environmental Assessment Act (1980) received Cabinet approval. The Act was applied at the procedural level through the Environmental Assessment Regulations (1984). The EIA review process in NL has been documented by Storey (1987) as one of the best in Canada. Over the course of 10 years, EIA in NL matured considerably. The process itself strengthened, and the EIA statements became more focused inasmuch as they reflected a better balance between physical and human environmental issues and greater cooperation between proponents, governments, and communities (Storey 1987). However, the public participation component of the process was occasionally weak. There were cases where the public was not actively involved in the review process even though they had been given information about a project. The submission of an environmental impact statement (EIS) was not always followed by public debate (Storey 1987); hence, the public often suspected that the EIA process was used to justify decisions rather than to examine feasible alternatives.

In a paper comparing the federal assessment process and the 10 provincial EIA processes, NL ranked as either 'strong' or 'excellent' on nine out of 10 criteria, including scope of application, definition of environment, and public participation (L. Smith 1991). After a period of public review, NL carried out significant revisions to its EIA legislation.[7] The aim of the Environmental Protection Act (2002) and the Environmental Assessment Regulations (2000) is a more disciplined and focused approach to the EIA process, as well as improved enforcement and public accountability.

Under the current regulations, the Government of Newfoundland and Labrador has the authority to reject a project at the outset if it is contrary to law or policy or is deemed not in the public interest. The minister can attach conditions to an undertaking upon its release at any stage, such as the proviso that proponents post a bond for site remediation. The legislation of 2000 and 2002 has established at the initial registration stage criteria upon which to base ministerial determinations.

Improvements have also been made to the public review component of the EIA process. The time allocated for public review, including mandatory public review at all major stages of the process, has been extended. The revisions also include conditions for the publication of EIA approvals, such as a public registry for information pertaining to EIAs. Another change in the legislation involves the exemption of some undertakings. In the past, all new undertaking had to be registered with the NL Department of the Environment. Under the current regulations, activities such as peat harvesting, small agricultural enterprises, tree farms, small-scale projects in special land-use areas (except those near salmon rivers), and projects considered to be relatively benign are exempt from the registration process with the minister. However, large amendments to prior approvals must still be registered. The Department of Forest Resources and Agrifoods now approves small amendments. The EIA process has also been simplified. In the past, the development of the EIS terms of reference (ToR) and the guidelines were separate stages; now they have been collapsed into one. Furthermore, proponents are no longer required to prepare draft ToRs for review and approval.

The NL Department of the Environment's Environmental Assessment Committee now prepares the EIS guidelines and submits them to the public and the proponent for review. The current legislation also acknowledges the substance of harmonization—the link between NL's and Canada's EIA processes, where both Acts are triggered by legislation.

Lastly, enforcement for infractions of legislation, regulations, or conditions has been toughened. The legislation now allows for higher penalties for infractions. In the past, the maximum fine for an offence by individuals and/or corporations was $10,000. Currently, the maximum penalty is $1 million for an offence by a corporation and $50,000 for an offence by an individual. The minister also has the authority to issue stop-work orders and/or remediation orders for those activities that fail to comply with the legislation, regulations, or conditions.

New Brunswick

In New Brunswick, EIA had its beginning in 1972 with the Lorneville Impact Study (NBDE 1975). The study considered the impacts associated with a supertanker port and a thermal energy generating station planned for the Lorneville area (see Figure 21.1). The project fell under the authority of both the federal and provincial governments. The Saint John River Basin Board commissioned the EIA, but provincial environment staff coordinated the study. The Lorneville project was one of a number of energy-related projects proposed in New Brunswick at that time. Others included a hydroelectric dam on the Green River in northwestern New Brunswick, the expansion of the Dalhousie thermal generating station, and the Point Lepreau nuclear generating station. However, project studies in New Brunswick seldom included all the elements of a comprehensive EIA (NBDE 1975). Public involvement was largely ineffectual in spite of widespread public interest. There were also concerns about the way the studies were supervised and carried out, which was usually by government employees (i.e., federal or provincial) (WGA 1980).

In 1974, a committee comprised of deputy ministers was struck to prepare a proposal for an EIA in New Brunswick. On 8 October 1975, the Cabinet established a comprehensive EIA policy rather than an Act specific to EIA. The intent of the policy was to make government decision makers and the public aware of the risks and trade-offs associated with an undertaking. However, under the policy, an EIA would only occur for major developments funded by the province of New Brunswick. The policy directed all major developments proposed by government or its agencies to carry out an EIA prior to any decision regarding their implementation. Environment design studies were also required to ensure that the potential adverse environmental effects associated with the project were minimized and the benefits enhanced. The proponent was responsible for funding the study, and the NB Department of the Environment assisted with the development of the guidelines. Proponents of private projects or activities, on the other hand, were not required to register their proposals.

In December 1985, the minister for municipal and cultural affairs and the environment released the draft EIA regulation for public review and comment. Following this, the Department of the Environment conducted an extensive consultation process with interested individuals and organizations regarding the draft regulation. Amendments to the Clean Air Act in the 1983 and 1985 legislative sessions provided a statutory framework for the new regulation. In 1987, EIA moved to full legislative status in New Brunswick with the adoption of the Clean Environment Act (1987) and the Environmental Impact Assessment Regulation 87-83.

The principal difference between Regulation 87-83 and the previous EIA policy is that registration and screening now capture all projects, including those proposed by municipalities and private developers, as well as those proposed by the Government of New Brunswick and its agencies. Regulation 87-83 also establishes comprehensive administrative procedures for assessing proposals and specific opportunities for public involvement at various stages of the EIA process. Basic features of the current EIA methodology include the following (WGA 1980):

(a) A focus on the environment and its systems (not just the project and its effluents)
(b) Consideration of socio-economic impacts on an equal scale with ecological impacts
(c) Concern for resource and land use
(d) Inclusion of public participation in the procedures

These amendments were the last significant revisions to the legislation. The changes reflect an effort on the part of the NB government to better capture and reflect the interests of individuals living in New Brunswick whose lives or well-being may be, or are, affected by new undertakings.

Prince Edward Island

Prior to the enactment of its current legislation, Prince Edward Island relied on administrative procedures established in 1973 under Executive Council Minute No. 16/73 to carry out EIA. Although EIA was not exercised on a comprehensive basis on the Island, proponents were still expected to prepare an environmental impact statement under the guidance of PEI's Environment Control Commission (Emond 1985). The commission, a body appointed by and accountable to the minister, remains responsible for reviewing most proposed projects (including modest projects like farm buildings), carrying out public meetings, and conducting educational assemblies. Other than the near environmental catastrophe (off the north shore of PEI) posed by the accidental sinking of the *Irving Whale* in the Gulf of St Lawrence in 1970, with its 24,000 tonnes of bunker 'C' oil, environmental concerns on Prince Edward Island have been fairly mundane (MacDonald 2000).[8] Issues such as bans on dumping, metal beverage containers, and soil erosion from agricultural practices have tended to dominate the environmental landscape.

In circumstances where an EA did take place, the requirement that it do so followed from federal/provincial cost-sharing arrangements (under the EARP). MacDonald coyly points out that anything of importance on Prince Edward Island during the 1970s 'either flowed from or through' the Island's Comprehensive Development Plan (2000, 303).[9]

A project of special significance to Islanders is the Northumberland Strait Crossing Project (officially known as the 'Confederation Bridge' or unofficially as the 'Fixed Link'; see Figure 21.1). During the development phase of the project, a controversy arose among some Islanders because Public Works Canada called for costed proposals for the project before the mandated draft 'Generic Initial Environmental Evaluation' was submitted for public review.[10] The point of contention was the minister's failure to request an assessment on the bridge itself, as opposed to other alternatives (e.g., a tunnel).

In response, the 'Friends of the Island' went before the Supreme Court of Canada to contest the general approach used in the evaluation and captured in the draft report (see *Canadian Environmental Law Review*, vol. 10, p. 204). The court subsequently ordered a public review of the draft; this was the first time that public consultation was ordered under the EARP.[11] However, the motivation driving the Friends of the Island was grounded mostly in economic issues, rather than biophysical issues (MacDonald 2000). The Confederation Bridge illustrates an all too familiar tension in EIA, namely economic interests competing against environmental interests.

In 1988, statutory provisions for EIA were established on PEI under the Environmental Protection Act (1988). The Act, which remains in place to date, leaves the requirement for EIA to the minister's discretion. Section 9(1) directs proponents to file a written proposal with the Department of the Environment to obtain ministerial approval before proceeding with an undertaking. The department also makes use of interim guidelines that outline its general approach to the EIA process. In contrast to the other jurisdictions, Prince Edward Island has yet to introduce new, or amend existing, legislation.

Nova Scotia

Environmental impact assessment in Nova Scotia was, at the outset, sanctioned under the general provisions of the province's environmental protection legislation, rather than under specific EIA legislation (Emond 1985). The Water Act and the Environmental Protection Act (S.N.S. 1973, c. 6) were used to administer EISs in order to preserve and protect the environment within the jurisdiction of Nova Scotia (Emond 1985). The environment includes the air, land, or parts thereof. The minister of the environment was responsible for administering the Act and determining whether an EIA was warranted. The proponent could be a government agency or department, a private corporation, or an individual proposing, controlling, or managing an undertaking. Projects proposed by individual farms and households were exempt from the EIA review process.

The Environmental Assessment Division of the NS Department of Environment and Labour was and remains responsible for screening and advising the minister on decisions relating to the project and environment. From 1973 to 1995, the minister also relied on the Environmental Control Council for direction. The council, appointed by the minister, was on all accounts 'an administrative body, which operated primarily as a fact-finding body' without executive authority; its role was to report directly to the minister (Harrison 1990). The council was responsible for reviewing EIA reports, carrying out public consultation (e.g., hearings), and recommending to the minister the approval (subject to conditions) or the rejection of a proposal.

Similar to today, the proponent at that time was responsible for initiating contact with the department regarding a proposed project. Further, undertakings were divided into two categories: Class 1 and Class 2. The minister had decision-making authority, and the Department of Environment and Labour was responsible for preparing study guidelines. The proponent, on the other hand, was responsible for preparing the EIS.

There are four avenues for public input: the proponent, the department, the Environmental Control Council, and a panel review. In Nova Scotia, the first public hearings involving the Control Council were held at Point Aconi, Cape Breton (see Figure 21.1). The purpose of the hearings was to discuss a 165-MW coal-fired generating station proposed by the Nova Scotia Power Corporation (Harrison 1990). The public review component of this assessment revealed that rules and regulations with regard to practice and procedures were essential for the council to exercise its powers appropriately at such hearings.

In 1986, the NS Department of Environment and Labour, in consultation with three levels of government (federal, provincial, and municipal), began drafting a new environmental assessment act with accompanying regulations (Harrison 1990). In 1989, the Environmental Assessment Act (1989) and the Environmental Assessment Regulations (1989) were formally established. The principles underlying the Act were 'the preservation and protection of the environment'. The legislation broadened the scope of the EIA process and provided greater opportunity for public consultation. Included in the Act was a broader definition of *environment*, as well as a definition for *undertaking* and *significant*. Under the Act, everyone proposing a project in Nova Scotia was required to register the proposal with the Department of Environment and Labour. Legislative provisions were also made for specific time frames at major stages of the EIA process. However, the time allocated for public review was limited to hearings. Table 21.1 outlines the time frames for the establishment of EIA policy, legislation, and regulations in Nova Scotia, as well as in the other three Atlantic provinces.

In 1995, Nova Scotia amended its Environmental Assessment Act. Under the revised Environment Assessment Act (S.N.S. 1994–95), the Control Council was disbanded and, in its place, the Environmental Assessment Board was established, with regulations specific to its activities. A significant part of the council's mandate was transferred to the Assessment Board. With the amended Act came broader opportunities for public review and input, along with the allocation of specific time frames for each phase of the EIA review process.

Current Institutional Framework: Requirements, Practices, and Procedures

Newfoundland and Labrador (NL)

The Environmental Assessment Division of the NL Department of the Environment is the responsible authority for administering the EIA process in NL. The division promotes the protection and prudent use of the environment through the review of proposed undertakings at the planning stage. EIA is legislated under Part IV of the Environmental Protection Act (2002). The EIA process is set out in the Environmental Assessment Regulations (2003). The aim of the legislation is to 'protect the environment and quality of life of the people of the province; and facilitate the wise management of the natural resources of the province' (*EPA*, 2002, c. E-14.2, s. 46). The goal of the assessment process is to include the various departmental agencies and the public at every stage of the process (see NSDEL 1995 [2003]).

1. The Requirement for Registration

In NL, individuals, private firms, and government agencies that propose a particular undertaking are directed to register the proposed project for evaluation (with the required fee) before the proponent may proceed with the final design of an undertaking. Part III of the regulations designates, but is not limited to, projects and activities requiring registration, as well as those that are exempt. The requirement for registration also includes any plan to modify, rehabilitate, abandon, or demolish an undertaking. Projects not listed that nonetheless can significantly impact

Table 21.1 Establishment of EIA policy, legislation, and regulations in Atlantic Canada

	1970s	1980s	1990s	2000s
NS	Established: Environmental Protection Act, chapter 6 of the Revised Statutes of Nova Scotia, 1973; Water Act, chapter 225, 1973; Guidelines on EIA under the |Environmental Protection Act, chapter 6; and Environmental Control Council	Established: Environmental Assessment Act (1989); and Environmental Assessment Regulations (1989)	Proclamation of revised: Environmental Assessment Act (1994–95); and Environmental Assessment Board Regulations (1994–95) Approval: Guidelines for Community Liaison Committee, 1993	Amendments to Environmental Assessment Regulations (2008)
NB	Adoption: specific EIA policy, 1975	Amendments to the Clean Environment Act, 1983 and 1985, and adoption of the Environmental Impact Assessment Regulation 87-83 (1987)		New Brunswick and Quebec sign environmental cooperation agreement, 2001
PEI	Administrative procedures for EIA legislated under Executive Council Minute 16/73, 1973	Established: Environmental Protection Act, 1988; Interim Guidelines—Environmental Assessment, 1988		Amendments to interim 1988 EIA guidelines and adoption of Environmental Assessment Fees Regulations (2005)
NL	Approval for drafting of an environmental assessment act, 1978	Established: Environmental Assessment Act, 1980; and Environmental Assessment Regulations, 1984		Established: Environmental Protection Act, S.N.L. 2002, c.E-14.2; and Environmental Assessment Regulations (2003)

the environment must also be registered. The NL Department of the Environment provides the proponent with an application form for registering a proposal. Once received, the application is assessed for completeness. If the application is found deficient, the minister may request additional information from the proponent before any further decision is made on the proposal.

2. The Minister's Screening Decision

Once the minister determines that the application is complete, the proposal is released to the public and concerned government agencies for review and input. A maximum of 30 days is allotted for examination of this material. The screening process seeks to establish the undertaking's interactions with the environment, the proponent's ability to conduct the undertaking in an environmentally sound manner, and the impact of unknown or experimental technology intended to be used during the undertaking. Forty-five days after the proposal is registered, the minister will advise the proponent on one of four options (see Figure 21.2). First, the minister can decide that the undertaking may proceed as outlined in the application, subject to any terms and conditions or other acts and regulations (i.e., provincial, municipal, and federal). Second, an environmental preview report (EPR) may be required to provide additional information not contained in the registration document to allow suitable appraisal of the project. Third, an environmental impact statement may be required to assess any significant environmental impact connected to the project that is indicated; an EIS is also required if there is significant public concern regarding the activity. Lastly, the project may be rejected if the associated environmental effects are determined to be unacceptable or not in the public interest. Cabinet makes the latter determination.

A public registry of all screening decisions made under the regulation is maintained and updated regularly by the Environmental Assessment Division. Once the undertaking is registered, the minister has seven days to announce the project. The minister must also make copies of the registration documentation available to the public for review, marking the initial opportunity for public involvement in the process. After the registration announcement, the public has a limit of 35 days to submit written comments on the project to the minister.

3. If an EPR or EIS Is Required

If an environmental preview report or environmental impact statement is warranted, the minister will first appoint an Assessment Committee comprised of technical experts from those provincial and federal departments that have an interest in the undertaking. The Assessment Committee is responsible for assisting in the preparation of the terms of reference for either an EPR or an EIS. In addition, the Assessment Committee reviews all EPR or EIS documents referred to it by the minister, as well as submitted public comments. Should the minister decide that an EPR or an EIS is warranted, the proponent would be required to carry out a study to assess the nature and significance of the potential impacts. The proponent is also responsible for the costs associated with the study and preparation of the EPR or EIS, as well as the costs associated with supplementary studies and revisions.

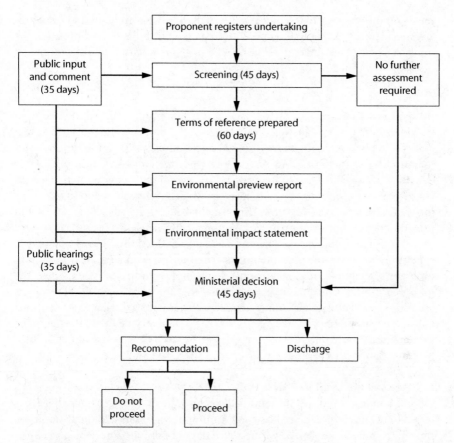

Figure 21.2   Newfoundland and Labrador EIA Process

Source: Storey 1987; also available at http://www.ucs.mun.ca/~kstorey/newfound.htm.

4. Developing the EPR or EIS Guidelines: A Public Procedure

The next step in the EIA process is the preparation of the guidelines for the EPR or EIS.[12] The Assessment Committee assists the proponent in preparing the draft terms of reference. The guidelines are based on registration-stage comments and meetings with the proponent, government agencies, and public groups. The guidelines serve as the terms of reference to assist the proponent in determining the significance of environmental effects while he or she carries out the respective study. The guidelines identify important environmental issues that the proponent must consider while assessing the impacts of a particular undertaking, and specify the general approach that the proponent should follow when conducting the EIA. The minister may direct the committee to amend or vary the guidelines before approving them. If the minister decides that the guidelines are acceptable, he or she will then submit them for public review. Forty days are allocated for public review of and comment on the guidelines. Interested persons may submit written comments on the proposal to the minister. Once the various comments and responses are considered, the final guidelines are prepared. Once they are completed, the minister releases the final guidelines to the proponent within 60 days of an EPR decision or within 120 days of an EIS decision.

5. Preparation of an EPR or EIS

If the minister decides that an EPR is warranted, the proponent will be required to address the main unanswered questions in the document as outlined in the guidelines. The content of an EPR is based on readily available information rather than on original fieldwork. An EIS, on the other hand, involves the study of key issues associated with the biophysical and socio-economic environmental effects connected to a project. It typically involves original research targeting the existing environment and anticipated environmental effects. Specifically, an EIS should provide a detailed description and evaluation of alternatives to the project, as well as methods for evaluating the accuracy of impact predictions.

During this phase, the proponent is required to carry out a public information program that should optimally include a meeting with local residents near the proposed project. The goal here is to ensure that the concerns of local residents are accurately identified and understood. Moreover, copies of all reports on original studies must be made available so that the public may be fully informed. The proponent is required to provide not fewer than seven days' notice of a scheduled meeting. Public comments are recorded and addressed in the EIS.

6. Review of the Draft EPR

Once the minister receives the completed environmental preview report, it is referred to the Environmental Assessment Committee for examination. In addition, the report must be publicized within seven days of ministerial receipt, and copies are to be made available to the public. Anyone who wishes to comment on the EPR has 35 days to submit their comments, in writing, to the minister after the announcement is made. Upon examination, the Assessment Committee will decide if the EPR is deficient, whether an EIS is required, or whether the undertaking should be released to the proponent.

The minister then decides, on the basis of the committee's recommendation, whether the EPR is complete and complies with the Act. He or she may decide that a subsequent EIS is not necessary and that the undertaking can go forward, subject to terms or conditions established by the minister. On the other hand, the minister may decide that the EPR fails to address environmental concerns adequately and hence that an EIS is warranted. Either way, the minister must notify the proponent of the status of the proposal no more than 45 days after ministerial receipt of an EPR (or amended/revised EPR) and then inform the public of the decision.

7. Conducting the Study and Preparing the EIS

As previously noted, an EIS is a more comprehensive information-gathering component of the EIA process than an EPR. If an EIS is required, the proponent must assess the nature and significance of the potential impacts in greater detail in order to remedy the noted deficiencies and thereby advance the EIA process. Completing an EIS study itself may require further investigative work. Once this work is completed and the EIS prepared and submitted, the review process is reactivated; that is, the minister refers the EIS to the review committee for examination.

Typically, the principal objective of an EIA study is to predict expected environmental impacts, should the project proceed. This is accomplished by gathering available resource information, conducting field investigations as required, and using scientific methods to evaluate potential interactions between the environment and activities associated with the undertaking. The study is expected to identify methods of enhancing the positive impacts and minimizing the negative impacts of the project. Information gathered during the study is first compiled in a draft EIA report. If the minister decides that further study is warranted, the proponent, should he or she wish to proceed with the undertaking, is required to prepare an EIS.

8. Component Studies

The minister may, in addition, require a proponent to prepare and submit one or more component studies in the preparation of the EIS. The purpose of a component study is to describe and provide data on specific aspects of the environment. For example, data may be required on elements of a valuable ecosystem (e.g., caribou, fish, or rare plants) that may be significantly affected by the project. When completed and submitted, the component studies are released to the review committee for examination. Although part of the EIS, the component studies have separate guidelines, public review and approval processes, and specific timelines.

9. Review of the Draft EIS

Once the draft report is submitted to the minister, it is turned over to the Environmental Assessment Committee, which has 35 days to carry out a detailed examination of the material. The committee's task is to determine whether the document has adequately addressed the issues raised in the study guidelines. Review of the draft EIS study is an interactive process that includes discussion between the committee and the proponent to clarify specific technical issues. On the advice of the committee and with ministerial approval, the public is then brought into a discussion of the environmental impacts described in the EIS.

10. Public Review and Comment on the EIS

Once the minister has received the environmental impact statement, he or she is required to announce the submission of the EIS within seven days and make copies of the EIS available to the public. Those individuals wanting to submit responses or comments on the document must do so in writing to the minister within 50 days after the EIS submission is announced. However, if the draft EIS does not adequately address all aspects of the guidelines, the proponent is advised of the deficiencies that need to be addressed and the public process is delayed.

On the other hand, if the minister accepts the EIS, a second and more comprehensive opportunity for public involvement in the EIA process occurs. The review committee prepares a 'general review statement' summarizing its comments on the document, and this statement, along with the final EIS, is released to the public for review and comment. A public meeting schedule is organized and announced through various media, stating the time and place for public discussions to be held concerning the proposal. Thirty days are allotted for this stage of the process.

11. Public Meetings
Public meetings to discuss the EIA generally take place at or near the proposed project location. The purpose of the meetings is to provide all interested parties with an opportunity to make comments, raise concerns, or ask questions for clarification about any matter covered in the EIA study. Public comments and briefs are submitted in writing either directly to the provincial Department of the Environment or to departmental staff at any public meeting held on the EIA. Following such meetings, an additional period of 15 days is set aside for members of the public to submit additional written comments on the proposal.

12. Public Hearings
In NL, provisions have also been made for public hearings regarding a project. A hearing is an interactive process involving a board, which is appointed by and is responsible to the minister, the proponent, and the public. Hearings have specific timelines for announcements and public review that can be extended if necessary. Copies of all written briefs are submitted to the board, whereas copies of hearing proceedings are submitted to the minister. All submitted documentation associated with the EIS is made available to the proponent as well as to interested members of the public.

13. The Final Decision
At the end of the public review, a summary of all written briefs is prepared and submitted by the review committee to the minister, including transcripts of public meetings and any additional comments received following the final public meeting. The summary is released publicly, and copies are sent to every person who actively participated in the meetings. At the same time, the full package of EIA study information, including the public participation summary, is forwarded to the minister for consideration.

Giving consideration to the review committee's recommendation, the minister notifies the Lieutenant-Governor in Council (Cabinet) whether the undertaking is contrary to a law or policy or, alternatively, is acceptable. The final decision to issue or deny an approval resides with the Lieutenant-Governor in Council. The minister must notify the proponent of this ultimate decision not more than 45 days after the determination has been made.

The approval may specify terms and conditions to which the proponent must adhere while implementing the project. However, if the proponent violates the terms and conditions, fails to disclose relevant facts, or submits inaccurate information, approval for the project can be suspended or revoked. Furthermore, if the proponent does not initiate the project within three years of receiving approval to proceed from the Lieutenant-Governor in Council, the release is void. Accordingly, if the proponent then wishes to proceed with the undertaking, he or she must once again register the project.

New Brunswick

In New Brunswick, the Department of the Environment and Local Government (DELG) is responsible for administering the EIA process in collaboration with other concerned regulatory agencies. For projects that may have significant adverse environmental impacts, the DELG, in cooperation with relevant regulatory agencies, prepares the guidelines for the EIA report and the formal review documents of the report. The EIA process does not replace the regulatory process; rather, the report is 'used to collect information required in regulatory processes' (Couch et al. 1981).

1. The Requirement for Registration
Individuals, private firms, and government agencies that propose a particular undertaking are directed under Regulation 87-83 to formally register the details of an undertaking with the minister of the environment and local government.[13] Schedule 'A' of Regulation 87-83 sets out criteria for the nature and scope of undertakings that must be registered—effectively, any plan to abandon, demolish, modify, or rehabilitate an existing project, including any project undertaken before the regulation came into force. After a project is registered, the proposal is screened to determine if it should undergo an EIA. This process must take place before any work on the actual project commences. The Department of the Environment and Local Government provides the proponent with an application form for registering a proposal. The form captures the details of the proposed project or activity. If the minister determines that the application is deficient, he or she may request additional information from the proponent before making a decision.

2. The Minister's Screening Decision
Once the minister has received enough detail regarding the project, the proposal is examined. The purpose of the screening process is to evaluate, on the basis of readily available information, the environmental issues associated with any activity connected to the proposed project. A maximum of 30 days is allocated for this stage of the screening process. Once the examination is complete, the proponent is notified whether the project will be released or whether an EIA is required (see Figure 21.3). If the minister determines that an EIA is warranted to establish the nature and significance of potential impacts through further study, the proponent receives written notice to this effect prior to the issuance of any notification or other public statement on the proposal. Alternatively, if the minister decides that an EIA is not necessary, he or she will notify the proponent in writing that the undertaking has been approved/released, subject to any terms or conditions established by the minister. A public register of all screening decisions made under Regulation 87-83 is maintained and updated regularly at the Project Assessment Branch of the DELG.

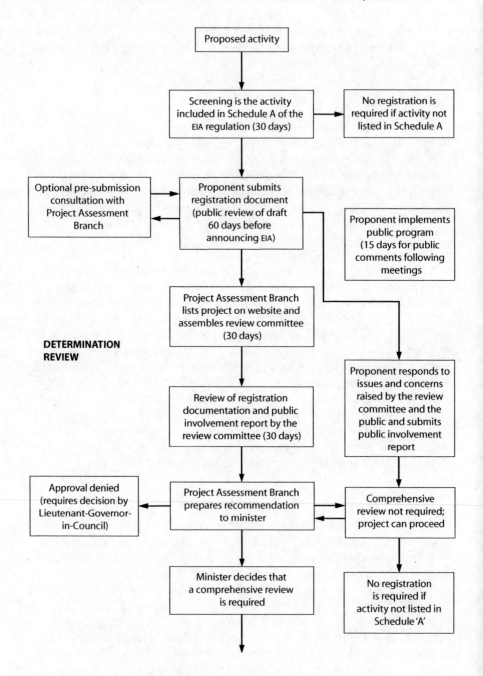

Figure 21.3a   New Brunswick EIA Process

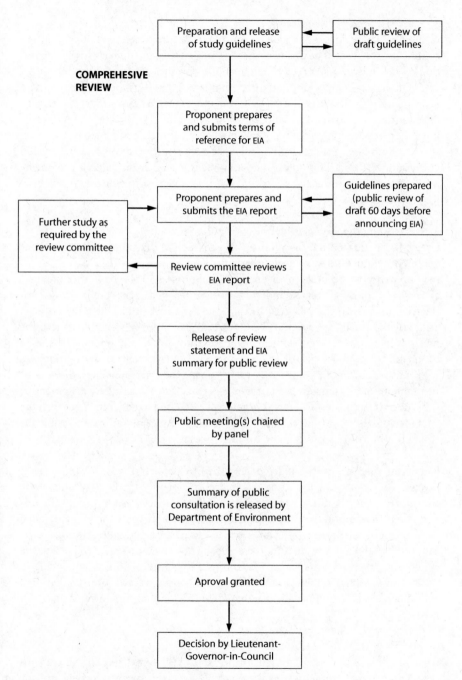

COMPREHESIVE REVIEW

Preparation and release of study guidelines → Public review of draft guidelines

Proponent prepares and submits terms of reference for EIA

Further study as required by the review committee → Proponent prepares and submits the EIA report ← Guidelines prepared (public review of draft 60 days before announcing EIA)

Review committee reviews EIA report

Release of review statement and EIA summary for public review

Public meeting(s) chaired by panel

Summary of public consultation is released by Department of Environment

Aproval granted

Decision by Lieutenant-Governor-in-Council

Figure 21.3b   New Brunswick EIA Process

Source: NBDE 2007.

3. If an EIA Study Is Required

If the minister decides that an EIA is warranted, the proponent will be required to conduct an EIA study if he or she wishes to proceed with the undertaking. Preparation of EIA study guidelines is the first step in this process. Guidelines for an EIA study identify the important environmental issues that must be considered in an assessment of the impacts of a particular development. They also specify the general approach a proponent should follow while conducting the EIA study. To assist the EIA process, the minister appoints a review committee comprised of technical specialists from various government agencies whose jurisdictions may be affected by the undertaking. This committee may include staff from federal, provincial, municipal, or regional agencies. Its first major task is to formulate draft guidelines for the EIA study.

4. Developing the Study Guidelines: A Public Procedure

Under Regulation 87-83, the minister must issue the draft guidelines for public comment within 60 days of announcing the EIA requirement. This marks the initial opportunity for public involvement in the process. Once the draft guidelines are available for public review, individuals and interested parties are free to submit their comments on the proposal to the Minister in writing. Thirty days are set aside for the public review process. Once public input has been considered, the minister issues the final guidelines for the EIA study to the proponent. The release of the guidelines must take place no more than 60 days after the draft guidelines were released for public comment. After receiving the final guidelines, the proponent is required to prepare the terms of reference for the study, which he or she submits to the minister. The terms of reference describe in detail the approach the study team will use during their investigations. The proponent is responsible for seeing the study to its fruition and for any associated costs.

5. Conducting the Study and Preparing the EIA Report

In New Brunswick, the EIA study is expected to identify methods of enhancing the positive impacts and minimizing the negative impacts resulting from the undertaking. In addition to considering impacts, the study must include a detailed description of the project, an evaluation of alternatives, and a description of the proposed methods for evaluating the accuracy of impact predictions. During the study process, proponents are also encouraged to consult with residents living in the area of potential impact to ensure that their concerns are accurately identified and understood. The information gathered during the study is compiled in a draft EIA report.

6. Review of the Draft EIA Report

After the draft report is submitted to the minister, the next step in the EIA process is a detailed examination of the report. A review committee carries out the examination to determine whether the document has adequately addressed the issues raised in the study guidelines. The review of the draft EIA study involves an interactive process between the review committee and the proponent (or his/her representatives), the purpose of which is to clarify specific technical issues surrounding the project.

When the examination is complete, the review committee advises the minister of its findings. If the minister is satisfied that the EIA report meets the criteria outlined in the guidelines, the report is released to the public for review. On the other hand, if the draft report is deficient with respect to any aspect of the guidelines, the public process is temporarily postponed. The minister advises the proponent of the report's deficiencies, and if the proponent wishes to proceed with the project, he or she is required to address these deficiencies, which may require additional investigative work. After the work is completed and the revised report meets the guidelines to the minister's satisfaction, the public process is reactivated. At this time, the proponent is required to submit 30 copies of the final report to the minister, in both official languages.

7. Public Review and Comment on the EIA Report
This stage marks the second opportunity for public involvement in the EIA process. This public review is more comprehensive than is the review of the draft guidelines. The Project Assessment Branch prepares a summary of the final report, and the review committee prepares a summary of its response to the proposal. Within 30 days of receiving the final report, the minister releases the combined documentation for public review and comment. The purpose of making the documentation public is to promote and assist public familiarity with the project. At the same time, the minister announces an agenda for the discussion of the EIA information through various media, including notification in the *Royal Gazette*. These notices will state the time and place for public discussions to be held concerning the proposal. Public meetings are scheduled following the release of the study information. The public is invited to submit written comments or briefs in response to the study. These comments are sent directly to the DELG or delivered to staff at the public meetings.

8. Public Meetings
Public meetings held to discuss an EIA generally take place near the area where the project is proposed to be located. The meetings provide all interested parties with an opportunity to comment, raise concerns, or ask questions for clarification about the substance of the EIA study. During the meetings, formality is generally kept to a minimum. Representatives of the proponent, the DELG, and other members of the EIA study review committee are required to be present at the meetings. The proceedings are recorded and a verbatim transcript is produced for subsequent study.

Following the last meeting, an additional period of 15 days is set aside for the public to submit further written comment on the proposal. At the end of this phase, a summary of public participation is prepared. This summary is based on the written briefs submitted to the minister, transcripts of public meetings, and any additional comments received following the final public meeting. This summary is released publicly, and copies are sent to every person who participated actively in the public discussion. At the same time, the full package of EIA study information, including the public participation summary, is forwarded to the minister for final consideration.

9. The Final Decision

Once the EIA process is complete, the minister submits a report and a recommendation regarding the direction of the proposal to the Lieutenant-Governor in Council. Based on the minister's report and recommendation, the Lieutenant-Governor in Council either issues or denies an approval for the undertaking. The approval may be subject to terms and conditions or other relevant legislation (e.g., the provisions of the Clean Environment Act or the Clean Air Act) to which the proponent must adhere while implementing the project. When an approval is issued, the minister may appoint a Monitoring Committee in order to track the progress of the undertaking, the success of any mitigation measures proposed in the EIA, and the overall impact of the project on the environment. The Lieutenant-Governor in Council may subsequently suspend or revoke an approval should the proponent violate the terms and conditions imposed for the project or should there be reason to believe that the proponent has failed to disclose relevant facts or has submitted inaccurate information.

Prince Edward Island

The responsible authority for administering the legislation with respect to the management, protection, and enhancement of the environment in Prince Edward Island is the minister of fisheries, aquaculture and environment. These tasks are achieved in part through EIA under the provisions of the Environmental Protection Act (1988). The PEI Department of Environment and Energy (PEDOE) is responsible for the day-to-day task of applying the EIA legislation.

1. Requirement for Registration

Under Schedule I of the interim guidelines, the proponent of any manufacturing undertaking is directed to register his or her proposal with the PEDOE before the project can commence. The Interim Guidelines outline the details that must be provided in the application at the time of submission. Projects that must be registered include 'any construction, industry, operation or other project or any alteration or modification of any existing undertaking which will or may' cause the emission or discharge of any contaminant into the environment (PEDOE 1989a). This requirement includes projects that may have an effect on rare and endangered species or on the biophysical environment, or that generate significant public concern about potential effects on the environment, whether that concern is real or perceived. The department provides the proponent with an application form for registering the proposal.

2. The Minister's Screening Decision

After registration, the proposed undertaking is screened to determine whether the project could have a significant effect on the environment and thus warrant an EIA (see Figure 21.4). The PEI Department of Environment and Energy uses an interdepartmental technical review committee to examine the proposal for completeness. The department may request additional information or ask for specific commitments from the proponent (e.g., environmental protection plans, environmental studies, mitigation measures) (see PEDOE 1989b). On the department's recommendation, the minister then decides whether the undertaking should be released, whether more information is required, or whether an EIA is warranted. The minister advises the proponent in writing of his or her determination.

3. Developing the Study Guidelines: A Public Procedure

If an EIA is required, PEDOE staff work with the proponent to develop specific impact assessment guidelines for the study. The guidelines identify the environmental issues that the study should consider and the approach the study should take. Further, the guidelines outline the nature and scope of the information required to resolve any potential concerns and issues associated with the project or activity. The public is also given notice that an EIA is required, the purpose being to familiarize the public with the proposal so that they may comment or raise concerns.

4. Conducting the Study and Preparing the EIA Report

When an EIA study is required, the proponent is responsible for conducting the study, preparing the environmental impact statement, and paying the associated costs. The principal objective of the study is to gather the necessary resource information through field investigation and scientific evaluations of potential interactions between the project and the environment. The interim guidelines suggest that an EIA study should describe the project and existing environment, find ways to enhance the project's positive impacts and minimize its negative impacts, describe methods for evaluating the accuracy of predictions, and review alternatives.

5. Review of the Draft EIA Report

When the assessment process has been completed, the draft EIA report is submitted to the minister. The minister then turns the report over to the technical review committee, who examine it for completeness. If the report is found deficient (i.e., lacking information based on the guidelines), the proponent may be required to carry out further investigations, should he or she wish to proceed.

6. Public Review and Comment on the EIA Report

If the PEDOE is satisfied that the potential environmental impacts of the undertaking have been addressed, the impact statement and technical review committee comments are submitted for public review. Information meetings may be held to provide the public with an opportunity to review the project, offer their com-

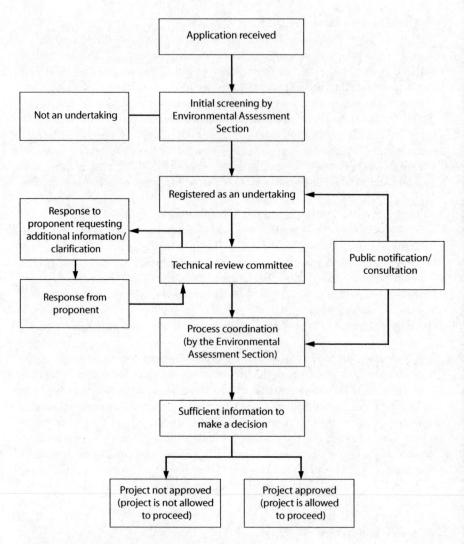

Figure 21.4 Prince Edward Island EIA Process

Source: PEDOE 2005.

ments, and raise any concerns. As yet, Prince Edward Island has not established specific timelines for public review.

7. Public Meetings
If there is a strong public interest in the project, the minister may appoint a Board of Inquiry to hold public meetings to discuss concerns and interests. Once the review process is complete, the public responses are submitted to the board and subsequent findings are made public.

8. The Final Decision

Public concerns are forwarded to the Department of Environment and Energy for consideration prior to any final decision on the undertaking. Should legitimate concerns be identified through the public consultation process that were not adequately addressed in the document, additional study may be required. Once the review process is completed, the minister decides whether to deny or to issue approval of the undertaking (subject to conditions or regulatory requirements or any other requirement imposed by the province of Prince Edward Island or a municipality).

Nova Scotia

In Nova Scotia, the Department of the Environment and Labour is the responsible authority for environmental assessment. The Environmental Assessment Branch of the department administers the EIA review process. The branch promotes the protection and prudent use of the environment through the review of proposed undertakings at the planning stage. EA is legislated by Part IV of the Environment Act (1995). The EA process is set out in the Environmental Assessment Regulations (1995) and the Environmental Assessment Board Regulations (1995).

1. The Requirement for Registration

The Environmental Assessment Regulations outline the criteria for the scope and nature of proposed undertakings that must be submitted for registration. Under section 3(1) of the Environmental Assessment Act, individuals, private firms, and government agencies that propose an undertaking are required to formally submit an application with the set fee and details as outlined in the guidelines or prescribed by the minister. Once the project is registered with the minister—that is, seven days after the registration fee is paid—the project is screened to determine whether an EA is warranted before any actual work on the project commences (see Figure 21.5a). The requirement for registration includes any plan to modify, rehabilitate, abandon, or demolish an undertaking. This registration requirement also applies to any undertaking registered prior to 17 March 1995.

Unless specified otherwise, the general guidelines set out the criteria for the content and scope of the information that the application should include. The minister may, if required, request additional information from the proponent before making a decision. If the application is not complete, the minister or administrator has 14 days following registration to notify the applicant in writing that additional information is required. An additional set fee must accompany the submission of supplementary information.

2. The Minister's Screening Decision

Once the registration document is deemed complete, the Environmental Assessment Division coordinates the document's review by the Environmental Assessment Branch and any other relevant provincial and federal agencies. The division advises the minister of the project's potential effects on the natural environment.

A maximum of 25 days is set aside for examination of the application. Based on that examination, the minister determines whether the undertaking is ranked as Class I or Class II. Approvals for Class I undertakings may include terms and conditions, a focus report, or an environmental assessment report (EAR). Class II undertakings require the proponent to carry out an EA (see Figure 21.5d). The purpose of the screening process is to identify any environmental issues associated with the proposed project and to ensure that the activity conforms to, but is not limited to, the guidelines, regulations, policies, and standards. The applicant's past performance in providing environmental protection for an activity or project is also considered.

A public register of all applications and screening decisions made under the regulations is maintained and updated regularly at the Environmental Assessment Division. After the application is considered complete, 30 days are set aside for public review and comment. The proponent is also required to place a notice publicizing the application in both a provincial newspaper with province-wide coverage and a local newspaper. Following the public review and submission to the minister, if more information is required, the minister or administrator must notify the proponent within 50 days of that decision. If the minister decides that more information is needed, or that a focus report is required, or that the project should be registered as a Class II undertaking, the administrator must notify the public within 12 days of that determination.

3. If a Focus Report Is Required
If a focus report is required, preparation of the study guidelines is the next stage of the process (see Figure 21.5a). The Assessment Division is responsible for preparing the required terms of reference for the focus report study. The administrator has 25 days to provide the proponent with the terms of reference in writing. If the project is to go forward, the proponent is required to carry out the study and prepare and submit the appropriate documents at his or her expense. Typically, the focus report is based on readily available information rather than on original research.

4. Developing the Focus Report Guidelines: A Public Procedure
Once the focus report is submitted to the minister, it is released to the public within 12 days. A period of 30 days is allocated for public review and comment. Twenty days after the final date for public review, the administrator must submit to the minister a summary of the comments received from the public and various government agencies. Taking the comments into account, the minister has 14 days to deliberate the project's status. If the focus report indicates that adverse effects or significant environmental effects associated with the undertaking are not likely to occur or are mitigable, the minister will approve the undertaking, subject to specified terms and conditions and other approvals required. However, if the focus report indicates that there may be adverse effects, or significant environmental effects, associated with the undertaking, the minister may determine that an EA is warranted. Alternatively, if the significant environmental effects are determined to be unacceptable, the minister may reject the undertaking.

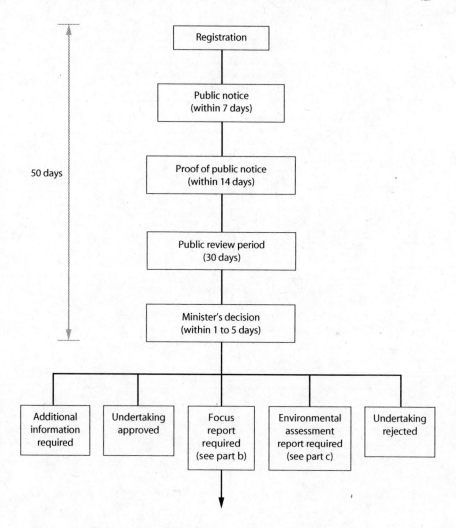

Figure 21.5a   Nova Scotia EIA Process, Class I Undertaking

Source: NSDEL 2008.

5. If an EA Report Is Required

If the minister decides that an environmental assessment is warranted, the proponent is required to carry out an EA study should he or she wish to proceed with the undertaking. Preparation of the EA terms of reference is the next step in this process (see Figure 21.5b). Following the minister's decision, the administrator has 25 days to prepare and provide the proponent with the terms of reference in writing. The terms of reference specify the important environmental issues that must be considered in assessing the potential impacts of the proposed undertaking as well as the general approach a proponent should follow while conducting the EA.

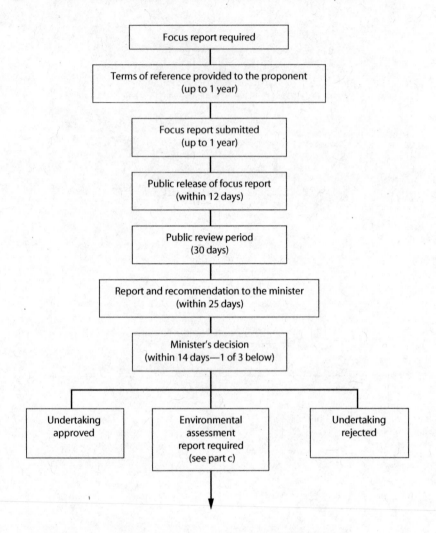

Figure 21.5b   Nova Scotia EIA Process, Class I Undertaking

Source: NSDEL 2001.

6. Developing the Study Guidelines: A Public Procedure

Once completed, the terms of reference for the EA study are then submitted for public review and comment. Forty days are set aside for this stage of the process. Five days after the public review, the administrator will advise the proponent of any comments received during the public process. The proponent then has 21 days to respond in writing to the comments. Once the latter deadline is reached, the administrator has 14 days to provide to the proponent, in writing, the final terms of reference for carrying out the EA study.

7. Conducting the Study for and Submitting the EA Report
After receiving the final terms of reference, the proponent has two years to submit a preliminary draft of the EA report. Once the draft report has been submitted to the minister, it is examined to see if it addresses the particulars specified in the terms of reference. If the report is deficient in any respect, the minister may request additional information. Alternatively, if the report is complete and accepted, the administrator has 12 days to advise the proponent of the minister's decision.

8. Public Review and Comment on the EA Report
In the case of a Class I undertaking that has not been referred to the Environmental Assessment Board, the public is duly notified and invited to comment on the EA report. The administrator has 12 days to publish a notice announcing the release of the report to the public and inviting public input. Following that notification, the public has 48 days to submit comment to the minister in writing. However, if the minister determines that the 48 days are insufficient, he or she can extend the review period, but the proponent must be advised of the same in writing.

9. Referral to the Environmental Assessment Board
For a Class I undertaking that has been referred to the Environmental Assessment Board and for all Class II undertakings (see Figure 21.5c and Figure 21.5d), the provisions for public notice and consultation are outlined in the Environmental Assessment Board Regulations (1995). The time frame for initial announcements and public review are the same as for EA reports not referred to the board. Once the EA report has been completed and submitted by the proponent, the minister has 10 days to refer the report, categorized as either a Class I or a Class II undertaking, to the board. The board, a body appointed by and accountable to the minister, is responsible for reviewing EA reports for all undertakings referred to it by the minister and reporting on potential environmental effects connected to the undertaking.

10. Public Meetings
The Environmental Assessment Board may be required to carry out public hearings or meetings as prescribed by the minister. It may also refer a dispute or issue connected to an undertaking to an alternative dispute resolution procedure under the provisions of the Act and regulations. Nonetheless, the final decision resides with the minister.

11. The Final Decision
Once the minister has received a summary of the board's comments and recommendations and/or the results of an alternative dispute resolution, he or she has 21 days to decide, and advise the proponent of, the following: 'whether the undertaking is approved subject to any other approval required by an enactment; is approved subject to such conditions as the Minister deems appropriate and any other approval required by an enactment; or is rejected' (NSDEL 1995). Before the

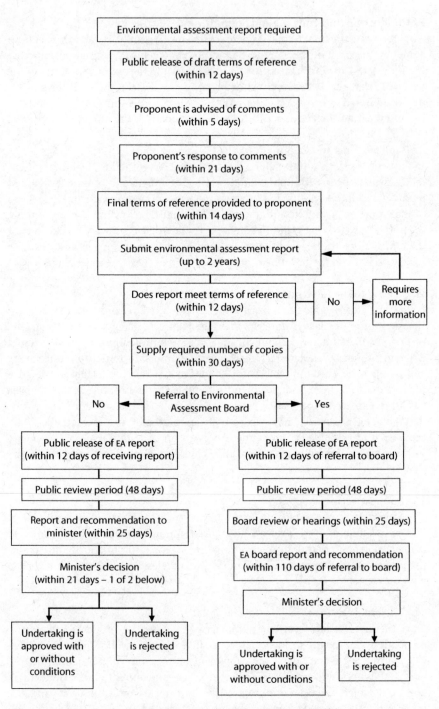

Figure 21.5c   Nova Scotia EIA Process, Class II Undertaking

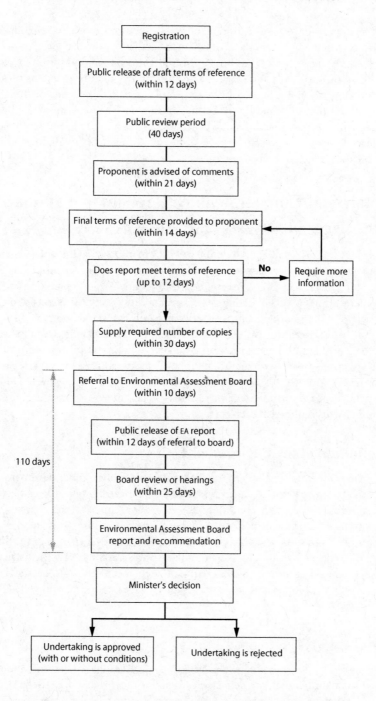

Figure 21.5d   Nova Scotia EIA Process, Class II Undertaking

application is approved, the proponent may be required to carry out a consultative process with the public in the area where the activity or the proposed project is to be located. Once a project has been approved, the proponent has two years to start work on the project. If necessary, the proponent may require an extension should he or she wish to proceed with the project. Under the provisions of the Act, the final decision must be publicized by the minister or the administrator in the *Royal Gazette,* in one newspaper having general circulation in the locality in which the undertaking concerned is to be located, and in one newspaper with province-wide circulation.

Analytical/Comparative: Differences and Similarities

EIA as a Planning Process and Impact Assessment

In Prince Edward Island and New Brunswick, environmental impact assessment is both a planning process and an impact assessment. Proponents proposing an undertaking in these jurisdictions are required to follow a logical planning process as a general strategy for predicting, lessening, and mitigating the effects of a project or activity. The assessment process in Prince Edward Island and New Brunswick is fluid; it advances and evolves in response to the interaction between the various participants and to the information available. Comparatively, in Nova Scotia and NL the process is highly structured. Proponents are required to evaluate potential impacts of a project or activity during the course of an EIA, as well as identify mitigative measures and assess the residual impacts of a project, in response to the requirements outlined at each stage of the process.

Definition of the Environment

The environmental assessment legislation in Nova Scotia defines the environment in broad terms. The environment includes the physical, biological, social-economic, cultural, and aesthetic conditions as well as the technical realm. The social and economic environments are captured in the definition insofar as the life of humans or communities is indirectly or directly affected by a change in the physical and biological environment. Newfoundland and Labrador, New Brunswick, and Prince Edward Island, in contrast, define the environment more narrowly to include only the biophysical components and related socio-economic effects of an activity.

Concept of Significance

The term *significant* is used extensively in EIA legislation, regulations, and guidelines to quantify/qualify the effects associated with a project or activity. However, Nova Scotia is the only jurisdiction that provides a specific definition for the term. According to the regulations, a *significant* impact is one that, 'with respect to an environmental effect, [is] an adverse impact in the context of its magnitude, geographic extent, duration, frequency, degree of reversibility, possibility of occurrence or any combination of the foregoing' (NSDEL 1995).

Application and Scope of the Environmental Assessment Act

EIA is applicable to both the public and private sectors in all Atlantic jurisdictions, including individuals, private firms, and government agencies or departments. The proponent can be an individual, corporation, or government department that owns, manages, or controls a project. The scope of the Act in all jurisdictions is captured by the definition of an *undertaking*. In Nova Scotia and New Brunswick, EIAs are applicable to all projects and activities, such as the modification, extension, abandonment, demolition, or rehabilitation of existing undertakings as well as new projects. They may also apply to policies, plans, or programs, but are subject to ministerial discretion. In NL and Prince Edward Island, EIA applies only to projects and activities.

Size of Projects Assessed

The size of a project or activity is not a condition for an EIA; rather, the magnitude of risk that a project or activity may affect the environment is the basis for determining whether an environmental assessment is required. Furthermore, should a project create considerable concern among the public, whether the risk is genuine or perceived, the minister may decide that the proponent must carry out an EIA. The projects that fall within the departmental lists are ranked on the magnitude of expected risk to the environment associated with a project.

Nova Scotia, New Brunswick, and NL provide proponents with specific lists that stipulate projects or activities that must be registered. In contrast to the other provinces, Nova Scotia has two lists under Schedule 'A': Class I for minor undertakings and Class II for major undertakings. The impacts generated by Class I projects are generally considered to be less significant than those generated by Class II projects and hence are often screened out early in the process. Proponents of Class II projects are by design required to carry out an EIA. The list includes, for example, nuclear facilities, highway construction that is 10 km or more in length and designed for four or more lanes of traffic, electric and hydroelectric generating facilities with production ratings of 10 MW or more. An electric generating facility utilizing wind energy as its sole power source is excluded under the regulations, regardless of its production rating. In Prince Edward Island, the Environmental Protection Act captures undertakings that may cause both minor and major impacts, irrespective of whether the projects are small or large.

Policy-Level Environmental Impact Assessments

Only two jurisdictions in Atlantic Canada have legislative provisions for policy-level EIAs: Nova Scotia and NL. In both jurisdictions, a committee may be appointed by the minister to carry out the assessment. Thus far, Prince Edward Island and

New Brunswick do not routinely carry out policy-level assessments. Nonetheless, the minister can request the Environmental Coordinating Committee, a formal board appointed by and responsible to the minister, to 'inquire any matter pertaining to the environment' on Prince Edward Island (PEDOE 1989a).

Cumulative Effects

In Atlantic Canada, no jurisdiction explicitly requires the proponent to deal with the cumulative effects of an undertaking. Nonetheless, this particular requirement is implied in various interpretive guidelines that have been produced to help practitioners carry out environmental assessments. The respective jurisdictions appear to take the view that it is the proponent's responsibility to determine if the cumulative effects of a project are of concern. In Nova Scotia, for example, one might infer from the definition of *significant* provided in the regulations that the proponent should assess the cumulative effects associated with a project. In other words, the adverse effects of an activity must be evaluated in the context of the magnitude, geographic extent, duration, and frequency of a project.

Alternatives to the Project

All jurisdictions in Atlantic Canada have requirements involving alternative methods of implementing projects (e.g., site or design). In New Brunswick and Nova Scotia, proponents are directed to examine alternative methods of implementing projects. In PEI and NL, however, proponents are explicitly required to examine genuine options—that is, alternatives that are functionally different from the project being proposed. For instance, if the proposal involves a highway, the proponent may be directed to examine alternative forms of transportation in terms of their effectiveness in minimizing environmental impacts (e.g., railway rather than trucking).

Approvals Granted under Environmental Impact Assessment

All provincial jurisdictions in Atlantic Canada, with the exception of Prince Edward Island, issue a formal EIA approval, licence, or permit. After an undertaking has received an environmental approval at a specific stage of the process, the proponent may require additional approvals from other federal or provincial agencies (e.g., industrial approvals). In Prince Edward Island, the process is less formal; in this province, a specialist advises other agencies to issue their approvals.

Provisions for Exceptions

With the exception of New Brunswick, the Atlantic jurisdictions make provisions in their EIA legislation for granting exemptions. In Prince Edward Island, there are defined thresholds or criteria in place that establish which projects or programs are not required to be registered. Such criteria include any project or activity that is smaller than a specified limitation with regard to time, area, length, volume, size, or output. The legislation in Nova Scotia and NL allows for discretionary exemptions that are granted by government. In NL, projects or activities that are small or perceived to be relatively benign—and thus exempt—include, for example, activities involving the wholesale trading, collection, and temporary storage of waste materials (e.g., hazardous) destined for recycling or located within an area designated for that purpose in a development plan (NL, *Environmental Assessment Regulations* [2003]). Comparatively, the exceptions in Nova Scotia include policies, plans, or programs developed after 17 March 1995 that will not directly or indirectly cause a significant environmental effect, or an undertaking that was registered pursuant to section 149 of the *Revised Statutes of Nova Scotia* (1989) but prior to the Environmental Assessment Act (S.N.S. 1994–95, c. 1) and the regulations made thereunder.

Public Involvement

Public involvement is accepted as the cornerstone of EIA in at least three of the four Atlantic jurisdictions; NL, New Brunswick, and Nova Scotia have explicit statutory requirements making public involvement a binding feature of the EIA process. Yet, the approach to public involvement in each jurisdiction differs, though marginally. Public involvement may be limited to public comment at various stages of the assessment process—after registration, during preparation of guidelines or terms of reference for EPR or EIS, and following the submission of the EPR or EIS. The applicable respective provincial Act or regulations establish a specific time frame for public involvement. For instance, the Public Participation Regulation enabled under the Clean Air Act in New Brunswick 'contains a number of information requirements and minimum time frames with regards to the public review of Class 1 Air Quality Approvals and draft Air Quality Objectives' (Public Participation Regulation—*Clean Air Act,* N.B. Reg. 2001-98). The procedure for public involvement in Nova Scotia and NL is established by the regulations but can be adjusted during the EIA process by the minister. In Prince Edward Island, public consultation as a feature of EIA is not a legislative requirement; nonetheless, the interim guidelines strongly encourage proponents to make themselves available to the public.

Filing an EIS or EPR

All jurisdictions require the proponent to file an environmental impact statement or environmental preview report should an examination of the project application reveal that such is warranted. NL has provisions for an additional step in its EIA process. This step entails the filing of an EPR and/or an EIS, if such is considered necessary. Moreover, proponents in NL may be required to carry out, prepare, and submit component studies as a section of the EIS.

Independent Review of Environmental Assessment by a Panel or Board

All jurisdictions in Atlantic Canada make use of an independent review committee that evaluates and reports on the completeness of EIA submissions when this is considered necessary. Prince Edward Island uses an in-house Technical Committee comprised of staff from relevant departments, whereas the other jurisdictions engage committees comprised of technical specialists and/or employees of the provincial and/or federal government. In general, the role of the review committee or Board of Inquiry is to examine EPRs or EISs and to help prepare guidelines for the respective study. In Nova Scotia, New Brunswick, and NL, the review committee is comprised of individuals drawn from the various government agencies whose jurisdictions may be affected by the undertaking. In NL, the committee must have a minimum of five members. In addition, the regulations explicitly stipulate that committee members must not own or have another interest in property located within or adjacent to the area proposed for that undertaking during the time of consideration.

In contrast to the other jurisdictions, Nova Scotia has an Environmental Assessment Board whose activities are subject to specific regulations. The board conducts public hearings, investigations, and/or studies. In Nova Scotia and NL, the review committee may also be involved in alternative dispute resolution mechanisms. In Prince Edward Island, like Nova Scotia, if the minister determines that public concern regarding a project is significant, he or she will appoint an Assessment Review Panel whose primary function is to conduct public hearings, an investigation, or a study with regard to the undertaking. In all jurisdictions, the minister or the Governor in Council (Cabinet) assigns powers to the board, panel, or committee.

Authority of Review Panel or Board

For each Atlantic jurisdiction, the role of the review panel or board is to appraise documents connected with an EIA process and to make recommendations. The minister or Governor in Council makes the final decision. The use of a review panel or board generally occurs in the special cases that may constitute highly controversial projects, new technologies, or a major commitment of natural resources or public funds. For example, large hydroelectric projects, nuclear waste disposal technology, and pulp and paper mills are generally referred to an independent review because they constitute a significant impact on the environment. The respective Governor in Council or minister in each Atlantic jurisdiction generally determines the authority and function of the review panel or board.

Formality of the Review Panel or Board Carrying Out the Review

In all Atlantic Canada jurisdictions, review panels or boards are formally constituted but are not required to follow the strict rules of law, procedure, or evidence such as is required of a court. Although the hearings are conducted in a structured manner (i.e., the process is quasi-judicial), efforts are made to keep them informal. Hence, the test of evidence used in law courts and legal procedures is not necessarily adhered to (e.g., evidence is not sworn and cross-examination of an individual witness is infrequent). Nonetheless, the process is interactive; adversarial views and opinions of various parties are tested and challenged by the proponent, lawyer, and the public before the hearing panel or board. Witnesses are examined, and rules of administrative justice prevail (e.g., fairness, impartiality). Generally, the proceedings of the hearing and all written briefs submitted to the board are made available to the proponent and to interested members of the public. However, none of the provincial jurisdictions have a formal funding system to support participants at hearings or during the panel/board review or the planning process.

Conflict Resolution Provision

Although conflict or alternative dispute resolution can potentially save time and avoid expense, only two jurisdictions in Atlantic Canada make such provisions: Nova Scotia and NL. Nova Scotia promotes the use of mediation and conflict resolution. Hence, an undertaking may be referred to an alternative dispute resolution mechanism at any stage of the assessment process. These provisions are actualized when the minister concludes that such techniques are appropriate for resolving a dispute or issue. In NL, on the other hand, alternative dispute resolution mechanisms are only used when the minister and the proponent disagree over the terms and conditions of an approval.

Health within Environmental Impact Assessment Legislation

The concept of human health is captured by different legislative acts and requirements in the Atlantic provinces.[14] In all jurisdictions, human life and/or well-being are included in the definition of environment, and the relevant legislation makes reference to the social, economic, recreational, cultural, and aesthetic conditions and factors that influence the life of humans in a community.

Government Policy, Stages, Processes, and Procedures

Screening Procedures and Process Initiation

Screening is a procedure by which a proposed project is assessed to determine the next step in the EIA process. All four jurisdictions subject project proposals to a screening process. The process is initiated when the proponent registers the proposal with the appropriate branch or division. In Nova Scotia, NL, and New Brunswick, staff and other interested government departments or agencies review submitted proposals. Based on those reviews, the appropriate branch or division will determine whether the project will likely have a significant effect on the environment. In Prince Edward Island, partial screening is the order of the day; an in-house committee screens all submitted proposals. If in the course of screening, however, the branch or division determines that the potential effects associated with the project are significant, an EIA may be warranted. These determinations are based on experiences with similar projects and other readily available information.

NL is the only jurisdiction that has legislated specific screening criteria to establish what must or may be considered in the assessment of a project proposal. In Nova Scotia, projects are screened to determine if they fall into the Class I or Class II category. Proponents of Class II projects are automatically required to carry out an EIA. Nova Scotia also has provisions for a 'one window' screening process for mining (e.g., metallic, peat), for which there are 'sector-specific guidelines'.

Scoping

Scoping is the process of identifying the key environmental issues associated with a project (e.g., groundwater or the aquatic habitat). Scoping runs the entire length of possibilities in Atlantic Canada, from no key issues to several central concerns. NL and New Brunswick have legislative provisions for a comprehensive approach to environmental assessments, including scoping. For example, in NL, when there is an indication of significant adverse environmental effects from and/or public concern over a proposed project, the minister may order an EIS. A comprehensive study would include a complete description of the project, including alternatives, original research on the existing environment, identification and evaluation of potentially significant environmental effects, evaluation of proposed mitigation measures to minimize harmful effects, and monitoring programs. The minister may decide that component studies should be carried out to make the EIS more comprehensive. To that end, data may be generated on valuable ecosystem elements, such as rare plants or caribou. Although component studies may be part of the EIS in Newfoundland and Labrador, they are separate documents with guidelines specific to each study.

Prince Edward Island, in contrast, takes a partial approach to scoping; environmental assessment focuses on key decision topics such as transportation or the fisheries. In New Brunswick, the scoping of significant issues is always considered part of a project assessment (e.g., impact on air quality). Although Nova Scotia and Prince Edward Island have no provisions in the legislation or regulations for scoping of significant issues, their respective legislation does allow the minister to exact additional information from the proponent regarding a project at major stages of the process.

Project Terms of Reference

The terms of reference (TOR) typically constitute the first formal stage in the approval of an EIA. The TOR can best be described as a study on how to do a study. In Nova Scotia, New Brunswick, and NL, the TOR for an undertaking is developed in consultation with the public, municipalities, and government agencies. The ToR is intended to outline how the EIA study will be carried out. Hence, the ToR for a project is the starting point for managing the EIA process and determining the adequacy of an EPR or EIS. Thus far, Nova Scotia, New Brunswick, and NL are the only jurisdictions where 'terms of reference' have been formally established under the respective provincial Acts. In Prince Edward Island, the approach is informal; department staff generate the partial terms of reference that are made available to the proponent. In Nova Scotia, staff of the Assessment Division prepare the draft ToR, which, upon review, is revised and released to the proponent. In New Brunswick, the Assessment Branch prepares the draft guidelines that are released to the proponent, who subsequently provides the minister with the final ToR. NL has recently collapsed the ToR and guidelines stages into one. Presently in NL, an Assessment Committee, comprised of technical experts from provincial and federal departments with an interest in the proposed project, prepares guidelines for the EPR or EIS. The preparation of the guidelines for an undertaking is collaborative. Those involved are the Assessment Committee, the proponent, government agencies, and the public. The guidelines also include issues identified in the public's comments on the registration document and at public meetings.

Specific Mitigation and Monitoring

All jurisdictions in Atlantic Canada require proponents to specify what mitigative measures they propose to draw on in order to reduce, eliminate, or control the adverse environmental effects of a project or activity, and to enhance the positive impacts of an undertaking. The mitigative measures (as a rule) are captured in the 'terms and conditions' generated for the project. Terms and conditions are typically part of most project approvals and hence constitute a legal and binding requirement on the proponent. All jurisdictions in Atlantic Canada operate according to the 'polluter pays principle', whereby anyone whose activities harm the environment is made responsible for remedying or rehabilitating the adversely affected area and for paying the costs of that action. In their EIA legislation, NL and Nova Scotia include the provision for the restitution of damage to the environment as a feature of mitigation, not simply a penalty for infractions of the Act or regulations or for non-compliance with terms and conditions. In Nova Scotia and NL, monitoring programs are put in place to ensure that the mitigation measures are effective. As a feature of the comprehensive review, NL additionally requires proponents to identify and evaluate the proposed mitigation measures and monitoring programs that will be used to minimize harmful effects connected to a project.

Filing an Environmental Impact Statement or Report

As required, proponents in all Atlantic jurisdictions are directed to file a document known as either an environmental impact statement or an environmental preview report. Each of the jurisdictions requires that guidelines be prepared that establish the content, organization, and level of detail to assist proponents in the preparation of an EIS or EPR. The proponent is responsible for the cost of the study, which, in most cases, is carried out by a team of consultants offering a variety of technical expertise.

The principal objective of an EIA study is to predict the impacts that may be expected should the project proceed. This is accomplished by gathering available resource information, conducting field investigations as required, and using scientific methods to evaluate potential interactions between the environment and the activities associated with the undertaking. An EIA study is expected to identify methods of enhancing the positive and minimizing the negative impacts associated with an undertaking.

In addition to considering impacts, the study should describe the project in detail and evaluate the alternatives or, as is the case in NL, describe the proposed methods for evaluating the accuracy of impact predictions. During the study process, proponents are also encouraged to consult with residents about potential impacts to ensure that their concerns are properly identified and understood. In Nova Scotia, a focus report may be required when a review of the registration document indicates that more information is required on specific aspects of the project. Alternatively, the proponent may be required to prepare and submit an EIS if any of the following conditions apply:

1.   The project falls within the Class II category.
2.   The focus report fails to address specific impacts.
3.   The effects of the project are determined to be significant or of significant public concern.

Review of EIAs by Government and the Public

All jurisdictions in Atlantic Canada require that the draft or revised EPR or EIS be released to government, its representatives, and agencies with an interest in the project, as well as the public, for review and comment after it has been submitted to the respective departmental division or branch. The comments received from the various sources are placed in the respective departmental libraries for public viewing. A key difference in the jurisdictions is the procedure each has adopted for notifying the public and giving them the opportunity to be involved in the initial phase of the process. In Nova Scotia and NL, the first opportunity for public involvement in the process occurs at the screening stage. Additional opportunities for public involvement occur at each major phase of the review process. Alternatively, in New Brunswick, the first opportunity for public review and input does not occur until the draft guidelines have been prepared. The second opportunity occurs following the submission of an EIA study. In Prince Edward Island, public involvement, hypothetically, may not occur at any stage of the EIA process.

Terms and Conditions of Approval

In Nova Scotia, New Brunswick, and Prince Edward Island, terms and conditions are an important feature of the approval granting process and are fully used in the administration of projects; they set out the monitoring, mitigation, and design strategies that are to be applied in order to reduce, eliminate, or improve the effects of the project. NL, in contrast, grants project approvals with partial terms and conditions.

Surveillance of Construction or Implementation

It is in the terms and conditions that the requirement for surveillance, monitoring, and periodic auditing and reporting can be found. In Atlantic Canada, only NL carries out surveillance of construction or implementation of a project to ensure compliance with the conditions of approval. In contrast, Nova Scotia, New Brunswick, and Prince Edward Island carry out surveillance on only some of the key project activities and their associated impacts on the environment (e.g., activities in close proximity to a water course or involving a rare or threatened species).

Compliance Monitoring of the Effects or Post-construction Evaluation

The purpose of monitoring is to ensure compliance, not only with EIA approval requirements but also with other environmental legislation. All jurisdictions require some monitoring of environmental effects. Typically, the proponent is responsible for carrying out compliance monitoring and reporting, as prescribed in an approval, the Act, or regulations. Reports are submitted to the appropriate legislated authority or, as a matter of routine, to local regulatory officials. The proponent is also responsible for reporting (as well as preventing and/or addressing) the release of a substance into the environment in amounts that exceed those authorized or prescribed in the approval, or as directed in the Act or regulations. Departmental monitoring is carried out on a partial or optional basis, but it is often public complaints that trigger an investigation or inspection. In NL, the minister has authority to appoint a monitoring committee to track the progress and the impact on the environment of a project, as well as the success of any mitigation measures required.

Periodic Audits of Approvals and the EIA Process

Environmental impact assessment in Atlantic Canada is an evolving process. In Nova Scotia and NL, there are provisions for periodic evaluation of the assessment process. New Brunswick, like Prince Edward Island, periodically undertakes partial reviews. Presently, NL is the only jurisdiction that carries out periodic audits of approvals.

Granting Approvals

All Atlantic jurisdictions apply a regime of environmental approvals and permits. Proponents are directed to register a proposed undertaking with their respective provincial Environmental Assessment Branch or Division for an approval rather than contact regional authorities directly. Following the assessment approval, other applicable federal, provincial, and municipal approvals and/or permits must be obtained; this is the typical process across the entire region. In New Brunswick, in contrast to the other jurisdictions, an undertaking is always subject to some form of assessment before a determination is made regarding the project.[15] Nova Scotia has taken steps to simplify the approval process. For example, the development of a cranberry operation on upland sites, as well as pits and quarries, requires a number of operating approvals. In the future, the required approvals will be included in a one-step decision-making process. Nonetheless, the one-step approval still requires that the public be given an opportunity to review and comment on proposed projects.

In Nova Scotia and Prince Edward Island, the minister has authority to grant or deny approvals for proposed undertakings. In NL and New Brunswick, the minister makes his or her recommendation regarding a proposal to the Governor in Council, and it is the Governor in Council that grants or denies approval of the project. In all jurisdictions, an approval may be subject to terms and conditions.

Timelines

New Brunswick, Nova Scotia, and NL have specific time frames for each major phase of the EIA process. Following the submission of EIA documentation, specific time allowances are made for public consultation, and the various boards or panels, administrators, and ministers must see that these requirements are met with respect to both the public and the proponent. Prince Edward Island, on the other hand, takes a more casual approach to the EIA process; time frames are not legislated or set out in its EIA guidelines.

Strategic-Level versus Policy-Level EIA

EIA in all Atlantic jurisdictions is regarded as a strategic mechanism. In keeping with the principles of sustainable development, EIA is a planning and decision-making tool used to support and promote the protection and conservation of the environment. In general, the goal of EIA is to promote better project planning by identifying and assessing, before an activity commences, the possible adverse effects that activity may have on the environment. In NL, the Act clearly expresses the intent of the EIA process, which is 'to facilitate the wise management of the natural resources of the province and to protect the environment and quality of life of the people of the province' (NL, *Environmental Protection Act*, S.N.L. 2002, c. E-14.2).

Climate Change

In the Atlantic jurisdictions, there are no explicit references to climate change in any provincial-level environmental legislation or regulations. References to the climate are limited to the quantity and quality of the air or the constituents thereof. Yet initiatives are being launched through innovative partnerships, such as ClimAdapt in Halifax, Nova Scotia, to incorporate climate change adaptation management frameworks into EIAs, municipal risk management processes, and industry infrastructure development practices.[16]

Harmonization as a Cookie Cutter Approach

The aim of harmonization is to standardize the type and scope of studies required to carry out an EIA, the format and presentation of the submissions provided in support of a registration application, and the methods used to evaluate submissions and prepare reports. It is not clear whether harmonization of provincial and federal EIA legislation is creating a 'cookie cutter' approach to the assessment process (i.e., a standardized process). The format and presentation of the submissions provided in support of a registration application and the methods used to evaluate submissions and prepare reports in NL, New Brunswick, and Prince Edward Island are strikingly similar. In each jurisdiction, the proponent is directed to describe the project, its geographic location, physical features, funding, construction and operating schedule/details, and any consultations and studies already completed.

In contrast to the other jurisdictions, Nova Scotia takes those minimal requirements a step further. In the 'Proponent's Guide to Environmental Assessment', Nova Scotia makes it clear that the information in the application must be sufficient to allow the minister to make a decision on the undertaking. This information includes public concerns regarding a project and the current condition of the 'valued environmental components' that are important to the stakeholders. The proponent is required to predict environmental effects and adverse effects relating to the environment, such as 'groundwater, surface water, flora and fauna, aquatic habitat and any other aspect of the environment' (NSDEL 2001). Other required studies are examinations of the socio-economic conditions of the locale, the physical and cultural heritage of the project area, or anything (e.g., structure or site) that may have historical, archaeological, paleontological, or architectural significance, as well as any effects that may impair or damage people's health and reasonable enjoyment of life or property (NSDEL 2001). Furthermore, the proponent must describe all measures used to avoid or mitigate any negative effects and to maximize any positive effects connected with the undertaking, including any monitoring programs and remedial actions that may be conducted throughout the life and the decommissioning of the undertaking. These requirements make the application more comprehensive but are significantly more costly from the proponent's perspective, particularly when one factors in fees required on registering the project. In the other jurisdictions, many of these requirements are found in later stages of the EIA process.

EIA Process Highly Specified or Unspecified

The EIA process in Nova Scotia and NL is highly specified. Each stage of the process is purposely laid out with clear timelines for, and expectations of, the various players in the EIA process. The EIA process in New Brunswick is less structured. Although there are specified time frames and expectations of its players at specific stages, at the decision stage there are no such time requirements. In Prince Edward Island, the EIA process is unstructured. There are no provisions in the legislation for time frames at each stage of the process. Moreover, the legislation fails to set out what is expected of the parties during the process. The approach is loosely defined; expectations and requirements are left to the discretion of the administrative staff or the minister.

Does the Legislation List or Capture the Role of the Agencies?

In NL and Nova Scotia, the respective legislation describes the role or duty that is binding on the minister and Lieutenant- or Governor in Council in administering the EIA process. The Act in Nova Scotia, in particular, clearly lists the all duties of the minister and the administrator of the Assessment Branch of the department. For example, the administrator is responsible for administering and directing the day-to-day operations of the Environmental Assessment Board, organizing all regular and special activities of the board, and supervising public hearings or review processes carried out by the board. Prince Edward Island's Act, on the other hand, leaves the duties of the minister to the direction of the Lieutenant-Governor in Council.

Penalties/Monitoring/Policing

As previously noted, all Atlantic jurisdictions operate under the polluter pays principle. Each jurisdiction has penalties for proponents who do not comply with the legislation, regulations, or terms and conditions supplied with a specific project approval. The relevant branch or division in each region is responsible for policing, compliance, and monitoring. Although projects are investigated and monitored, the respective jurisdictions rely largely on the public to report breaches of the Act or regulations.

Cooperation Agreements with the Government of Canada

All Atlantic jurisdictions allow for cooperation agreements with the Government of Canada. Nonetheless, such arrangements are typically determined on a project-by-project basis, with no general agreement covering all or most undertakings where both federal and provincial governments are involved. For example, the recently rejected Whites Point quarry project in Digby County, NS (see Figure 21.1) was the focus of a Joint Panel Review and Hearings. The proposed project involved the construction of a basalt quarry processing facility and marine terminal located on Digby Neck. For this project, the federal minister of the environment and the minister of the environment and labour for Nova Scotia established a Joint Review Panel to supervise the EIA process and arrange for participant funding. The federal and provincial departments worked cooperatively to prepare the draft guidelines regarding the scope of the EIS. Once public input had been received and considered, the final EIS guidelines were prepared and released to the proponent. After the proponent prepared the EIS and distributed the EIS to the public, panel hearings were convened to hear public commentary on the project or any document or activity related to the project. Analogous Joint Panel Review EIA processes are ongoing in New Brunswick and Newfoundland.

Fee Structure

All the Atlantic jurisdictions have introduced registration fees to cover a portion of the cost of administering the regulations and performing the assessment of undertakings. Each jurisdiction has established its own fixed fee structure. The fee arrangement outlines the time that the fee must be paid and the specific fee that the proponent must pay, depending on the type of project or service provided.

In Newfoundland and Labrador, for example, the Department of the Environment (effective 1998) applies a fixed fee to each significant stage of the EIA process. Within this non-refundable fixed fee structure, the fees range from $200 for project registration to $13,000 for an EA statement.[17] The province also charges for such things as above normal operating costs for an EIA, costs associated with Environment Assessment Committee site visits, and Joint Panel Review and Hearings processes. Proponents who are exempt from the fees include: (a) a municipality; (b) a band or a council of a band as defined by the Indian Act (Canada); (c) a Canadian charitable organization registered under the Income Tax Act (Canada); (d) an individual who proposes to carry out an undertaking for the purposes of constructing a residence; and so on. Similar fee exemptions are evident in New Brunswick and Nova Scotia.

In Nova Scotia, the fixed fees range from $578.98 for an approval transfer to $15,067.40 for a Category 1 EA registration. The fixed fee structure applies to all undertakings in Nova Scotia identified in Schedule 'A' of the regulations.[18] In contrast to the other jurisdictions, the province now requires (effective 6 August 2008) that the registration fees be paid seven days in advance of registering the project with Nova Scotia Environment. The province also charges for such things as above normal operating costs associated with Environmental Assessment Board reviews and hearings and joint federal-provincial environmental assessment panels.

While the fees in Nova Scotia are non-refundable, in Prince Edward Island a portion of the fees may be refunded if the project cost decreases during the assessment process. Alternatively, if the project cost increases during the assessment process, the proponent must pay an additional EA application fee. The registration fee structure in Prince Edward Island (effective 21 May 2005) ranges from $100 for a project estimated to cost under $200,000 to $10,000 for a project estimated to cost $2,500,000 and over. The applicable projects include an array of undertakings, from the expansion of commercial livestock operations to the construction of windpower generators. However, large commercial livestock operations are not subject to the EIA process in either Nova Scotia or New Brunswick because they are subject to provisions made in other regulatory legislation.

New Brunswick introduced a non-refundable fixed fee structure (effective 1 April 2005) for registering an undertaking, submitting supplementary EIA documentation, and modifying an existing approval. As in Nova Scotia, the fixed fee structure in New Brunswick applies to all undertakings identified in Schedule 'A' of the regulations. The fixed fee ranges from $5,000 for a Category I EA registration to $1,000 for a Category III registration, depending on the complexity of the project and the associated potential impacts. There are two key differences between the four jurisdictions: (1) the timing of fee payment; and (2) the basis of fee payment—that is, the cost (PE), schedule (NS and NB), or stage of the EIA (NL).

Legal Provisions for EIA Cooperation

Nova Scotia, New Brunswick, and Newfoundland have legal provisions in place that allow the respective provincial minister (with the approval of the Lieutenant-Governor in Council or the Governor in Council in Nova Scotia and NL) to enter into agreements with other jurisdictions with regard to EIA—the Government of Canada, a province, or a territory. Nova Scotia has provisions under the Act that allow the Governor in Council to transfer the administration and control of an EIA to an individual or other government jurisdiction (federal, provincial, or municipal). New Brunswick has provisions in place that allow agreements with jurisdictions in the United States; whereas, New Brunswick has a bilateral agreement with the Province of Quebec regarding environmental concerns.

Principles for Process Harmonization

All Atlantic jurisdictions are signatories to the Canada-Wide Accord on Environmental Harmonization. Under the agreement, each jurisdiction works cooperatively with the Government of Canada to carry out a single assessment when a project requires an assessment by both levels of government. The principles that underlie harmonization include the polluter-pays principle, the precautionary principle, and recognition that 'pollution prevention is the preferred approach to environmental protection' (CCME 1991). All jurisdictions in Atlantic Canada—NL under the Environmental Protection Act (2002), New Brunswick under the Clean Environment Act (S.N.B., c. C-6), Nova Scotia under the Environment Act (1994–95), and Prince Edward Island under the Environmental Protection Act (1988)—adhere to the principles of the Canada-Wide Accord. Nonetheless, only Newfoundland and Nova Scotia explicitly refer to the principles in their legislation.

Brief Critique/Analysis (Strengths and Weaknesses)

One of the problems with the EIA process is the complexity of some of the information in the EIA documents submitted by proponents. Some information, such as that found in component studies, is often exceedingly technical. Thus, the information is inaccessible to a large sector of the public that may not be familiar with the technology or language. In the spirit of transparency, perhaps standardization in the format and presentation of the submissions provided in support of a registration application, as well as standardization in the methods used to evaluate submissions and prepare reports, would help resolve this concern. Another limitation in Atlantic Canada may be the lack of participant funding at the provincial level. Currently, those who wish to participate in, and/or contribute to, the EIA process have to rely on their own financial resources.

Positive changes are occurring in the Atlantic region, such as greater access to EIA information and enhanced opportunities for public involvement. In addition, as evidenced in NL and Nova Scotia, provisions that require proponents to post a bond for site remediation/reclamation go some distance in lessening the impact of development. Perhaps opportunities for public involvement throughout the life a project, particularly during remediation/reclamation, would enhance the (positive) contribution a project makes. The Liaison Committees now being employed in Nova Scotia are exemplars of long-term public involvement.

Current Status (Any Pending or Recent Changes)

Newfoundland has not amended its EIA legislation since 2002, and it does not plan to do so in the foreseeable furture.[19] Alternatively, Nova Scotia has recently amended its Environmental Assessment Regulations (2008) to improve the clarity of its regulations. The province has also extended the overall timeline for Class I undertakings from 25 days to 50 days, and the public comment period from 10–14 days to 30 days to give the public more time for commentary. These timeline changes will bring Nova Scotia's EIA regulations more in line with other jurisdictions. The following are some of the other more notable changes (for a complete list, see NSDEL 2008):

- Natural gas processing plants are to be included in Class I projects to capture new industries involving liquefied natural gas and sour gas.
- Energy projects are reclassified so that they will require a more comprehensive review if they have the potential to produce higher levels of greenhouse gases and air pollutants.
- Tidal power projects capable of producing at least 2 MW of energy will now trigger a Class 1 environmental assessment.

Although New Brunswick has no formal plans to revise its EIA legislation, the assessment branch implements internal administrative changes on an ongoing basis to further greater consistency across projects. To improve the clarity of the province's regulations, the Department of the Environment recently published A Guide to Environmental Impact Assessment in New Brunswick (2007). This publication summarizes the main components of New Brunswick's EIA process and the requirements of Regulation 87-83. The Department of the Environment, Energy, and Forestry in Prince Edward Island has not only published its 'EIA Guidelines' to advance the clarity of its regulations but also made two notable changes to its EIA regulations. First, the province has introduced requirements for public consultation, and second, it has established standards for public notification/consultation, thereby bringing the Island's EIA regulations more in line with those of the other Atlantic jurisdictions.

Recent or Ongoing Comprehensive EIAs

In New Brunswick, there are currently two ongoing comprehensive EIAs. The first is for the proposed construction and operation of a new petroleum refinery by Irving Oil Company, Limited, capable of processing up to 300,000 barrels per day of crude oil and producing a variety of petroleum products. The second is for the proposed construction and operation of an industrial landfill and related facilities and infrastructure by J.D. Irving, Limited, for the disposal of such wastes as lime mud, fly ash, grits, and dregs generated at the Irving Pulp and Paper Mill. At the time of this writing, the final guidelines that the respective proponents must follow in conducting the EIA have been issued.

Two projects in New Brunswick involving comprehensive EIAs have recently been granted approval. The first was the proposed modifications to the Petitcodiac River causeway between Moncton and Riverview. The technical review committee for this project provided a harmonized provincial and federal review for the EIA. The second project was the proposed removal of the Eel River dam, located in Restigouche County, south of the Town of Dalhousie. The Eel River dam, built in 1963 by the Town of Dalhousie to provide an industrial water source for the area, has resulted in a variety of changes in the hydrology of the Eel River watershed. The review for this project also involved a harmonized provincial and federal EIA process. The objectives of the project are to (1) provide a long-term solution to fish passage; (2) establish conditions that will lead to the natural re-establishment of salt marsh wetlands upstream of the current dam location; and (3) improve the habitat for soft-shelled clams and other shellfish upstream and downstream of the dam (NBDE 2006).

Newfoundland has had a number of comprehensive EIAs involving such large projects as the Voisey's Bay mine/mill project. The project, which was granted approval in 1999, included the construction and operation of a mine and mill to extract and ship ore containing nickel, cobalt, and copper from Voisey's Bay. The EIA process included a Joint Review Panel and Hearings. However, Aboriginal groups were highly critical of the EIA process associated with the project. The Innu Nation wanted to protect the integrity of the EIA process, a process established collaboratively by Canada, Newfoundland, the Innu Nation, and Inuit people (Fry 1998).[20] The Innu Nation appealed to the courts to stop the Newfoundland minister of environment from issuing any permits or approvals to advance the infrastructure proposed by the Voisey's Bay Nickel Company.

Currently, Newfoundland has a number of ongoing comprehensive EIAs, two of which involve a Joint Review Panel. A Joint Review Panel was formed for the

EIA of the Lower Churchill hydroelectric generation project. The proposed project involves the construction and operation of two hydroelectric power generating facilities on the lower section of the Churchill River at Gull Island and Muskrat Falls in Labrador by Newfoundland and Labrador Hydro. The final guidelines for the preparation of the environmental impact statement were released on 15 July 2008. A Joint Review Panel was also formed for a comprehensive EIA, recently released (29 April 2008), of the Newfoundland and Labrador refinery project. This EIA involves the proposed construction and operation of an oil refinery at the head of Placentia Bay by the Newfoundland and Labrador Refining Corporation. Production at the refinery is planned to start in late 2010 or early 2011, with initial production in the order of 300,000 barrels per day.

Prince Edward Island, by contrast, has not seen a comprehensive EIA carried out on the Island since the construction of the Confederation Bridge, whereas Nova Scotia has experienced the completion of two comprehensive EIAs involving a Joint Review Panel and Hearings. The first was the proposed Sydney Tar Ponds and Coke Ovens sites remediation project. In January 2007, the minister of environment and labour for Nova Scotia announced that the Sydney Tar Ponds Agency had been granted an EIA approval to move forward with the clean-up. With the decision came the appointment of a three-member Remediation Monitoring Oversight Board to monitor and report on how the Department of Environment and Labour is performing as the primary regulator of the Sydney Tar Ponds and Coke Ovens clean-up project (Joint Panel Review 2006). The decision brought with it a degree of closure to a long-standing environmental controversy involving a 31-hectare steel plant and coke ovens site, where there had been a number of unsuccessful attempts to clean up 700,000 tonnes of contaminated sediments. The decision also brought to a close a 10-year public consultation process involving the Joint Action Group and a three-week public participation process involving panel hearings (Palen et al. 2004).[21] The second comprehensive EIA was the Joint Review Panel and Hearings for the Whites Point quarry and marine terminal development. In November 2007, the minister announced his decision to deny Bilcon of Nova Scotia approval to construct, operate, and decommission a large basalt quarry, processing facility, ship loading facility, and marine terminal at Whites Point, Digby County, Nova Scotia, for the export of aggregate to New Jersey. The minister based his decision on the Review Panel's opinion that the 'burdens outweigh the benefits and that it would not be in the public interest to proceed with the Whites Point Quarry and Marine Terminal development' (Joint Review Panel 2007). In their report, the panel members made seven interrelated recommendations relating to the implementation of a comprehensive coastal zone management policy or plan for Nova Scotia, including a moratorium on new approvals for development along the North Mountain, amendment of the province's EIA regulations requiring an EIA of quarry projects of any size, revisions to Transport Canada's ballast water regulations, and so forth (for all of the recommendations, see JRPR 2007). The Joint Review Panel and Hearings in this case demonstrated that—contrary to widely held public opinion—the final decision is not a forgone conclusion.

Conclusions

Since its infancy, EIA in Atlantic Canada has experienced significant change. The provinces of Nova Scotia, Prince Edward Island, Newfoundland and Labrador, and New Brunswick are beginning to work together more cooperatively, building partnerships with other provinces, the Government of Canada, and governments in the United States. Atlantic Canada continues to attend to transboundary matters and the coordination of legislative and regulatory initiatives in all its jurisdictions. A central aim of the jurisdictions in Atlantic Canada continues to be to provide access to information on proposed projects and facilitate effective public participation in the formulation of decisions affecting the environment, as well as to provide the public with opportunities to participate in the review of legislation. Moreover, new regulations and policies that provide for a responsive, effective, fair, timely, and efficient administrative and regulatory system are part of that aim. As a feature of this evolution, public input in the EIA process can have greater positive effects on environmental health—and human health and well-being—than will strictly punitive measures. EIA, as an instrument whose conceptual machinery continues to change, may be one of humanity's greatest resources to lessen the impact that human activity visits on the environment.

Notes

1. New Brunswick uses the term *environmental impact assessment* (EIA) rather than *environmental assessment* (EA), in contrast to the other jurisdictions. In this chapter, I will use the words synonymously.

2. In 1971, Statistics Canada identified Prince Edward Island as the most economically disadvantaged province in Canada, surpassing Newfoundland.

3. Further references in this chapter to Newfoundland and Labrador will be denoted as NL.

4. *Navigable Waters Protection Act Established by the Revised Statutes of Canada*, S.C. 1985, c. N-22, accessed on 16 February 2001 at http://www.tc.gc.ca/actsregs/nwpa/english/nwpa.html.

5. Most projects being proposed in the 1970s and 1980s in Atlantic Canada were energy related, either for the production (e.g., hydroelectric) or the distribution (e.g., transmission lines) of energy. Hydroelectric generating stations were typical of the developments involving EIAs being proposed in the Atlantic region to meet the increasing demand for electricity. Others include the Hinds Lake project in central Newfoundland, which started construction in 1977; the Upper Salmon River project on the south coast, which started construction in 1979; and the Cat Arm project on the northwest coast, which started in 1981. In the mid-1970s, Newfoundland Light and Power constructed an oil-fired electric generating station at Holyrood. 'Heritage Society of Newfoundland', accessed on 31 August 2003 at http:// www.heritage.nf.ca/society/hydro.html.

6. Premier Hatfield announced that the project would proceed regardless of the outcome of any public meeting about the project. See Emond 1978.

7. Public meetings to review proposed legislation were held at nine locations throughout NL in October 2001. The Government of Newfoundland and Labrador acknowledged

that 'public input is an essential step in developing progressive legislation.' Ministerial News Release, NLIS 5, Government of Newfoundland and Labrador, 19 September 2001.

8. In the same year, the 11,000-tonne tanker *Arrow*, while transporting bunker 'C' oil to Nova Scotia, struck a submerged reef on its approach to the Strait of Canso, subsequently spilling 50,000 barrels of oil into Chedabucto Bay. The sinking of the *Arrow* is considered to be one of the worst ecological disasters in Canadian waters. The impact on the environment was immense, as the oil covered miles of beach and ruined the local fishery. Clean-up took two years and cost about $4,500,000. *Halifax Chronicle-Herald,* 8 June 2002; *National Post,* 8 June 2002.

9. Dr G. Edward MacDonald, currently a history professor at the University of Prince Edward Island and previously employed at the Prince Edward Island Museum and Heritage Foundation, points out that the post–Second World War history of Prince Edward Island is of two epochs: 'before the Plan' and 'after it' (MacDonald 2000).

10. The draft was both a feasibility study and an environmental risks-and-benefits assessment. Under the EARP, the minister was responsible for ensuring that the draft report underwent public scrutiny.

11. The triggers that initiated the EIA process for the Confederation Bridge were public funds from federal sources committed to the project and the Navigable Waters Protection Act. Navigable Waters Protection Act Established by the Revised Statutes of Canada, S.C. 1985, c. N-22, accessed on 16 February 2003 at http://www.tc.gc.ca/actsregs/nwpa/english/nwpa.html.

12. The EPR guidelines serve to help identify the relevant environmental issues that should be considered when assessing the impacts of a particular development. They also specify the general approach a proponent should follow while conducting an EPR.

13. The following section on EIA in New Brunswick has been adapted from NBDE 2002.

14. No jurisdiction makes any specific mention of Aboriginal health within the context of environmental impact assessment.

15. Conversation with staff at NB Assessment Branch of the Department of the Environment and Local Government, November 2003.

16. Climate Change Adaptation Network (ClimAdapt) is a Halifax-based private sector–driven environmental network established in 2001. ClimAdapt was established to provide innovative climate change adaptation expertise in Canada and internationally. See ClimAdapt at http://www.climadapt.com/networkmembers.html.

17. Government of Newfoundland and Labrador, Department of Environment, available at http://www.govt.nf.ca/env/default.asp.

18. For the environmental assessment fee schedule in Nova Scotia, see http://www.gov.ns.ca/nse/ea/docs/EAFeeSchedule.pdf.

19. The information in this section with respect to possible changes in legislation and regulations is drawn from discussions with the respective director of the division/branch or departmental staff.

20. Critics of the project argued that the EIA panel members are not qualified to assess the proposed Voisey's Bay mine in light of the norm of cultural integrity of the Innu.

21. The Joint Action Group (JAG) was a highly empowered community group, which included local citizens, government representatives, and technical experts; its mandate

was to reach a community consensus on safe, technically sound cleanup options. The JAG lasted 7.5 years and its members put in more than 100,000 volunteer hours in 950 public meetings, with over 1,700 residents taking part in workshops.

Current EIA Legislation

Newfoundland and Labrador. *Environmental Protection Act*, S.N.L. 2002, c. E-14.2. Government of Newfoundland and Labrador, Newfoundland and Labrador, Department of Environment. Environmental Assessment Division. St John's, NL (2002). Accessed on 24 May 2004 at http://www.gov.nf.ca/hoa/statutes/e14-2.htm.

Newfoundland and Labrador. *Environmental Assessment Regulations*, 54/03. Government of Newfoundland and Labrador, Newfoundland and Labrador, Department of Environment. Environmental Assessment Division. St John's, NL (2003). Accessed on 24 May 2004 at http://www.gov.nf.ca/hoa/regulations/rc030054.htm#13.

New Brunswick. *The Clean Air Act*, O.C. 87–558. Environmental Impact Section. New Brunswick Department of the Environment and Local Government. Fredericton, NB (1987). Accessed on 24 May 2004 at http://www.canlii.org/nb/sta/csnb/20030912/s.n.b.c.c-6/.

New Brunswick. *Environmental Impact Assessment Regulations* 87-83. New Brunswick Department of the Environment and Local Government. Fredericton, NB. (1987). Accessed on 24 May 2004 at http://www.canlii.org/nb/regu/crnb/20030912/n.b.reg.87-83/.

Nova Scotia. *Nova Scotia Environmental Assessment Act*, S.N.S. 1994–95, c.1. Nova Scotia Department of the Environment and Labour. Environmental Assessment Division. Halifax, NS (1995). Accessed on 24 May 2004 at http://www.gov.ns.ca/legi/legc/statutes/environ1.htm.

Nova Scotia. *Environmental Assessment Regulations*, s. 49, 1994–95. Nova Scotia Department of the Environment and Labour. Environmental Assessment Division. Halifax, NS (1995). Accessed on 24 May 2004 at http://www.gov.ns.ca/just/regulations/REGS/envassmt.htm.

Nova Scotia. *Environmental Assessment Board Regulations*, s. 49, 1994–95. Nova Scotia Department of the Environment and Labour. Environmental Assessment Division. Halifax, NS (1995). Accessed on 24 May 2004 at http://www.gov.ns.ca/just/regulations/REGS/env2795.htm.

Prince Edward Island. *Environmental Protection Act,* c. E-9, 1988. Prince Edward Island Department of Fisheries, Aquaculture and Environment, Department of Environment and Energy. Charlottetown, PEI. Accessed on 24 May 2004 at http://www.gov.pe.ca/law/statutes/pdf/e-09.pdf.

References

CCME (Canadian Council of Ministers of the Environment). 1991. *Guide to the Canada-Wide Accord on Environmental Harmonization.* Retrieved 27 October 2003 from http:/www.ccme.ca/assests/pdf/guide_to_accord_e.pdf.

CCREM (Canadian Council of Resource and Environment Ministers). 1978. *Canadian environmental impact assessment processes: A discussion paper for the Canadian Council of Resource and Environment Ministers.* Edmonton: Canadian Council of Resource and Environment Ministers Environmental Impact Assessment Task Force.

Couch, W.J., J.F. Herity, and R.E. Munn. 1981. *Environmental impact assessment in Canada.* Vol. 6. Ottawa: Federal Environmental Assessment Review Office.

Emond, P. 1978. *Environment assessment law in Canada.* Toronto: Emond-Montgomery Ltd.

———. 1985. The legal framework of environmental impact assessment in Canada and application of the legislation. In J.B.R. Whitney and V.W. Maclaren (Eds), *Environmental impact assessment: The Canadian experience.* Environmental Monograph No. 5, 53–73. Toronto: Institute for Environmental Studies.

Guernsey, J., R. Dewar, S. Weerasinghe, S. Kirkland, and P. Veugelers. 2000. Incidence of cancer in Sydney and Cape Breton County, Nova Scotia 1979–1997. *Canadian Journal of Public Health* 91 (4): 285–92.

Harrison, J.L. 1990. *Environmental impact assessment in Nova Scotia.* Winnipeg: 14th Annual Assembly: Canadian Environmental Advisory Councils.

Joint Review Panel. 2006. *Environmental Assessment Report on the Proposed Sydney Tar Ponds and Coke Ovens Sites Remediation Project.* Ottawa ON; Halifax, NS: Public Works and Government Services Canada.

———. 2007. *Environmental assessment of the Whites Point quarry and marine terminal project.* Halifax, NS. Accessed on 11 September 2008 at http://www.gov.ns.ca/nse/ea/whitespointquarry/WhitesPointQuarryFinalReport.pdf.

Lang, R., and A. Armour. 1977. The process of environmental assessment: Making it work in Canada. Paper presented at the Environmental Assessment in Canada: Processes and Approaches Conference, Guelph, ON.

MacDonald, E. 2000. *If you're stronghearted: Prince Edward Island in the twentieth century.* Charlottetown: Prince Edward Island Museum and Heritage Foundation.

NBDE (New Brunswick Department of the Environment and Local Government). 1975. *Environmental impact assessment in New Brunswick.* Fredericton: Environment New Brunswick.

———. 2002. *Environmental Impact Assessment in New Brunswick.* Fredericton: New Brunswick Department of the Environment and Local Government.

———. 2006. *Summary of the Environmental Impact Assessment for the Removal of the Eel River Dam.* Fredericton: New Brunswick Department of the Environment and Local Government. Accessed on 13 September at http://www.gnb.ca/0009/0377/0002/EelRiverDamSummary-FINAL-E.pdf.

———. 2007. *A Guide to Environmental Impact Assessment in New Brunswick.* Fredericton: New Brunswick Department of the Environment and Local Government. Accessed on 11 September 2008 at http://www.gnb.ca/0009/0377/0002/11-04-e.pdf.

NSDEL (Nova Scotia Department of Environment and Labour). 1995. *Environmental Assessment Regulations,* as amended up to O.I.C. 2003–67 (28 February 2003), N.S. Reg. 44/2003. Accessed on 24 May 2004 at http://www.gov.ns.ca/just/regulations/REGS/envassmt.htm.

———. 2001. *A proponent's guide to environmental assessment.* Halifax: Nova Scotia Department of Environment and Labour.

———. 2008. *Amendments to Nova Scotia Environmental Assessment Regulations* (a summary), as amended on 6 August 2008. Accessed on 11 September 2008 at http://www.gov.ns.ca/nse/ea/docs/EARegulationAmendmentsGuide.pdf.

Palen, H.R.J., P.N. Duinker, B. Gilbert, J.D. Levy, M.D. MacLeod, and H.A. Van Wilgenburg. 2004. *Community-based processes in the context of contaminated sites: Lessons from the Joint Action Group for Environmental Clean-Up of the Muggah Creek Watershed.* Halifax, NS: Environment Canada, Atlantic Region.

PEDOE (Prince Edward Island Department of the Environment). 1989a. *Environmental impact assessment in Prince Edward Island: Interim guidelines.* Charlottetown: Prince Edward Island Department of the Environment.

————. 1989b. *Environmental impact assessment in Prince Edward Island: Interim report.* Charlottetown: Prince Edward Island Department of the Environment.

Pushchak, R. 1985. The political and institutional context for environmental impact assessment in Ontario. In J.B.R. Whitney and V.W. Maclaren (Eds), *Environmental impact assessment: The Canadian experience.* Environmental Monograph No. 5, 75–87. Toronto: Institute for Environmental Studies.

Rainham, D. 2002. Risk communication and public response to industrial chemical contamination in Sydney, Nova Scotia: A case study. *Journal of Environmental Health* 65 (5): 25, 26–32, 34.

Sadler, B. 1979. Public participation and the planning process intervention and integration. *Plan Canada* 19:8–12.

Smith, L. 1991. Canada's changing impact assessment provision. *Environmental Impact Assessment Review* 11 (1): 5–9.

————. 1984. Public participation in policy making: The state-of-art in Canada. *Geoforum* 15 (2): 253–9.

Storey, K. 1987. Environmental impact assessment in Newfoundland: A review. *Operational Geographer* 12 (1): 31–4.

WGA (Washburn and Gillis Associates). 1980. A review of environmental impact assessment in New Brunswick: A report to the New Brunswick Department of the Environment. Saint John: Washburn and Gillis Associates Ltd.

Contributors

The Editor

KEVIN S. HANNA is an associate professor of geography and environmental studies at Wilfrid Laurier University. His research and teaching focus on land-use planning, impact assessment, and integrated approaches to environment and natural resource management.

The Authors

DEREK R. ARMITAGE is an assistant professor of geography and environmental studies at Wilfrid Laurier University. His areas of research are community-based natural resource management and impact assessment. He has also worked on projects for public- and private-sector organizations, including Fisheries and Oceans Canada, the World Bank, and the Inter-American Development Bank.

DOUGLAS BAKER is a professor and Chair of Urban and Regional Planning in the School of Urban Development at the Queensland University of Technology. His research focuses on infrastructure and land-use planning, performance-based planning, and integrated environmental management.

DARREN BARDATI is director of Environmental Studies at the University of Prince Edward Island. He served in the Department of Environmental Studies and Geography at Bishop's University, Quebec, between 1996 and 2008. His research projects examine the creation of local-expert knowledge partnerships, source water protection at the watershed scale, and coastal community adaptation to climate change.

RYAN W. BARRY is a technical advisor to the Nunavut Impact Review Board. His current research interests include issues related to impact assessment in arctic marine areas and cumulative effects assessment theory and practice.

MARIE ANN BOWDEN is a professor of law at the University of Saskatchewan. Her research interests include environmental law, public-interest advocacy, water law, and property law. She is the editor of the *Canadian Journal of Environmental Law & Policy* and co-author, with Marjorie Benson, of *Understanding Property: A Guide to Canada's Property Law*.

ROGER CREASEY is an adjunct professor in the Faculty of Environmental Design at the University of Calgary. He also practises impact assessment and environmental science as the manager of Ecosystem Management at Shell Canada.

ALAN DIDUCK is an assistant professor and the director of environmental studies at the University of Winnipeg. His research and teaching focus on environmental law and on public involvement and social learning in resource and environmental management.

ANN MARIE FARRUGIA-UHALDE is a natural heritage specialist with the Town of Richmond Hill in the Toronto region. Her work is in the management of terrestrial natural heritage resources and community involvement in natural heritage restoration.

PATRICIA FITZPATRICK is in the Department of Geography at the University of Winnipeg. Her research focuses on the links between impact assessment, public involvement, and participant learning.

LEN GERTLER (1923–2005) was one of Canada's leading city and regional planners. He was the founding director of the School of Urban and Regional Planning at the University of Waterloo, which he joined in 1966. Following his retirement from teaching, he became vice-chair of the Ontario Environmental Review Tribunal. His research and writing focused on regional planning, urban policy, and the Canadian environment. One of his most significant and lasting achievements was to head the planning study, commissioned by the province of Ontario, that led to the protection of the Niagara Escarpment by the Government of Ontario and its subsequent designation as a UNESCO World Biosphere Reserve.

ROBERT B. GIBSON is in the Department of Environment and Resource Studies at the University of Waterloo. His research and writing have centred on decision-making successes and failures in environmental planning, assessment, and regulation in various Canadian jurisdictions and on the emerging design and practice of sustainability assessment.

SONYA GRACI is a member of the Rogers School of Management at Ryerson University in Toronto. She has also worked for the Ontario Ministry of the Environment in environmental assessment and on the development of non-regulatory partnerships.

SOPHIA C.R. GRANCHINHO is a technical advisor to the Nunavut Impact Review Board. Her main responsibilities involve reviews of proposed projects and the monitoring and evaluation of the impacts of projects.

JILL HARRIMAN-GUNN is in the Department of Geography at the University of Saskatchewan. Her research focuses on strategic impact assessment.

LYN HARTLEY has worked in environmental management in Canada's North for the past 15 years. She teaches at Yukon College and consults on projects for a wide range of communities and organizations.

R. JAMIE HERRING is a member of the Department of Development Sociology at Cornell University. His research centres on the role of environmental assessment in resource conservation as well as on the interconnections between war and the environment.

DAN KELLAR is in the Department of Geography and Environmental Management at the University of Waterloo. His research interests centre on impact assessment, adaptation, and the impacts of climate change of high-elevation environments and their communities and economies.

KENTON LOBE teaches in the Department of International Development Studies at the Canadian Mennonite University. He is a graduate of the Natural Resources Institute at the University of Manitoba, where his research focused on community-based resource management of small-scale fisheries in South Asia.

BRUCE MITCHELL is a professor of geography and associate provost of Academic and Student Affairs at the University of Waterloo. His research focuses on integrated water resource management in both developed and developing countries.

BRAM F. NOBLE is an assistant professor of geography at the University of Saskatchewan. His teaching and research interests are resource management, environmental decision making, and the impacts of resource development.

MEAGAN NOONAN is in the Geography and Environmental Studies Program at Wilfrid Laurier University. She specializes in Canadian resource and environmental policy.

RON PUSHCHAK is a professor of occupational and public health at the School of Urban and Regional Planning, Ryerson University. He is also the director of Ryerson's Program in Environmental Applied Science and Management. His research and teaching centre on environmental planning, impact assessment, risk assessment, and decision making in siting hazardous facilities.

ERIC RAPAPORT is an assistant professor in the School of Environmental Planning at the University of Northern British Columbia. His research and teaching centre on regional infrastructure planning, impact assessment, and integrated resource management.

WILLIAM A. ROSS, a professor of environmental science in Environmental Design at the University of Calgary, has studied and participated in the professional practice of environmental impact assessment around the world since 1973.

J. JEFFREY RUSK is the director of technical services at the Nunavut Impact Review Board, located in Cambridge Bay, Nunavut. His research focuses on the development of decision-support tools for impact assessment and land-use planning practitioners.

MURRAY B. RUTHERFORD is in the Resource and Environmental Management Program at Simon Fraser University, Vancouver. His research focuses on the human dimensions of environmental policy and planning. He has studied ecosystem-based management, carnivore conservation policy, water planning, and environmental impact assessment.

A. JOHN SINCLAIR is a professor at the Natural Resources Institute, University of Manitoba. His research and teaching focus on community involvement and learning through resource and environmental decision making, particularly environmental assessment.

D. SCOTT SLOCOMBE is a professor in the Department of Geography and Environmental Studies at Wilfrid Laurier University. His work focuses on ecosystem-based management, sustainability, and protected areas, with a geographic emphasis on western and northern Canada and Australia.

HENDRICUS A. VAN WILGENBURG conducts research on public participation in environmental impact assessment and risk assessment. He also teaches environmental and agricultural ethics at the Nova Scotia Agricultural College and prepares impact assessments.

BERT WEICHEL teaches physical geography and environmental science at the University of Saskatchewan. He also works as a consultant in the areas of land-use planning and natural resource management.

Index